Preliminary Edition

Calculus in Context

I

James Callahan, Smith College
David Cox, Amherst College
Kenneth Hoffman, Hampshire College
Donal O'Shea, Mount Holyoke College
Harriet Pollatsek, Mount Holyoke College
Lester Senechal, Mount Holyoke College

Advisory Committee of the Five College Calculus Project
Peter Lax, New York University, Courant Institute, Chairman
John Neuberger, The University of North Texas
John Truxal, State University of New York, Stony Brook
Gilbert Strang, MIT
Solomon Garfunkel, COMAP, Inc.
Barry Simon, California Institute of Technology

W. H. FREEMAN AND COMPANY
NEW YORK

The principal investigators acknowledge the following sources of funding for the project: National Science Foundation, The Pew Memorial Trust, Sloan Foundation, IBM, and the Five Colleges, Inc.

ISBN: 0-7167-2537-1

Printed in the United States of America

2 3 4 5 6 7 8 9 0 HC 9 9 8 7 6 5 4

Contents

List of Programs

Preface

Our point of view   We believe that calculus can be for our students what it was for Euler and the Bernoullis: a language and a tool for exploring the whole fabric of science. We also believe that much of the mathematical depth and vitality of calculus lies in these connections to the other sciences—the mathematical questions that arise are compelling in part because the answers matter to other disciplines as well. In developing a calculus curriculum, we found it helpful to spell out our **starting points**, our **curricular goals**, our **functional goals**, and our view of the **impact of technology**. Our starting points are a summary of what calculus is really about. Our curricular goals are what we aim to convey about the subject in the course. Our functional goals describe the attitudes and behaviors we hope our students will adopt in using calculus to approach scientific and mathematical questions. We emphasize that what is missing from these lists is as significant as what appears. In particular, we did *not* begin by asking what parts of the traditional course to include or discard.

Starting Points

- Calculus is fundamentally a way of dealing with functional relationships that occur in scientific and mathematical contexts. The techniques of calculus must be subordinate to an overall view of the underlying questions.
- Technology radically enlarges the range of questions we can explore and the ways we can answer them. Computers and graphing calculators are much more than tools for teaching the traditional calculus.

- The concept of a dynamical system is central to science. Therefore, differential equations belong at the center of calculus, and technology makes this possible *at the introductory level.*
- The process of successive approximation is a key tool of calculus, even when the outcome of the process—the limit—cannot be explicitly given in closed form.

Curricular Goals

- Develop calculus in the context of scientific and mathematical questions.
- Treat systems of differential equations as fundamental objects of study.
- Construct and analyze mathematical models.
- Use the method of successive approximations to define and solve problems.
- Develop geometric visualization with hand-drawn and computer graphics.
- Give numerical methods a more central role.

Functional Goals

- Encourage collaborative work.
- Empower students to use calculus as a language and a tool.
- Make students comfortable tackling large, messy, ill-defined problems.
- Foster an experimental attitude towards mathematics.
- Help students appreciate the value of approximate solutions.
- Develop the sense that understanding concepts arises out of working on problems, not simply from reading the text and imitating its techniques.

Impact of Technology

- Differential equations can now be solved numerically, so they can take their rightful place in the introductory calculus course.
- The ability to handle data and perform many computations allows us to explore examples containing more of the messiness of real problems.

- As a consequence, we can now deal with credible models, and the role of modelling becomes much more central to our subject.
- In particular, introductory calculus (and linear algebra) now have something more substantial to offer to life and social scientists, as well as to physical scientists, engineers and mathematicians.
- The distinction between pure and applied mathematics becomes even less clear (or useful) than it may have been.

By studying the text you can see, quite explicitly, how we have pursued the curricular goals. In particular, every one of those goals is addressed within the very first chapter. It begins with questions about describing and analyzing the spread of a contagious disease. A model is built, and the model is a system of coupled non-linear differential equations. We then begin a numerical assault on those equations, and the door is opened to a solution by successive approximations.

Our implementation of the functional goals is less obvious, but it is still evident. For instance, the text has many more words that the traditional calculus book—it is a book to be *read*. Also, the exercises make unusual demands on students. Most exercises are not just variants of examples that have been worked in the text. In fact, the text has rather few simple "template" examples.

It will also become apparent to you that the text reflects substantial **shifts in emphasis** in comparison to the traditional course. Here are some of the most striking:

HOW THE EMPHASIS SHIFTS:

INCREASE	DECREASE
concepts	techniques
geometry	algebra
graphs	formulas
brute force	elegance
numerical solutions	closed-form solutions

Euler's method is a good example of what we mean by "brute force". It is a general method of wide applicability. Of course when we use it to solve a differential equation like $y'(t) = t$, we are using a sledgehammer to crack a peanut. But at least the sledgehammer *does* work. Moreover, it works with coconuts (like $y' = y(1 - y/10)$), and it will eventually even knock down a house (like $y' = \cos(t^2)$). Of course, students also see the elegant special methods that can be invoked to solve $y' = t$ and $y' = y(1 - y/10)$ (separation of variables and partial fractions are discussed in chapter 11), but they understand that they are fortunate indeed when a real problem will succumb to these special methods.

Our curriculum is not aimed at a special clientele. On the contrary, we think that calculus is one of the great bonds that unifies science, and all students should have an opportunity to see how the language and tools of calculus help forge that bond. We emphasize, though, that this is not a "service" course or calculus "with applications," but rather a course rich in mathematical ideas that will serve *all* students well, including mathematics majors. The student population in the first-semester course is especially diverse. In fact, since many students take only one semester, we have aimed to make the first six chapters stand alone as a reasonably complete course. In particular, we have tried to present contexts that would be more or less broadly accessible. The emphasis on the physical sciences is clearly greater in the later chapters; this is deliberate. By the second semester, our students have gained skill and insight that allows them to tackle this added complexity.

Teacher's manual   Working toward our curricular and functional goals has stretched us as well as our students. Teaching in this style is substantially different from the calculus courses most of us have learned from and taught in the past. Therefore we have prepared a manual based on our experiences and those of colleagues at other schools with specific suggestions for use of the text. Prospective teachers are urged to consult it.

Origins   The Five College Calculus Project has a singular history. It begins almost thirty years ago, when the Five Colleges were only Four: Amherst, Mount Holyoke, Smith, and the large Amherst campus of the University of Massachusetts. These four resolved to create a new institution which would be a site for educational innovation at the undergraduate level; by 1970, Hampshire College was enrolling students and enlisting faculty.

Early in their academic careers, Hampshire students grapple with primary sources in all fields—in economics and ecology, as well as in history and literature. But journal articles don't shelter their readers from home truths: if a mathematical argument is needed, it is used. In this way, students in the life and social sciences found, sometimes to their surprise and dismay, that they needed to know calculus if they were to master their chosen fields. However, the calculus they needed was not, by and large, the calculus that was actually being *taught*. The journal articles dealt directly with the relation between quantities and their rates of change—in other words, with differential equations.

Confronted with a clear need, those students asked for help. By the mid-1970s, Michael Sutherland and Kenneth Hoffman were teaching a course for those students. The core of the course, was calculus, but calculus as it is *used* in contemporary science. Mathematical ideas and techniques grew out of scientific questions. Given a process, students had to recast it as a model; most often, the model was a set of differential equations. To solve the differential equations, they used numerical methods implemented on a computer.

The course evolved and prospered quietly at Hampshire. More than a decade passed before several of us at the other four institutions paid some attention to it. We liked its fundamental premise, that differential equations belong at the center of calculus. What astounded us, though, was the revelation that differential equations could really *be* at the center.

This book is the result of our efforts to translate the Hampshire course for a wider audience. The typical student in calculus has not been driven to study calculus in order to come to grips with his or her own scientific questions—as those pioneering students had. If calculus is to emerge organically in the minds of the larger student population, a way must be found to involve that population in a spectrum of scientific and mathematical questions. Hence, calculus *in context*. Moreover, those contexts must be understandable to students with no special scientific training, and the mathematical issues they raise must lead to the central ideas of the calculus—to differential equations, in fact.

Coincidentally, the country turned its attention to the undergraduate science curriculum, and it focused on the calculus course. The National Science Foundation created a program to support calculus curriculum development. To carry out our plans we requested funds for a five-year project; we were fortunate to receive the only multi-year curriculum development grant awarded in the first year of the NSF program. This text is the outcome of our five-year effort.

Advice to students   In a typical high school math text, each section has a "technique" which you practice in a series of exercises very like the examples in the text. This book is different. In this course you will be learning to use calculus both as a tool and as a language in which you can think coherently about the problems you will be studying. As with any other language, a certain amount of time will need to be spent learning and practicing the formal rules. For instance, the conjugation of *être* must be almost second nature to you if you are to be able to read a novel—or even a newspaper—in French. In calculus, too, there are a number of manipulations which must become automatic so that you can focus clearly on the content of what is being said. It is important to realize, however, that becoming good at these manipulations is not the goal of learning calculus any more than becoming good at declensions and conjugations is the goal of learning French.

Up to now, most of the problems you have learned to solve in math classes have had definite answers such as "17", or "the circle with radius 1.75 and center at (2,3)", or the like, and such definite answers are satisfying (and even comforting). However, many interesting and important questions, like "How far is it to the planet Pluto," or "How many people are there with sickle-cell anemia," or "What are the solutions to the equation $x^5 + x + 1 = 0$" can't be answered exactly. Instead, we have ways of finding **approximations** of the answers, and the more time and/or money we are willing to expend, the better our approximations may be. While many calculus problems do have exact answers, such problems often tend to be special or atypical in some way. Therefore, while you will be learning how to deal with these "nice" problems, you will also be developing ways of making good approximations to the solutions of the less well-behaved (and more common!) problems.

The computer or the graphing calculator is a tool that that you will need for this course, along with a clear head and a willing hand. We don't assume that you know anything about this technology ahead of time. Everything necessary is covered completely as we go along.

You can't learn mathematics simply by reading or watching others. The only way to really internalize the material is to work on problems yourself. It is by grappling with the problems that you will come to see what it is you do understand, and to see where your understanding is incomplete or fuzzy.

One of the most important mathematical skills you can develop is that of playing with problems. Don't simply shut your mind down when you come to the end of an assigned problem. These problems have been designed not so much to capture the essence of calculus as to prod your thinking, to get you wondering about the concepts being explored. See if you can think up and answer variations on the problem. Does the problem suggest other questions? The ability to ask good questions of your own is at least as important as being able to answer questions posed by others.

We encourage you to work with others on the exercises. Two or three of you of roughly equal ability working on a problem will often accomplish much more than would any of you working alone. You will stimulate one another's imaginations, combine differing pieces of a problem into a greater whole, and keep up each other's spirits in the frustrating times. This is particularly effective if you first spend time individually working on the material. Many students find it helpful to schedule a regular time to get together to work on problems.

Above all, take time to pause and admire the beauty and power of what you are learning. Aside from its utility, calculus is one of the most elegant and richly structured creations of the human mind and deserves to be profoundly admired on those grounds alone. Enjoy!

Acknowledgments   Certainly none of this would have been possible without the support of the National Science Foundation and of Five Colleges, Inc. We particularly want to thank Louise Raphael who, as the first director of the calculus program at the National Science Foundation, had faith in us and recognized the value of what had already been accomplished at Hampshire when we began our work. Five College Coordinators Conn Nugent and Lorna Peterson have supported and encouraged our efforts, and Five College treasurer and business manager Jean Stabell assisted us in countless ways throughout the project.

We are very grateful to the members of our Advisory Board: to Peter Lax, for his faith in us and his early help in organizing and chairing the Board; to Solomon Garfunkel, for his advice on politics and publishing; to John Neuberger for his passionate convictions; to Barry Simon, for using our text and giving us his thoughtful and imaginative suggestions for improving it; to Gilbert Strang, for his support of a radical venture; to John Truxal, for his detailed commentaries and insights into the world of engineering.

Among our colleagues, James Henle of Smith College deserves special thanks. Besides his many contributions to our discussions of curriculum and pedagogy, he developed the computer programs that have been so valuable for our teaching: Graph, Slinky, Superslinky, and Tint. Jeff Gelbard and Fred Henle ably extended Jim's programs to the Mac-Intosh and to DOS Windows and X Windows. All of this software is (or soon will be) available on anonymous ftp at emmy.smith.edu. Mark Peterson, Robert Weaver, and David Cox also developed software that has been used by our students.

Several of our colleagues made substantial contributions to our frequent editorial conferences and helped with the writing of early drafts. We offer thanks to David Cohen, Robert Currier and James Henle at Smith; David Kelly at Hampshire; and Frank Wattenberg at the University of Massachusetts. Mary Beck, who is now at the University of Virginia, gave heaps of encouragement and good advice as a co-teacher of the earliest version of the course at Smith. Anne Kaufmann, an Ada Comstock Scholar at Smith, assisted us with extensive editorial reviews from the student perspective.

We appreciate the contributions of our colleagues who participated in numerous debriefing sessions at semester's end and gave us comments on the evolving text. We thank George Cobb, Giuliana Davidoff, Alan Durfee, Janice Gifford, Mark Peterson, Margaret Robinson and Robert Weaver at Mount Holyoke; Michael Ablertson, Ruth Haas, Mary Murphy, Marjorie Senechal, Patricia Sipe, and Gerard Vinel at Smith. We learned, too, from the reactions of our colleagues in other disciplines who participated in faculty workshops on Calculus in Context.

We profited a great deal from the comments and reactions of early users of the text. We extend our thanks to Marian Barry at Aquinas College, Peter Dolan and Mark Halsey at Bard College, Donald Goldberg and his colleagues at Occidental College, Benjamin Levy at Beverly High School, Joan Reinthaler at Sidwell Friends School, Keith Stroyan at the University of Iowa, and Paul Zorn at St. Olaf College. Later users who have helped us are Judith Grabiner and Jim Hoste at Pitzer College, Allen Killpatrick and his colleagues at the University of the Redlands, and Barry Simon at CalTech.

A dissemination grant from the NSF funds regional workshops for faculty planning to adopt Calculus in Context. We are grateful to Donald Goldberg and to Henry Warchall of the University of North Texas for coordinating the first two of these workshops.

We owe a special debt to our students over the years, especially those who assisted us in teaching, but also those who gave us the benefit of their thoughtful reactions to the course and the text. Seeing what they were learning encouraged us at every step.

Chapter 1

A Context for Calculus

Calculus gives us a language to describe how quantities are related to one another, and it gives us a set of computational and visual tools for exploring those relationships. Usually, we want to understand how quantities are related in the context of a particular problem—it might be in chemistry, or public policy, or mathematics itself. In this chapter we take a single context—an infectious disease spreading through a population—to see how calculus emerges and how it is used.

§1. The Spread of Disease

Making a Model

Many human diseases are contagious: you "catch" them from someone who is already infected. Contagious diseases are of many kinds. Smallpox, polio, and plague are severe and even fatal, while the common cold and the childhood illnesses of measles, mumps, and rubella are usually relatively mild. Moreover, you can catch a cold over and over again, but you get measles only once. A disease like measles is said to "confer immunity" on someone who recovers from it. Some diseases have the potential to affect large segments of a population; they are called *epidemics* (from the Greek words *epi*, upon + *demos*, the people.) *Epidemiology* is the scientific study of these diseases.

Some properties of contagious diseases

An epidemic is a complicated matter, but the dangers posed by contagion—and especially by the appearance of new and uncontrollable diseases—compel us to learn as much as we can about the nature of

1

epidemics. Mathematics offers a very special kind of help. First, we can try to draw out of the situation its essential features and describe them mathematically. This is calculus as *language*. We substitute an "ideal" mathematical world for the real one. This mathematical world is called a **model**. Second, we can use mathematical insights and methods to analyze the model. This is calculus as *tool*. Any conclusion we reach about the model can then be interpreted to tell us something about the reality.

The idea of a mathematical model

To give you an idea how this process works, we'll build a model of an epidemic. Its basic purpose is to help us understand the way a contagious disease spreads through a population—to the point where we can even predict what fraction falls ill, and when. Let's suppose the disease we want to model is like measles. In particular,

- it is mild, so anyone who falls ill eventually recovers;

- it confers permanent immunity on every recovered victim.

In addition, we will assume that the affected population is large but fixed in size and confined to a geographically well-defined region. To have a concrete image, you can imagine the elementary school population of a big city.

At any time, that population can be divided into three distinct classes:

Susceptible: those who have never had the illness and can catch it;

Infected: those who currently have the illness and are contagious;

Recovered: those who have already had the illness and are immune.

The quantities that our model analyzes

Suppose we let S, I, and R denote the number of people in each of these three classes, respectively. Of course, the classes are all mixed together throughout the population: on a given day, we may find persons who are susceptible, infected, and recovered in the same family. For the purpose of organizing our thinking, though, we'll represent the whole population as separated into three "compartments" as in the diagram at the top of the next page.

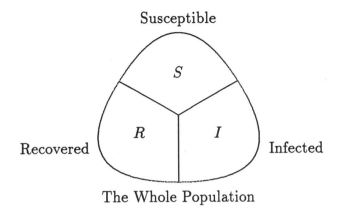

The goal of our model is to determine what happens to the numbers S, I, and R over the course of time. Let's first see what our knowledge and experience of childhood diseases might lead us to expect. When we say there is a "measles outbreak," we mean that there is a relatively sudden increase in the number of cases, and then a gradual decline. After some weeks or months, the illness virtually disappears. In other words, the number I is a **variable**; its value changes over time. One possible way that I might vary is shown in the following graph.

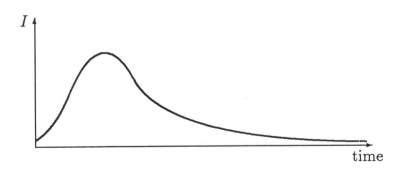

During the course of the epidemic, susceptibles are constantly falling ill. Thus we would expect the number S to show a steady decline. Unless we know more about the illness, we cannot decide whether everyone eventually catches it. In graphical terms, this means we don't know whether the graph of S levels off at zero or at a value above zero. Finally, we would expect more and more people in the recovered group as time passes. The graph of R should therefore climb from left to right. The graphs of S and R might take the forms shown on the next page.

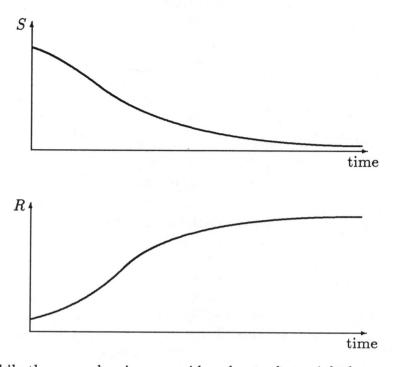

While these graphs give us an idea about what might happen, they raise some new questions, too. For example, because there are no scales marked along the axes, the first graph does not tell us how large I becomes when the infection reaches its peak, nor when that peak occurs. Likewise, the second and third graphs do not say how rapidly the population either falls ill or recovers. A good model of the epidemic should give us graphs like these and it should also answer the quantitative questions we have already raised—for example: When does the infection hit its peak? How many susceptibles eventually fall ill?

A Simple Model for Predicting Change

Suppose we know the values of S, I, and R today; can we figure out what they will be tomorrow, or the next day, or a week or a month from now? Basically, this is a problem of predicting the future. One way to deal with it is to get an idea how S, I, and R are *changing*. For example, suppose the city's Board of Health reports that the measles infection has been spreading at the rate of 470 new cases per day for the last several days. If that rate continues to hold steady and we start with 20,000 susceptible children, then we can expect 470 fewer susceptibles with each passing day. The immediate future would then look like this:

days after today	accumulated number of infections	remaining number of susceptibles
0	0	20 000
1	470	19 530
2	940	19 060
3	1410	18 590
⋮	⋮	⋮

Of course, these numbers will be correct only if the infection continues to spread at its present rate of 470 persons per day. If we want to follow S, I, and R into the future, our example suggests that we should pay attention to the **rates** at which these quantities change. To make it easier to refer to them, let's denote the three rates by S', I', and R'. For example, in the illustration above, S is changing at the rate $S' = -470$ persons per day. We use a minus sign here because S is *decreasing* over time. If S' stays fixed we can express the value of S after t days by the following formula:

$$S = 20000 + S' \cdot t = 20000 - 470\,t \text{ persons.}$$

Knowing rates, we can predict future values

Check that this gives the values of S found in the table when $t = 0$, 1, 2, or 3. How many susceptibles does it say are left after 10 days?

Our assumption that $S' = -470$ persons per day amounts to a mathematical characterization of the susceptible population—in other words, a model! Of course it is quite simple, but it led to a formula that told us what value we could expect S to have at any time t.

The equation $S' = -470$ is a model

The model will even take us backwards in time. For example, two days ago the value of t was -2; according to the model, there were

$$S = 20000 - 470 \times -2 = 20940$$

susceptible children then. There is an obvious difference between going backwards in time and going forwards: we already know the past. Therefore, by letting t be negative we can generate values for S that can be checked against health records. If the model gives good agreement with *known* values of S we become more confident in using it to predict *future* values.

Predictions
depend on the
initial value, too

To predict the value of S using the rate S' we clearly need to have a starting point—a known value of S from which we can measure changes. In our case that starting point is $S = 20000$. This is called the **initial value** of S, because it is given to us at the "initial time" $t = 0$. To construct the formula $S = 20000 - 470\,t$, we needed to have an initial value as well as a rate of change for S.

In the following pages we will develop a more complex model for all three population groups that has the same general design as this simple one. Specifically, the model will give us information about the rates S', I', and R', and with that information we will be able to predict the values of S, I, and R at any time t.

The Rate of Recovery

Our first task will be to model the recovery rate R'. We look at the process of recovering first, because it's simpler to analyze. An individual caught in the epidemic first falls ill and then recovers—recovery is just a matter of time. In particular, someone who catches measles has the infection for about fourteen days. So if we look at the entire infected population today, we can expect to find some who have been infected less than one day, some who have been infected between one and two days, and so on, up to fourteen days. Those in the last group will recover today. In the absence of any definite information about the fourteen groups, let's assume they are the same size. Then 1/14-th of the infected population will recover today:

$$\text{today the change in the recovered population} = \frac{I \text{ persons}}{14 \text{ days}}.$$

There is nothing special about today, though; I has a value at any time. Thus we can make the same argument about any other day:

$$\text{every day the change in the recovered population} = \frac{I \text{ persons}}{14 \text{ days}}.$$

This equation is telling us about R', the rate at which R is changing. We can write it more simply in the form

The first piece of
the S-I-R model

$$R' = \frac{1}{14}I \text{ persons per day.}$$

We call this a **rate equation**. Like any equation, it links different quantities together. In this case, it links R' to I. The rate equation for R is the first part of our model of the measles epidemic.

> Are you uneasy about our claim that 1/14-th of the infected population recovers every day? You have good reason to be. After all, during the first few days of the epidemic almost no one has had measles the full fourteen days, so the recovery rate will be much less than $I/14$ persons per day. About a week before the infection disappears altogether there will be no one in the early stages of the illness. The recovery rate will then be much greater than $I/14$ persons per day. Evidently our model is not a perfect mirror of reality!
>
> Don't be particularly surprised or dismayed by this. Our goal is to gain insight into the workings of an epidemic and to suggest how we might intervene to reduce its effects. So we start off with a model which, while imperfect, still captures some of the workings. The simplifications in the model will be justified if we are led to inferences which help us understand how an epidemic works and how we can deal with it. If we wish, we can then refine the model, replacing the simple expressions with others that mirror the reality more fully.

Notice that the rate equation for R' does indeed give us a tool to predict future values of R. For suppose today 2100 people are infected and 2500 have already recovered. Can we say how large the recovered population will be tomorrow or the next day? Since $I = 2100$,

$$R' = \frac{1}{14} \times 2100 = 150 \text{ persons per day}.$$

Thus 150 people will recover in a single day, and twice as many, or 300, will recover in two. At this rate the recovered population will number 2650 tomorrow and 2800 the next day.

These calculations assume that the rate R' holds steady at 150 persons per day for the entire two days. Since $R' = I/14$, this is the same as assuming that I holds steady at 2100 persons. If instead I varies during the two days we would have to adjust the value of R' and, ultimately, the future values of R as well. In fact, I *does* vary over time. We shall see this when we analyze how the infection is transmitted. Then, in chapter 2, we'll see how to make the adjustments in the values of R' that will permit us to predict the value of R in the model with as much accuracy as we wish.

Other diseases. What can we say about the recovery rate for a contagious disease other than measles? If the period of infection of the new illness is k days, instead of 14, and if we assume that $1/k$ of the infected people recover each day, then the new recovery rate is

$$R' = \frac{I \text{ persons}}{k \text{ days}} = \frac{1}{k}I \text{ persons per day}.$$

If we set $b = 1/k$ we can express the recovery rate equation in the form

$$R' = bI \text{ persons per day.}$$

The constant b is called the **recovery coefficient** in this context.

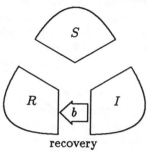

recovery

Let's incorporate our understanding of recovery into the compartment diagram. For the sake of illustration, we'll separate the three compartments. As time passes, people "flow" from the infected compartment to the recovered. We represent this flow by an arrow from I to R. We label the arrow with the recovery coefficient b to indicate that the flow is governed by the rate equation $R' = bI$.

The Rate of Transmission

Since susceptibles become infected, the compartment diagram above should also have an arrow that goes from S to I and a rate equation for S' to show how S changes as the infection spreads. While R' depends only on I, because recovery involves only waiting for people to leave the infected population, S' will depend on both S and I, because transmission involves contact between susceptible and infected persons.

Here's a way to model the transmission rate S'. First, consider a single susceptible person on a single day. On average, this person will contact only a small fraction, p, of the infected population. For example, suppose there are 5000 infected children, so $I = 5000$. We might expect only a couple of them—let's say 2—will be in the same classroom with our "average" susceptible. So the fraction of contacts is $p = 2/I = 2/5000 = .0004$. The 2 contacts themselves can be expressed as $2 = (2/I) \cdot I = pI$ contacts per day per susceptible.

Contacts are proportional to both S and I

To find out how many daily contacts the *whole* susceptible population will have, we can just multiply the average number of contacts per susceptible person by the number of susceptibles: this is $pI \cdot S = pSI$.

Not all contacts lead to new infections; only a certain fraction q do. The more contagious the disease, the larger q is. Since the number of daily contacts is pSI, we can expect $q \cdot pSI$ new infections per day (i.e., to convert contacts to infections, multiply by q). This becomes aSI if we define a to be the product qp.

Recall, the value of the recovery coefficient b depends only on the illness involved. It is the same for all populations. By contrast, the value of a depends on the general health of a population and the level of social interaction between its members. Thus, when two different populations experience the same illness, the values of a could be different. One strategy for dealing with an epidemic is to alter the value of a. Quarantine does this, for instance; see the exercises.

Since each new infection decreases the number of susceptibles, we have the rate equation for S:

$$S' = -aSI \text{ persons per day}.$$

The minus sign here tells us that S is decreasing when S and I are positive. We call a the **transmission coefficient**.

Just as people flow from the infected to the recovered compartment when they recover, they flow from the susceptible to the infected when they fall ill. To indicate the second flow let's add another arrow to the compartment diagram. Because this flow is due to the transmission of the illness, we will label the arrow with the transmission coefficient a. The compartment diagram now reflects all aspects of our model.

Here is the
second piece of
the S-I-R model

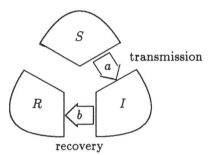

We haven't talked about the units in which to measure a and b. They must be chosen so that any equation in which a or b appears will balance. Thus, in $R' = bI$ the units on the left are persons/day; since the units for I are persons, the units for b must be 1/(days). The units in $S' = -aSI$ will balance only if a is measured in 1/(person-day).

The reciprocals have more natural interpretations. First of all, $1/b$ is the number of days a person needs to recover. Next, note that $1/a$ is measured in person-days (i.e., persons × days), which are the natural units in which to measure exposure. Here is why. Suppose you contact 3 infected persons for each of 4 days. That gives you the same exposure to the illness that you get from 6 infected persons in 2 days—both give 12 "person-days" of exposure. Thus, we can interpret $1/a$ as the level of exposure of a typical susceptible person.

Completing the Model

The final rate equation we need—the one for I'—reflects what is already clear from the compartment diagram: every loss in I is due to a gain in R, while every gain in I is due to a loss in S.

$$
\begin{aligned}
S' &= -aSI \\
I' &= aSI - bI \\
R' &= bI
\end{aligned}
$$

Here is the
complete S-I-R
model

If you add up these three rates you should get the overall rate of change
of the whole population. The sum is zero. Do you see why?

You should not draw the conclusion that the only use of rate equations is to model an epidemic.
Rate equations have a long history, and they have been put to many uses. Isaac Newton (1642–
1727) introduced them to model the motion of a planet around the sun. He showed that the
same rate equations could model the motion of the moon around the earth and the motion of an
object falling to the ground. Newton created calculus as a tool to analyze these equations. He
did the work while he was still an undergraduate—on an extended vacation, curiously enough,
because a plague epidemic was raging at Cambridge!

Today we use Newton's rate equations to control the motion of earth satellites and the
spacecraft that have visited the moon and the planets. We use other rate equations to model
radioactive decay, chemical reactions, the growth and decline of populations, the flow of electricity
in a circuit, the change in air pressure with altitude—just to give a few examples. You will have
an opportunity in the following chapters to see how they arise in many different contexts, and
how they can be analyzed using the tools of calculus.

The following diagram summarizes, in a schematic way, the relation
between our model and the reality it seeks to portray.

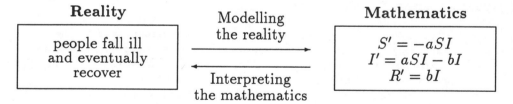

The model is
part of
mathematics; it
only
approximates
reality

The diagram calls attention to several facts. First, the model is a part
of mathematics. It is distinct from the reality being modelled. Second, the model is based on a simplified interpretation of the epidemic.
As such, it will not match the reality exactly; it will be only an **approximation**. Thus, we cannot expect the values of S, I, and R that
we calculate from the rate equations to give us the exact sizes of the
susceptible, infected, and recovered populations. Third, the connection
between reality and mathematics is a two-way street. We have already
travelled one way by constructing a mathematical object that reflects
some aspects of the epidemic. This is model-building. Presently we
will travel the other way. First we need to get mathematical answers
to mathematical questions; then we will see what those answers tell us
about the epidemic. This is interpretation of the model. Before we
begin the interpretation, we must do some mathematics.

Analyzing the Model

Now that we have a model we shall analyze it as a mathematical object. We will set aside, at least for the moment, the connection between the mathematics and the reality. Thus, for example, you should not be concerned when our calculations produce a value for S of 44,446.6 persons at a certain time—a value that will never be attained in reality. In the following analysis S is just a numerical quantity that varies with t, another numerical quantity. Using only mathematical tools we must now extract from the rate equations the information that will tell us just how S and I and R vary with t.

We already took the first steps in that direction when we used the rate equation $R' = I/14$ to predict the value of R two days into the future (see page 7). We assumed that I remained fixed at 2100 during those two days, so the rate $R' = 2100/14 = 150$ was also fixed. We concluded that if $R = 2500$ today, it will be 2650 tomorrow and 2800 the next day.

A glance at the full S-I-R model tells us those first steps have to be modified. The assumption we made—that I remains fixed—is not justified, because I (like S and R) is continually changing. As we shall see, I actually increases over those two days. Hence, over the same two days, R' is not fixed at 150, but is continually increasing also. That means that R' becomes larger than 150 during the first day, so R will be larger than 2650 tomorrow.

Rates are continually changing; this affects the calculations

The fact that the rates are continually changing complicates the mathematical work we need to do to find S, I, and R. In chapter 2 we will develop tools and concepts that will overcome this problem. For the present we'll assume that the rates S', I', and R' stay fixed for the course of an entire day. This will still allow us to produce reasonable estimates for the values of S, I, and R. With these estimates we will get our first glimpse of the predictive power of the S-I-R model. We will also use the estimates as the starting point for the work in chapter 2 that will give us precise values. Let's look at the details of a specific problem.

The Problem. Consider a measles epidemic in a school population of 50,000 children. The recovery coefficient is $b = 1/14$. For the transmission coefficient we choose $a = .00001$, a number within the range used in epidemic studies. We suppose that 2100 people are currently infected and 2500 have already recovered. Since the total population is 50,000, there must be 45,400 susceptibles. Here is a summary of the problem in mathematical terms:

Rate equations:

$$S' = -.00001SI$$
$$I' = .00001SI - I/14$$
$$R' = I/14$$

Initial values: when $t = 0$,

$$S = 45400 \qquad I = 2100 \qquad R = 2500$$

Tomorrow. From our earlier discussion, $R' = 2100/14 = 150$ persons per day, giving us an estimated value of $R = 2650$ persons for tomorrow. To estimate S we use

$$S' = -.00001\,SI = -.00001 \times 45400 \times 2100 = -953.4 \text{ persons/day.}$$

Hence we estimate that tomorrow

$$S = 45400 - 953.4 = 44446.6 \text{ persons.}$$

Since $S + I + R = 50000$ and we have $S + R = 47096.6$ tomorrow, a final subtraction gives us $I = 2903.4$ persons. (Alternatively, we could have used the rate equation for I' to estimate I.)

The fractional values in the estimates for S and I remind us that the S-I-R model describes the behavior of the epidemic only approximately.

Several days hence. According to the model, we estimate that tomorrow $S = 44446.6$, $I = 2903.4$, and $R = 2650$. Therefore, from the new I we get a new approximation for the value of R' tomorrow; it is

$$R' = \frac{1}{14}I = \frac{1}{14} \times 2903.4 = 207.4 \text{ persons/day.}$$

Hence, two days from now we estimate that R will have the new value $2650 + 207.4 = 2857.4$. Now follow this pattern to get new approximations for S' and I', and then use those to estimate the values of S and I two days from now.

The pattern of steps that just carried you from the first day to the second will work just as well to carry you from the second to the third. Pause now and do all these calculations yourself. See exercises 15 and 16 on page 22. If you round your calculated values of S, I, and R to the nearest tenth, they should agree with those in the following table.

Stop and do the
calculations

Estimates for the first three days

t	S	I	R	S'	I'	R'
0	45 400.0	2100.0	2500.0	−953.4	803.4	150.0
1	44 446.6	2903.4	2650.0	−1290.5	1083.1	207.4
2	43 156.1	3986.5	2857.4	−1720.4	1435.7	284.7
3	41 435.7	5422.1	3142.1			

Yesterday. We already pointed out, on page 5, that we can use our models to go *backwards* in time, too. This is a valuable way to see how well the model fits reality, because we can compare estimates that the model generates with health records for the days in the recent past.

To find how S, I, and R change when we go one day into the future we multiplied the rates S', I', and R' by a time step of $+1$. To find how they change when we go one day into the past we do the same thing, except that we must now use a time step of -1. According to the table above, the rates at time $t = 0$ (i.e., today) are

$$S' = -953.4 \qquad I' = 803.4 \qquad R' = 150.0.$$

Therefore we estimate that, one day ago,

$$
\begin{aligned}
S &= 45400 + (-953.4 \times -1) &= 45400 + 953.4 &= 46353.4 \\
I &= 2100 + (803.4 \times -1) &= 2100 - 803.4 &= 1296.6 \\
R &= 2500 + (150.0 \times -1) &= 2500 - 150.0 &= 2350.0
\end{aligned}
$$

Just as we would expect with a spreading infection, there are more susceptibles yesterday than today, but fewer infected or recovered. In the exercises for §3 you will have an opportunity to continue this process, tracing the epidemic many days into the past. For example, you will be able to go back and see when the infection started—that is, find the day when the value of I was only about 1.

Go forward a day
and then back
again

There and back again. What happens when we start with tomorrow's values and use tomorrow's rates to go back one day—back to today? We should get $S = 45400$, $I = 2100$, and $R = 2500$ once again, shouldn't we? Tomorrow's values are

$$
\begin{array}{lll}
S = 44446.6 & I = 2903.4 & R = 2650.0 \\
S' = -1290.5 & I' = 1083.1 & R' = 207.4
\end{array}
$$

To go backwards one day we must use a time step of -1. The predicted values are thus

$$
\begin{array}{rcll}
S & = & 44446.6 + (-1290.6 \times -1) & = & 45737.2 \\
I & = & 2903.4 + (1083.1 \times -1) & = & 1820.3 \\
R & = & 2650.0 + (207.4 \times -1) & = & 2442.6
\end{array}
$$

These are *not* the values that we had at the start, when $t = 0$. In fact, it's worth noting the difference between the original values and those produced by "going there and back again."

	original value	there and back again	difference
S	45400	45737.2	337.2
I	2100	1820.3	−279.7
R	2500	2442.6	−57.4

Do you see why there are differences? We went forward in time using the rates that were current at $t = 0$, but when we returned we used the rates that were current at $t = 1$. Because these rates were different, we didn't get back where we started. These differences do not point to a flaw in the model; the problem lies with the way we are trying to extract information from the model. As we have been making

estimates, we have assumed that the rates don't change over the course of a whole day. We already know that's not true, so the values that we have been getting are not exact. What this test adds to our knowledge is a way to measure just *how* inexact those values are—as we do in the table above.

In chapter 2 we will solve the problem of rough estimates by recalculating all the quantities ten times a day, a hundred times a day, or even more. When we do the computations with shorter and shorter time steps we will be able to see how the estimates improve. We will even be able to see how to get values that are mathematically exact!

Delta notation. This work has given us some insights about the way our model predicts future values of S, I, and R. The basic idea is very simple: *determine how S, I, and R change.* Because these changes play such an important role in what we do, it is worth having a simple way to refer to them. Here is the notation that we will use:

$$\Delta x \text{ stands for a } \textbf{change} \text{ in the quantity } x$$

The symbol "Δ" is the Greek capital letter *delta*; it corresponds to the Roman letter "D" and stands for **difference**.

Delta notation gives us a way to refer to changes of all sorts. For example, in the table above, between day 1 and day 3 the quantities t and S change by

$$\Delta t \;=\; 2 \quad \text{days}$$
$$\Delta S \;=\; -3010.9 \quad \text{persons.}$$

We sometimes refer to a change as a **step**. For instance, in this example we can say there is a "t step" of 2 days, and an "S step" of -3010.9 persons. In the calculations that produced the table at the top of the page we "stepped into" the future, a day at a time. Finally, delta notation gives us a concise and vivid way to to describe the relation between rates and changes. For example, if S changes at the constant rate S', then under a t step of Δt, the value of S changes by

$$\Delta S = S' \cdot \Delta t.$$

Δ stands for a change, a difference, or a step

Using the computer as a tool. Suppose we wanted to find out what happens to S, I, and R after a month, or even a year. We need only repeat—30 times, or 365 times—the three rounds of calculations we used to go three days into the future. The computations would take time, though. The same is true if we wanted to do ten or one hundred rounds of calculations per day—which is the approach we'll take in chapter 2 to get more accurate values. To save our time and effort we will soon begin to use a computer to do the repetitive calculations.

A computer does calculations by following a set of instructions called a **program**. Of course, if we had to give a million instructions to make the computer carry out a million steps, there would be no savings in labor. The trick is to write a program with just a few instructions that can be repeated over and over again to do all the calculations we want.

The usual way to do this is to arrange the instructions in a **loop**. To give you an idea what a loop is, we'll look at the S-I-R calculations. They form a loop. We can see the loop by making a flow chart.

The flow chart. We'll start by writing down the three steps that take us from one day to the next:

Step I Given the current values of S, I, and R, we get current S', I', and R' by using the rate equations.

$$
\begin{aligned}
S' &= -aSI \\
I' &= aSI - bI \\
R' &= bI
\end{aligned}
$$

Step II Given the current values of S', I', and R', we find the changes ΔS, ΔI, and ΔR over the course of a day by using the equations

$$\Delta S \text{ persons} = S' \frac{\text{persons}}{\text{day}} \times 1 \text{ day}$$

$$\Delta I \text{ persons} = I' \frac{\text{persons}}{\text{day}} \times 1 \text{ day}$$

$$\Delta R \text{ persons} = R' \frac{\text{persons}}{\text{day}} \times 1 \text{ day}.$$

Step III Given the current values of ΔS, ΔI, and ΔR, we find the new values of S, I, and R a day later by using the equations

$$
\begin{aligned}
\text{new } S &= \text{current } S + \Delta S \\
\text{new } I &= \text{current } I + \Delta I \\
\text{new } R &= \text{current } R + \Delta R.
\end{aligned}
$$

Each step takes three numbers as **input**, and produces three new numbers as **output**. Note that the output of each step is the input of the next, so the steps flow together. The diagram below, called a **flow chart**, shows us how the steps are connected.

The flow chart
forms a loop

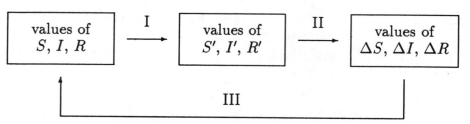

The calculations form a **loop**, because the output of step III is the input for step I. If we go once around the loop, the output of step III gives us the values of S, I, and R *on the following day*. The steps do indeed carry us into the future.

Each step involves calculating three numbers. If we count each calculation as a single instruction, then it takes nine instructions to carry the values of S, I, and R one day into the future. To go a million days into the future, we need add only one more instruction: "Go around the loop a million times." In this way, a computer program with only ten instructions can carry out a million rounds of calculations!

Later in this chapter (§3) you will find a real computer program that lists these instructions (for three days instead of a million, though). Study the program to see which instructions accomplish which steps. In particular, see how it makes a loop. Then run the program to check that the computer reproduces the values you already computed by hand. Once you see how the program works, you can modify it to get further information—for example, you can find out what happens to S, I, and R thirty days into the future. You will even be able to plot the graphs of S, I, and R.

A computer program will carry out the three steps

> Rate equations have always been at the heart of calculus, and they have been analyzed using mechanical and electronic computers for as long as those tools have been available. Now that small powerful computers have begun to appear in the classroom, it is possible for beginning calculus students to explore interesting and complex problems that are modelled by rate equations. Computers are changing how mathematics is done and how it is learned.

Analysis without a computer. A computer is a powerful tool for exploring the *S-I-R* model, but there are many things we can learn about the model without using a computer. Here is an example.

According to the model, the rate at which the infected population grows is given by the equation

$$I' = .00001\, SI - I/14 \quad \text{persons/day.}$$

In our example, $I' = 803.4$ at the outset. This is a positive number, so I increases initially. In fact, I will continue to increase as long as I' is positive. If I' ever becomes negative, then I decreases. So let's ask the question: when is I' positive, when is it negative, and when is it zero? By factoring out I in the last equation we obtain

$$I' = I \left(.00001\, S - \frac{1}{14} \right) \quad \text{persons/day.}$$

Consequently $I' = 0$ if either

$$I = 0 \qquad \text{or} \qquad .00001\,S - \frac{1}{14} = 0.$$

The first possibility $I = 0$ has a simple interpretation: there is no infection within the population. The second possibility is more interesting; it says that I' will be zero when

$$.00001\,S - \frac{1}{14} = 0 \qquad \text{or} \qquad S = \frac{100000}{14} \approx 7142.9.$$

If S is *greater* than 100000/14 and I is positive, then you can check that the formula

$$I' = I\left(.00001\,S - \frac{1}{14}\right) \text{ persons/day}$$

tells us I' is positive—so I is increasing. If, on the other hand, S is *less than* 100000/14, then I' is negative and I is decreasing. So $S = 100000/14$ represents a **threshold**. If S falls below the threshold, I decreases. If S exceeds the threshold, I increases. Finally, I reaches its peak when S equals the threshold.

The threshold determines whether there will be an epidemic

The presence of a threshold value for S is purely a mathematical result. However, it has an interesting interpretation for the epidemic. As long as there are at least 7143 susceptibles, the infection will spread, in the sense that there will be more people falling ill than recovering each day. As new people fall ill, the number of susceptibles declines. Finally, when there are fewer than 7143 susceptibles, the pattern reverses: each day more people will recover than will fall ill.

If there were fewer than 7143 susceptibles in the population *at the outset*, then the number of infected would only decline with each passing day. The infection would simply never catch hold. The clear implication is that the noticeable surge in the number of cases that we associate with an "epidemic" disease is due to the presence of a large susceptible population. If the susceptible population lies below a certain threshold value, a surge just isn't possible. This is a valuable insight—and we got it with little effort. We didn't need to make lengthy calculations or call on the resources of a computer; a bit of algebra was enough.

Exercises

Reading a Graph

The graphs on pages 3 and 4 have no scales marked along their axes, so they provide only *qualitative* information. The graphs below do have scales, so you can now answer *quantitative* questions about them. For example, on day 20 there are about 18,000 susceptible people. Read the graphs to answer the following questions. (Note: $S + I + R$ is *not* constant in this example.)

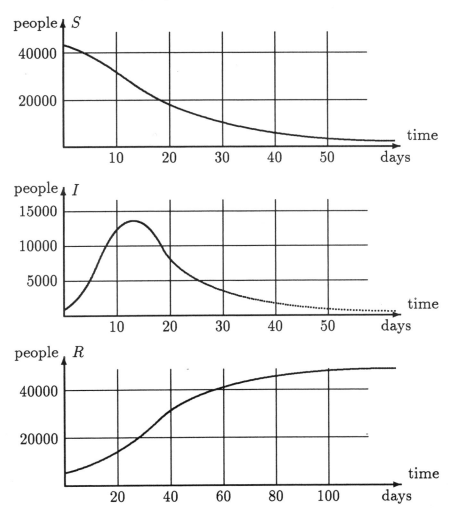

1.   When does the infection hit its peak? How many people are infected at that time?

2.   Initially, how many people are susceptible? How many days does it take for the susceptible population to be cut in half?

3.   How many days does it take for the recovered population to reach 25,000? How many people *eventually* recover? Where did you look to get this information?

4.   On what day is the size of the infected population increasing most rapidly? When is it decreasing most rapidly? How do you know?

5.   How many people caught the illness at some time during the first 20 days? (Note that this is not the same as the number of people who are infected on day 20.) Explain where you found this information.

6.   Copy the graph of R as accurately as you can, and then superimpose a sketch of S on it. Notice the time scales on the *original* graphs of S and R are different. Describe what happened to the graph of S when you superimposed it on the graph of R. Did it get compressed or stretched? Was this change in the horizontal direction or the vertical?

A Simple Model

These questions concern the rate equation $S' = -470$ persons per day that we used to model a susceptible population on pages 4–6.

7.   Suppose the initial susceptible population was 20,000 on Wednesday. Use the model to answer the following questions.

a)   How many susceptibles will be left ten days later?

b)   How many days will it take for the susceptible population to vanish entirely?

c)   How many susceptibles were there on the previous Sunday?

d)   How many days before Wednesday were there 30,000 susceptibles?

Mark Twain's Mississippi

The Lower Mississippi River meanders over its flat valley, forming broad loops called ox-bows. In a flood, the river can jump its banks and cut off one of these loops, getting shorter in the process. In his book *Life on the Mississippi* (1884), Mark Twain suggests, with tongue in cheek, that some day the river might even vanish! Here is a passage that shows us some of the pitfalls in using rates to predict the future and the past.

> In the space of one hundred and seventy six years the Lower
> Mississippi has shortened itself two hundred and forty-two miles.
> That is an average of a trifle over a mile and a third per year.
> Therefore, any calm person, who is not blind or idiotic, can
> see that in the Old Oölitic Silurian Period, just a million years
> ago next November, the Lower Mississippi was upwards of one
> million three hundred thousand miles long, and stuck out over
> the Gulf of Mexico like a fishing-pole. And by the same token
> any person can see that seven hundred and forty-two years from
> now the Lower Mississippi will be only a mile and three-quarters
> long, and Cairo [Illinois] and New Orleans will have joined their
> streets together and be plodding comfortably along under a sin-
> gle mayor and a mutual board of aldermen. There is something
> fascinating about science. One gets such wholesale returns of
> conjecture out of such a trifling investment of fact.

Let L be the length of the Lower Mississippi River. Then L is a variable
quantity we shall analyze.

8.   According to Twain, what is the **rate** at which L is changing, in
miles per year? Is this rate *positive* or *negative*? Explain.

9.   Twain wrote his book in 1884. Suppose the Mississippi that Twain
wrote about had been 1100 miles long; how long would it have become
in 1990?

[Answer: By 1990, the river would have become $141\frac{1}{3}$ miles shorter;
therefore, it would be only $958\frac{2}{3}$ miles long.]

10.   Twain does not tell us how long the Lower Mississippi was in 1884
when he wrote the book, but he does say that 742 years later it will be
only $1\frac{3}{4}$ miles long. How long must the river have been when he wrote
the book?

11.   Suppose t is the number of years since 1884. Write a formula that
describes how much L has changed in t years. Your formula should
complete the equation

the change in L in t years $= \ldots$.

12.   From your answer to question 10, you know how long the river was in 1884. From question 11, you know how much the length has changed t years after 1884. Now write a formula that describes how long the river is t years later.

[Answer: $L = 991\frac{1}{12} - 1\frac{1}{3}t$ miles.]

13.   Use your formula to find what L was a million years ago. Does your answer confirm Twain's assertion that the river was "upwards of 1,300,000 miles long" then?

14.   Was the river ever 1,300,000 miles long; will it ever be $1\frac{3}{4}$ miles long? (This is called a **reality check**.) What, if anything, is wrong with the "trifling investment of fact" which led to such "wholesale returns of conjecture" that Twain has given us?

The Measles Epidemic

We consider once again the specific rate equations

$$
\begin{aligned}
S' &= -.00001\,SI \\
I' &= .00001\,SI - I/14 \\
R' &= I/14
\end{aligned}
$$

discussed in the text on pages 10–15. We saw that at time $t = 1$,

$$S = 44446.6 \qquad I = 2903.4 \qquad R = 2650.0\,.$$

15.   Calculate the current rates of change S', I', and R' when $t = 1$, and then use these values to determine S, I, and R one day later.

16.   In the previous question you found S, I, and R when $t = 2$. Using these values, calculate the rates S', I', and R' and then determine the new values of S, I, and R when $t = 3$. See the table on page 13.

17.   **Double the time step.** Go back to the starting time $t = 0$ and to the initial values

$$S = 45400 \quad I = 2100 \quad R = 2500.$$

Recalculate the values of S, I, and R at time $t = 2$ by using a time step of $\Delta t = 2$. You should perform only a single round of calculations, and use the rates S', I', and R' that are current at time $t = 0$.

[Answer: $S = 43493.2$, $I = 3706.8$, $R = 2800.0$.]

18. **There and back again.** In the text we went one day into the future and then back again to the present. Here you'll go forward two days from $t = 0$ and then back again. There are two ways to do this: with a time step of $\Delta t = \pm 2$ (as in the previous question), and with a pair of time steps of $\Delta t = \pm 1$.

a) ($\Delta t = \pm 2$). Using the values of S, I, and R at time $t = 2$ that you just got in the previous question, calculate the rates S', I', and R'. Then using a time step of $\Delta t = -2$, estimate new values of S, I, and R at time $t = 0$. How much do these new values differ from the original values 45,400, 2100, 2500?

b) ($\Delta t = \pm 1$). Now make a new start, using the values

$$
\begin{array}{lll}
S \;=\; 43156.1 & I \;=\; 3986.5 & R \;=\; 2857.4 \\
S' \;=\; -1720.4 & I' \;=\; 1435.7 & R' \;=\; 284.7
\end{array}
$$

that occur when $t = 2$ if we make estimates with a time step $\Delta t = 1$. (These values come from the table on page 13.) Using two rounds of calculations with a time step of $\Delta t = -1$, estimate another set of new values for S, I, and R at time $t = 0$. How much do these new values differ from the original values 45,400, 2100, 2500?

c) Which process leads to a *smaller* set of differences: a single round of calculations with $\Delta t = \pm 2$, or two rounds of calculations with $\Delta t = \pm 1$? Consequently, which process produces better estimates—in the sense in which we used to measure estimates on page 14?

19. **Quarantine.** One of the ways to treat an epidemic is to keep the infected away from the susceptible; this is called quarantine. The intention is to reduce the chance that the illness will be transmitted to a susceptible person. Thus, quarantine alters the *transmission coefficient.*

a) Suppose a quarantine is put into effect that cuts in half the chance that a susceptible will fall ill. What is the new transmission coefficient?

b) On page 18 it was determined that whenever there were fewer than 7143 susceptibles, the number of infected would decline instead of grow.

We called 7143 a *threshold* level for S. Changing the transmission coefficient, as in part (a), changes the threshold level for S. What is the new threshold?

[Answer: The new threshold level is $200000/14 \approx 14286$.]

c) Suppose we start with $S = 45,400$. Does quarantine eliminate the epidemic, in the sense that the number of infected immediately goes down from 2100, without ever showing an increase in the number of cases?

[Answer: Quarantine does *not* eliminate the epidemic, because S is still above the new threshold level. In other words, I will increase, at least as long as S stays above the threshold.]

d) Since the new transmission coefficient is not small enough to guarantee that I never goes up, can you find a smaller value that *does* guarantee I never goes up? Continue to assume we start with $S = 45400$.

e) Suppose the initial susceptible population is 45,400. What is the *largest* value that the transmission coefficient can have and still guarantee that I never goes up? What level of quarantine does this represent? That is, do you have to reduce the chance that a susceptible will fall ill to one-third of what it was with no quarantine at all, to one-fourth, or what?

Other Diseases

20.   Suppose the spread of an illness similar to measles is modelled by the following rate equations:

$$
\begin{aligned}
S' &= -.00002\,SI \\
I' &= .00002\,SI - .08\,I \\
R' &= .08\,I
\end{aligned}
$$

Note: the initial values $S = 45400$, etc. that we used in the text do not apply here.

a) Roughly how long does someone who catches this illness remain infected? Explain your reasoning.

b) How large does the susceptible population have to be in order for the illness to take hold—that is, for the number of cases to increase? Explain your reasoning.

c) Suppose 100 people in the population are currently ill. According to the model, how many (of the 100 infected) will recover during the next 24 hours?

d) Suppose 30 *new* cases appear during the same 24 hours. What does that tell us about S'?

[Answer: The susceptible population is now decreasing at the rate of 30 persons per day. Thus $S' = -30$.]

e) Using the information in parts (c) and (d), can you determine how large the current susceptible population is?

[Answer: Currently, $S = 15000$ persons.]

21. a) Construct the appropriate S-I-R model for a measles-like illness that lasts for 4 days. It is also known that a typical susceptible person meets only about 0.3% of infected population each day, and the infection is transmitted in only one contact out of six.

b) How small does the susceptible population have to be for this illness to fade away without becoming an epidemic?

[Answer: When S drops below 500 the infection rate I' becomes negative, so the illness fades away.]

22.   Consider the general S-I-R model for a measles-like illness:

$$
\begin{aligned}
S' &= - aSI \\
I' &= aSI - bI \\
R' &= bI
\end{aligned}
$$

a) The threshold level for S—below which the number of infected will only decline—can be expressed in terms of the transmission coefficient a and the recovery coefficient b. What is that expression?

b) Consider two illnesses with the same transmission coefficient a; assume they differ only in the length of time someone stays ill. Which one has the lower threshold level for S? Explain your reasoning.

What Goes Around Comes Around

Some relatively mild illnesses, like the common cold, return to infect you again and again. For a while, right after you recover from a cold, you are immune. But that doesn't last; after some weeks or months, depending on the illness, you become susceptible again. This means there is now a flow from the recovered population to the susceptible. These exercises ask you to modify the basic *S-I-R* model to describe an illness where immunity is temporary.

23.   Draw a compartment diagram for such an illness. Besides having all the ingredients of the diagram on page 9, it should depict a flow from R to S. Call this **immunity loss**, and use c to denote the coefficient of immunity loss.

24.   Suppose immunity is lost after about six weeks. Show that you can set $c = 1/42$ per day, and explain your reasoning carefully. A suggestion: adapt the discussion of recovery in the text.

25.   Suppose this illness lasts 5 days and it has a transmission coefficient of .00004 in the population we are considering. Suppose furthermore that the total population is fixed in size (as was the case in the text). Write down rate equations for S, I, and R.

26.   We saw in the text that the model for an illness that confers permanent immunity has a threshold value for S in the sense that when S is above the threshold, I increases, but when it is below, I decreases. Does *this* model have the same feature? If so, what is the threshold value?

[Answer: There is a threshold at $S = 5000$.]

27.   For a mild illness that confers permanent immunity, the size of the recovered population can only grow. This question explores what happens when immunity is only temporary.

a) Will R increase or decrease if

$$S = 45400 \qquad I = 2100 \qquad R = 2500 \, ?$$

b) Suppose we shift 20000 susceptibles to the recovered population (so $S = 25400$ and $R = 22500$), leaving I unchanged. Will R increase or decrease?

[Answer: In (a), R increases, but in (b) it decreases.]

c) Using a total population of 50,000, give two other sets of values for S, I, and R that lead to a decreasing R.

d) In fact, the relative sizes of I and R determine whether R will increase or decrease. Show that

$$\text{if} \quad I > \tfrac{5}{42}R \quad \text{then } R \text{ will increase;}$$
$$\text{if} \quad I < \tfrac{5}{42}R \quad \text{then } R \text{ will decrease.}$$

Explain your argument clearly. A suggestion: consider the rate equation for R'.

28. **The steady state.** Any illness that confers only temporary immunity can appear to linger in a population forever. You may not always have a cold, but someone does, and eventually you catch another one. ("What goes around comes around.") Individuals gradually move from one compartment to the next. When they return to where they started, they begin another cycle.

Each compartment (in the diagram you drew in question 23) has an *inflow* and an *outflow*. It is conceivable that the two exactly balance, so that the *size* of the compartment doesn't change (even though its individual occupants do). When this happens for all three compartments simultaneously, the illness is said to be in a **steady state**. In this question you explore the steady state of the model we are considering. Recall that the total population is 50,000.

a) What must be true if the inflow and outflow to the I compartment are to balance?

[Answer: For I not to change, S must be 5000.]

b) What must be true if the inflow and outflow to the R compartment are to balance?

[Answer: For R not to change, it must be true that $I = \tfrac{5}{42}R$.]

c) If neither I nor R is changing, then the model must be at the steady state. Why?

d) What is the value of S at the steady state?

e) What is the value of R at the steady state? A suggestion: you know $R + I = 50000 -$ (the steady state value of S). You also have a connection between I and R at the steady state.

[Answer: R must satisfy the equation $\tfrac{47}{42}R = 45000$; thus $R \approx 40213$.]

§2.   The Mathematical Ideas

A number of important mathematical ideas have already emerged in our study of an epidemic. In this section we pause to consider them, because they have a "universal" character. Our aim is to get a fuller understanding of what we have done so we can use the ideas in other contexts.

We often draw out of a few particular experiences a lesson that can be put to good use in new settings. This process is the essence of mathematics, and it has been given a name—abstraction— which means literally "drawing from." Of course abstraction is not unique to mathematics; it is a basic part of the human psyche.

Functions

Functions and
their notation

A **function** describes how one quantity depends on another. In our study of a measles epidemic, the relation between the number of susceptibles S and the time t is a function. We write $S(t)$ to denote that S is a function of t. We can also write $I(t)$ and $R(t)$ because I and R are functions of t, too. We can even write $S'(t)$ to indicate that the rate S' at which S changes over time is a function of t. In speaking, we express $S(t)$ as "S of t" and $S'(t)$ as "S prime of t."

You can find functions everywhere. The amount of postage you pay for a letter is a function of the weight of the letter. The time of sunrise is a function of what day of the year it is. The crop yield from an acre of land is a function of the amount of fertilizer used. The position of a car's gasoline gauge (measured in centimeters from the left edge of the gauge) is a function of the amount of gasoline in the fuel tank. On a polygraph ("lie detector") there is a pen that records breathing; its position is a function of the amount of expansion of the lungs. The volume of a cubical box is a function of the length of a side. The last is a rather special kind of function because it can be described by an algebraic formula: if V is the volume of the box and s is the length of a side, then $V(s) = s^3$.

Most functions are *not* described by algebraic formulas, however. For instance, the postage function is given by a set of verbal instructions and the time of sunrise is given by a table in an almanac. The relation between a gas gauge and the amount of fuel in the tank is determined simply by making measurements. There is no algebraic formula that tells us how the number of susceptibles, S, depends upon t, either.

Instead, we find $S(t)$ by carrying out the steps in the flow chart on page 14 until we reach t days into the future.

In the function $S(t)$ the variable t is called the **input** and the variable S is called the **output**. In the sunrise function, the day of the year is the input and the time of sunrise is the output. In the function $S(t)$ we think of S as *depending on* t, so t is also called the independent variable and S the dependent variable. The set of values that the input takes is called the **domain** of the function. The set of values that the output takes is called the **range**.

The idea of a function is one of the central notions of mathematics. It is worth highlighting:

> **A function is a rule that specifies how the value of one variable, the input, determines the value of a second variable, the output.**

Notice that we say *rule* here, and not *formula*. This is deliberate. We want the study of functions to be as broad as possible, to include all the ways one quantity is likely to be related to another in a scientific question.

Some technical details. It is important not to confuse an expression like $S(t)$ with a product; $S(t)$ does *not* mean $S \times t$. On the contrary, the expression $S(1.4)$, for example, stands for the output of the function S when 1.4 is the input. In the epidemic model we then interpret it as the number of susceptibles that remain 1.4 days after today.

We have followed the standard practice in science by letting the single letter S designate both the *function*—that is, the *rule*—and the *output* of that function. Sometimes, though, we will want to make the distinction. In that case we will use two different symbols. For instance, we might write $S = f(t)$. Then we are still using S to denote the output, but the new symbol f stands for the function rule.

The symbols we use to denote the input and the output of a function are just names; if we change them, we don't change the function. For example, here are three ways to describe the same function g:

$$g \quad : \quad \text{multiply the input by 5, then subtract 3}$$
$$g(x) = 5x - 3$$
$$g(u) = 5u - 3.$$

A function has input and output

Graphs

A graph is a
function rule
given visually

A graph describes a function in a visual form. Sometimes—as with a seismograph or a lie detector, for instance—this is the *only* description we have of a particular function. The usual arrangement is to put the input variable on the horizontal axis and the output on the vertical— but it is a good idea when you are looking at a particular graph to take a moment to check; sometimes, the opposite convention is used! This is often the case in geology and economics, for instance.

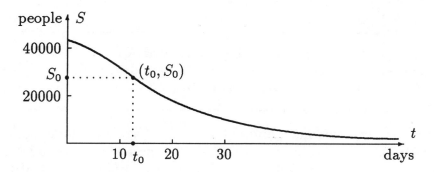

Sketched above is the graph of a function $S(t)$ that tells how many susceptibles there are after t days. Given any t_0, we "read" the graph to find $S(t_0)$, as follows: from the point t_0 on the t-axis, go vertically until you reach the graph; then go horizontally until you reach the S-axis. The value S_0 at that point is the output $S(t_0)$. Here t_0 is about 13 and S_0 is about 27,000; thus, the graph says that $S(13) \approx 27000$, or about 27,000 susceptibles are left after 13 days.

Linear Functions

If y depends on
x, then Δy
depends on Δx

Changes in input and output. Suppose y is a function of x. Then there is some rule that answers the question: What is the value of y for any given x? Often, however, we start by knowing the value of y for a particular x, and the question we really want to ask is: How does y respond to *changes* in x? We are still dealing with the same function—just looking at it from a different point of view. This point of view is important; we use it to analyze functions (like $S(t)$, $I(t)$, and $R(t)$) that are defined by rate equations.

The way Δy depends on Δx can be simple or it can be complex, depending on the function involved. The simplest possibility is that Δy and Δx are **proportional**:

$$\Delta y = m \cdot \Delta x, \quad \text{for some constant } m.$$

The defining
property of a
linear function

Thus, if Δx is doubled, so is Δy; if Δx is tripled, so is Δy. A function whose input and output are related in this simple way is called a **linear function**, because the graph is a straight line. Let's take a moment to see why this is so.

The graph. The graph consists of certain points (x, y) in the x, y-plane. Our job is to see how those points are arranged. Fix one of them, and call it (x_0, y_0). Let (x, y) be any other point on the graph. Draw the line that connects this point to (x_0, y_0), as we have done in the figure at the right. Now set

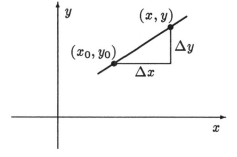

$$\Delta x = x - x_0 \qquad \Delta y = y - y_0.$$

As the figure shows, the slope of this line is $\Delta y / \Delta x = m$. Recall that m is a constant; thus, if we pick a new point (x, y), the slope of the connecting line won't change.

Since (x, y) is an arbitrary point on the graph, what we have shown is that **every point on the graph lies on a line of slope m through the point (x_0, y_0).** But there is only one such line—and all the points lie on it! That line must be the graph.

> A linear function is one that satisfies $\Delta y = m \cdot \Delta x$; its graph is a straight line whose slope is m.

Rates, slopes, and multipliers. The interpretation of m as a slope is just one possibility; there are two other interpretations that are equally important. To illustrate them we'll use Mark Twain's vivid description of the shortening of the Lower Mississippi River (see page 21). This will also give us the chance to see how a linear function emerges in context.

Twain says "the Lower Mississippi has shortened itself ... an average of a trifle over a mile and a third per year." Suppose we let L denote the length of the river, in miles, and t the time, in years. Then L

depends on t, and Twain's statement implies that L is a *linear* function of t—in the sense in which we have just defined a linear function. Here is why. According to our definition, there must be some number m which makes $\Delta L = m \cdot \Delta t$. But notice that Twain's statement has exactly this form if we translate it into mathematical language. Convince yourself that it says

Stop and do the translation

$$\Delta L \text{ miles} = -1\tfrac{1}{3} \frac{\text{miles}}{\text{year}} \times \Delta t \text{ years}.$$

Thus we should take m to be $-1\tfrac{1}{3}$ miles per year.

The role of m here is to convert one quantity (Δt years) into another (ΔL miles) by multiplication. All linear functions work this way. In the defining equation $\Delta y = m \cdot \Delta x$, multiplication by m converts Δx into Δy. Any change in x produces a change in y that is m times as large. For this reason we give m its second interpretation as a **multiplier**.

It is easier to understand why the usual symbol for *slope* is m—instead of s—when you see that a slope can be interpreted as a multiplier.

It is important to note that, in our example, m is not simply $-1\tfrac{1}{3}$; it is $-1\tfrac{1}{3}$ *miles per year*. In other words, m is the **rate** at which the river is getting shorter. All linear functions work this way, too. We can rewrite the equation $\Delta y = m \cdot \Delta x$ as a ratio

$$m = \frac{\Delta y}{\Delta x} = \text{the rate of change of } y \text{ with respect to } x.$$

For these reasons we give m its third interpretation as a **rate of change**.

For a linear function satisfying $\boldsymbol{\Delta y = m \cdot \Delta x}$, the coefficient $\boldsymbol{m}$ is rate of change, slope, and multiplier.

We already use y' to denote the rate of change of y, so we can now write $m = y'$ when y is a linear function of x. In that case we can also write

$$\Delta y = y' \cdot \Delta x.$$

This expression should recall a pattern very familiar to you. (If not, change y to S and x to t!) It is the fundamental formula we have been using to calculate future values of S, I, and R. We can approach the relation between y and x the same way. That is, if y_0 is an "initial value" of y, when $x = x_0$, then *any* value of y can be calculated from

$$y = y_0 + y' \cdot \Delta x \quad \text{or} \quad y = y_0 + m \cdot \Delta x.$$

Units. Suppose x and y are quantities that are measured in specific units. If y is a linear function of x, with $\Delta y = m \cdot \Delta x$, then m must have units too. Since m is the multiplier that "converts" x into y, the units for m must be chosen so they will convert x's units into y's units. In other words,

If x and y have units, so does m

$$\text{units for } y = \text{units for } m \times \text{units for } x.$$

This implies

$$\text{units for } m = \frac{\text{units for } y}{\text{units for } x}.$$

For example, the multiplier in the Mississippi River problem converts years to miles, so it must have units of miles per year. The rate equation $R' = bI$ in the S-I-R model is a more subtle example. It says that R' is a linear function of I. Since R' is measured in persons per day and I is measured in persons, we must have

$$\text{units for } b = \frac{\text{units for } R'}{\text{units for } I} = \frac{\frac{\text{persons}}{\text{days}}}{\text{persons}} = \frac{1}{\text{days}}.$$

Formulas for linear functions. The expression $\Delta y = m \cdot \Delta x$ declares that y is a linear function of x, but it doesn't quite tell us what y itself *looks like* directly in terms of x. In fact, there are several equivalent ways to write the relation $y = f(x)$ in a formula, depending on what information we are given about the function.

● **The initial-value form.** Here is a very common situation: we know the value of y at an "initial" point—let's say $y_0 = f(x_0)$—and we know the rate of change—let's say it is m. Then the graph is the straight line of slope m that passes through the point (x_0, y_0). The formula for f is

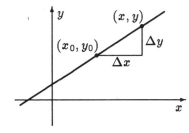

$$y = y_0 + \Delta y = y_0 + m \cdot \Delta x = y_0 + m(x - x_0) = f(x).$$

What you should note particularly about this formula is that it expresses y in terms of the initial data x_0, y_0, and m—as well as x. Since that data consists of a point (x_0, y_0) and a slope m, the initial-value formula is also referred to as the **point-slope form** of the equation of a line. It may be more familiar to you with that name.

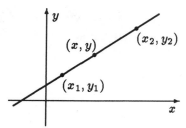

• **The interpolation form.** This time we are given the value of y at *two* points—let's say $y_1 = f(x_1)$ and $y_2 = f(x_2)$. The graph is the line that passes through (x_1, y_1) and (x_2, y_2), and its slope is therefore

$$m = \frac{y_2 - y_1}{x_2 - x_1}.$$

Now that we know the slope of the graph we can use the point-slope form (taking (x_1, y_1) as the "point", for example) to get the equation. We have

$$y = y_1 + m(x - x_1) = y_1 + \frac{y_2 - y_1}{x_2 - x_1}(x - x_1) = f(x).$$

Notice how, once again, y is expressed in terms of the initial data—which consists of the two points (x_1, y_1) and (x_2, y_2).

The process of finding values of a quantity between two given values is called **interpolation**. Since our new expression does precisely that, it is called the interpolation formula. (Of course, it also finds values outside the given interval.) Since the initial data is a pair of points, the interpolation formula is also called the **two-point formula** for the equation of a line.

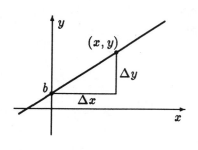

• **The slope-intercept form.** This is a special case of the intial-value form that occurs when the initial $x_0 = 0$. Then the point (x_0, y_0) lies on the y-axis, and it is frequently written in the alternate form $(0, b)$. The number b is called the **y-intercept**. The equation is

$$y = mx + b = f(x).$$

In the past you may have thought of this as *the* formula for a linear function, but for us it is only one of several. You will find that we will use the other forms more often.

Functions of Several Variables

Language and notation. Many functions depend on more than one variable. For example, sunrise depends on the day of the year but it also depends on the latitude (position north or south of the equator) of the observer. Likewise, the crop yield from an acre of land depends on the amount of fertilizer used, but it also depends on the amount of rainfall, on the composition of the soil, on the amount of weeding done—to mention just a few of the other variables that a farmer has to contend with.

A function can have several input variables

The rate equations in the *S-I-R* model also provide examples of functions with more than one input variable. The equation

$$I' = .00001\, SI - I/14$$

says that we need to specify both S and I to find I'. We can say that

$$F(S, I) = .00001\, SI - I/14$$

is a function whose input is the **ordered pair** of variables (S, I). In this case F is given by an algebraic formula. While many other functions of several variables also have formulas—and they are extremely useful—not all functions do. The sunrise function, for example, is given by a two-way table (see page 163) that shows the time of sunrise for different days of the year and different latitudes.

As a technical matter it is important to note that the input variables S and I of the function $F(S, I)$ above appear in a particular *order*, and that order is part of the definition of the function. For example, $F(1, 0) = 0$, but $F(0, 1) = -1/14$. (Do you see why? Work out the calculations yourself.)

Parameters. Suppose we rewrite the rate equation for I', replacing .00001 and 1/14 with the general values a and b:

$$I' = aSI - bI.$$

This makes it clear that I' depends on a and b, too. But note that a and b are not variables in quite the same way that S and I are. For example, a and b will vary if we switch from one disease to another or from one population to another. However, they will stay fixed while we consider a particular disease in a particular population. By contrast, S

and I will *always* be treated as variables. We call a quantity like a or b a **parameter**.

To emphasize that I' depends on the parameters as well as S and I, we can write I' as the output of a new function

$$I' = I'(S, I, a, b) = aSI - bI$$

Some functions depend on parameters

whose input is the set of *four* variables (S, I, a, b), in that order. The variables S, I, and R must also depend on the parameters, too, and not just on t. Thus, we should write $S(t, a, b)$, for example, instead of simply $S(t)$. We implicitly used the fact that S, I, and R depend on a and b when we discovered there was a threshold for an epidemic (page 18). In exercise 22 of §1 (page 25), you made the relation explicit. In that problem you show I will simply decrease over time (i.e., there will be no "burst" of infection) if

$$S < \frac{b}{a}.$$

There are even more parameters lurking in the S-I-R problem. To uncover them, recall that we needed *two* pieces of information to estimate S, I, and R over time:

1)   the rate equations;
2)   the initial values S_0, I_0, and R_0.

We used $S_0 = 45400$, $I_0 = 2100$, and $R_0 = 2500$ in the text, but if we had started with other values then S, I, and R would have ended up being different functions of t. Thus, we should really write

$$S = S(t, a, b, S_0, I_0, R_0)$$

to tell a more complete story about the inputs that determine the output S. Most of the time, though, we do *not* want to draw attention to the parameters; we usually write just $S(t)$.

Further possibilities. Steps I, II, and III on page 16 are also functions, because they have well-defined input and output. They are unlike the other examples we have discussed up to this point because they have more than one output variable. You should see, though, that there is nothing more difficult going on here.

In our study of the S-I-R model it was natural not to separate functions that have one input variable from those that have several.

This is the pattern we shall follow in the rest of the course. In particular, we will want to deal with parameters, and we will want to understand how the quantities we are studying depend on those parameters.

The Beginnings of Calculus

While functions, graphs, and computers are part of the general fabric of mathematics, we can also abstract from the *S-I-R* model some important aspects of the calculus itself. The first of these is the idea of a **rate of change**. In this chapter we just assumed the idea was intuitively clear. However, there are some important questions not yet answered; for example, how do you deal with a quantity whose rate of change is itself always changing? These questions, which lead to the fundamental idea of a **derivative**, are taken up in chapter 3.

Rate equations—more commonly called **differential equations**—lie at the very heart of calculus. We will have much more to say about them, because many processes in the physical, biological, and social realms can be modelled by rate equations. In our analysis of the *S-I-R* model, we used rate equations to estimate future values by assuming that rates stay fixed for a whole day at a time. The discussion called "there and back again" on page 14 points up the shortcomings of this assumption. In chapter 2 we will develop a procedure, called Euler's method, to address this problem. In chapter 4 we will return to differential equations in a general way, equipped with Euler's method and the concept of the derivative.

How the next three chapters are connected

Exercises

Functions and Graphs

1. Sketch the graph of each of the following functions. Label each axis, and mark a scale of units on it. For each line that you draw, indicate
  i) its slope;
  ii) its y-intercept;
  iii) its x-intercept (where it crosses the x-axis).

a) $y = -\frac{1}{2}x + 3$

c) $5x + 3y = 12$

b) $y = (2x - 7)/3$

2.   Graph the following functions. Put labels and scales on the axes.
a) $V = .3Z - 1$;   b) $W = 600 - P^2$.

3.   Sketch the graph of each of the following functions. Put labels and scales on the axes. For each graph that you draw, indicate
   i) its y-intercept;
   ii) its x-intercept(s).
For part (d) you will need the **quadratic formula**

$$x = \frac{-b \pm \sqrt{b^2 - 4ac}}{2a}$$

for the roots of the **quadratic equation** $ax^2 + bx + c = 0$.
a) $y = x^2$ c) $y = (x + 1)^2$
b) $y = x^2 + 1$ d) $y = 3x^2 + x - 1$

The next four questions refer to these functions:

$$
\begin{aligned}
c(x, y) &= 17 & &\text{a constant function} \\
j(z) &= z & &\text{the identity function} \\
r(u) &= 1/u & &\text{the reciprocal function} \\
D(p, q) &= p - q & &\text{the difference function} \\
s(y) &= y^2 & &\text{the squaring function} \\
Q(v) &= \frac{2v + 1}{3v - 6} & &\text{a rational function} \\
H(x) &= \begin{cases} 5 & \text{if } x < 0 \\ x^2 + 2 & \text{if } 0 \le x < 6 \\ 29 - x & \text{if } 6 \le x \end{cases} & & \\
T(x, y) &= r(x) + Q(y) & &
\end{aligned}
$$

4.   Determine the following values:

$$
\begin{array}{cccc}
c(5, -3) & j(17) & c(a, b) & j(u^2 + 1) \\
j(c(3, -5)) & s(1.1) & r(1/17) & Q(0) \\
Q(2) & Q(3/7) & D(5, -3) & D(-3, 5) \\
H(1) & H(7) & H(4) & H(H(H(-3))) \\
r(s(-4)) & r(Q(3)) & Q(r(3)) & T(3, 7)
\end{array}
$$

5.   True or false. Give reasons for your answers: if you say true, explain why; if you say false, give an example that shows why it is false.

a) For every non-zero number x, $r(r(x)) = j(x)$.

b) If $a > 1$, then $s(a) > 1$.

c) If $a > b$, then $s(a) > s(b)$.

d) For all real numbers a and b, $s(a + b) = s(a) + s(b)$.

e) For all real numbers a, b, and c, $D(D(a, b), c)) = D(a, D(b, c))$.

[Answer: (c), (d), and (e) are false.]

6.   Find all numbers x for which $Q(x) = r(Q(x))$.

7.   The **natural domain** of a function f is the largest possible set of real numbers x for which $f(x)$ is defined. For example, the natural domain of $r(x) = 1/x$ is the set of all non-zero real numbers.

a) Find the natural domains of Q and H.

b) Find the natural domains of $P(z) = Q(r(z))$; $R(v) = r(Q(v))$.

[Answer: The natural domain of R consists of all real numbers except $-1/2$ and 2.]

c) What is the natural domain of the function $W(t) = \sqrt{\dfrac{1 - t^2}{t^2 - 4}}$?

[Answer: t must satisfy either $-2 < t \leq -1$ or $1 \leq t < 2$.]

Computer Graphing

The purpose of these exercises is to give you some experience using a "graphing package" on a computer. This is a program that will draw the graph of a function $y = f(x)$ whose formula you know. You must type in the formula, using the following symbols to represent the basic arithmetic operations:

to indicate	type
addition	+
subtraction	-
multiplication	*
division	/
an exponent	^

The caret " ^ " appears above the "6" on a keyboard (Shift-6). Here is an example:

to enter:	type:
$\dfrac{7x^5 - 9x^2}{x^3 + 1}$	(7*x^5 - 9*x^2)/(x^3 + 1)

The parentheses you see here are important. If you do not include them, the computer will interpret your entry as

$$7x^5 - \frac{9x^2}{x^3} + 1 = 7x^5 - \frac{9}{x} + 1 \neq \frac{7x^5 - 9x^2}{x^3 + 1}.$$

In some graphing packages, you do not need to use * to indicate a multiplication. If this is true for the package you use, then you can enter the fractional expression above in the somewhat simpler form

$$(7x^5 - 9x^2)/(x^3 + 1).$$

To do the following exercises, follow the specific instructions for the graphing package you are using.

8.   Graph the function $f(x) = .6x + 2$ on the interval $-4 \leq x \leq 4$.

a) What is the y-intercept of this graph? What is the x-intercept?

b) Read from the graph the value of $f(x)$ when $x = -1$ and when $x = 2$. What is the difference between these y values? What is the difference between the x values? According to these differences, what is the slope of the graph? According to the *formula*, what is the slope?

9.   Graph the function $f(x) = 1 - 2x^2$ on the interval $-1 \leq x \leq 1$.

a) What is the y-intercept of this graph? The graph has two x-intercepts; what are they?

[Answer: The x-intercepts are at $\pm 1/\sqrt{2} \approx .7071068$.]

You can also find an x-intercept using the computer. The idea is to **magnify** the graph near the intercept until you can determine as many decimal places in the x coordinate as you want. For a start you should be able to see that the graph on your computer monitor crosses the x-axis somewhere between .6 and .8. Regraph $f(x)$ on the interval $.6 \leq x \leq .8$. You should then be able to determine that the x-intercept lies between .70 and .71. This means $x = .7\ldots$; that is, you know the location of the x-intercept to one decimal place of accuracy.

b) Regraph $f(x)$ on the interval $.70 \leq x \leq .71$ to get two decimal places of accuracy in the location of the x-intercept. Continue this process until you have 7 places of accuracy. What is the x-intercept? [Answer: $x = .7071067812\ldots.$]

The circular functions Graphing packages "know" the familiar functions of trigonometry. Trigonometric functions are qualitatively different from the functions in the preceding problems. Those functions are defined by algebraic formulas, so they are called **algebraic functions**. The trigonometric functions are defined by explicit "recipes," but *not* by algebraic formulas; they are called **transcendental functions**. For calculus, we always use the definition of the trigonometric functions as **circular functions**. This definition begins with a unit circle centered at the origin. Given the input number t, locate a point P on the circle by tracing an arc of length t along the circle from the point $(1, 0)$. If t is positive, trace the arc counterclockwise; if t is negative, trace it clockwise. Because the circle has radius 1, the arc of length t subtends a central angle of **radian** measure t.

The circular (or trigonometric) functions $\cos t$ and $\sin t$ are defined as the coordinates of the point P,

$$P = (\cos t, \sin t).$$

The other trigonometric functions are defined in terms of the sine and cosine:

$$\tan t = \sin t / \cos t \qquad \sec t = 1 / \cos t$$
$$\cot t = \cos t / \sin t \qquad \csc t = 1 / \sin t$$

Notice that when t is a positive acute angle, the circle definition agrees with the right triangle definitions of the sine and cosine:

$$\sin t = \frac{\text{opposite}}{\text{hypotenuse}} \quad \text{and} \quad \cos t = \frac{\text{adjacent}}{\text{hypotenuse}}.$$

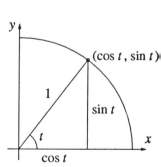

However, the circle definitions of the sine and cosine have the important advantage that they produce functions whose domains are the set of *all* real numbers. (What are the domains of the tangent, secant and cosecant functions?) In calculus, not only are the circle definitions invariably used, but also angles are always measured in radians. To convert between radians and degrees, notice that the circumference of a unit circle is 2π, so the radian measure of a semi-circular arc is half of this, and thus we have

$$\pi \text{ radians } = 180 \text{ degrees.}$$

Graphing packages "know" the trigonometric functions in exactly this form: circular functions with the input variable given in radians. The following exercises let you review the trigonometric functions and explore some of the possibilities using computer graphing.

10.   Graph the function $f(x) = \sin(x)$ on the interval $-2 \leq x \leq 10$.

a) What are the x-intercepts of $\sin(x)$ on the interval $-2 \leq x \leq 10$? Determine them to two decimal places accuracy.

b) What is the largest value of $f(x)$ on the interval $-2 \leq x \leq 10$? Which value of x makes $f(x)$ largest? Determine x to two decimal places accuracy.

[Answer: $x = 1.57....$]

c) Regraph $f(x)$ on the very small interval $-.001 \leq x \leq .001$. Describe what you see. Can you determine the slope of this graph?

11.   Graph the function $f(x) = \cos(x)$ on the interval $0 \leq x \leq 14$. On the same screen graph the *second* function $g(x) = \cos(2x)$.

a) How far apart are the x-intercepts of $f(x)$? How far apart are the x-intercepts of $g(x)$?

b) The graph of $g(x)$ has a pattern that repeats. How wide is this pattern? The graph of $f(x)$ also has a repeating pattern; how wide is *it*?

c) Compare the graphs of $f(x)$ and $g(x)$ to one another. In particular, can you say that one of them is a stretched or compressed version of the other? Is the compression (or stretching) in the vertical or the horizontal direction?

d) Construct a *new* function $f(x)$ whose graph is exactly the same shape as the graph of $g(x) = \cos(2x)$ (that is, it should have the same proportions), but make the graph of $f(x)$ twice as tall as the graph of $g(x)$. [A suggestion: either deduce what $f(x)$ should be, or make a guess. Then test your choice on the computer. If your choice doesn't work, think how you might modify it, and then test your modifications the same way.]

12.   The aim here is to find a solution to the equation $\sin x = \cos(3x)$. There is no purely *algebraic* procedure to solve this equation. Because the sine and cosine are not defined by *algebraic* formulas, this should not be particularly surprising. (Even for algebraic equations, there are only a few very special cases for which there are formulas like the quadratic formula. In chapter 5 we will look at a method for solving equations when formulas can't help us.)

a) Graph the two functions $f(x) = \sin(x)$ and $g(x) = \cos(3x)$ on the interval $0 \le x \le 1$.

b) Find a solution of the equation $\sin(x) = \cos(3x)$ that is accurate to six decimal places.

[Answer: $x = .392699\ldots$.]

c) Find *another* solution of the equation $\sin(x) = \cos(3x)$, accurate to four decimal places. Explain how you found it.

13.   Use a graphing program to make a sketch of the graph of each of the following functions. In each case, make clear the domain and the range of the function, where the graph crosses the axes, and where the function has a maximum or a minimum.

$$a)\ F(w) = (w - 1)(w - 2)(w - 3) \qquad b)\ Q(a) = \frac{1}{a^2 + 5}$$

$$c)\ E(x) = x + \frac{1}{x} \qquad\qquad d)\ e(x) = x - \frac{1}{x}$$

$$e)\ g(u) = \sqrt{\frac{u - 1}{u + 1}} \qquad\qquad f)\ M(u) = \frac{u^2 - 2}{u^2 + 2}$$

14.   Graph on the same screen the following three functions:

$$f(x) = 2^x \qquad g(x) = 3^x \qquad h(x) = 10^x.$$

Use the interval $-2 \le x \le 1.2$.

a) Which function has the largest value when $x = -2$?

b) Which is climbing most rapidly when $x = 0$?

c) Magnify the picture at $x = 0$ by resetting the size of the interval to $-.0001 \le x \le .0001$. Describe what you see. Estimate the slopes of the three graphs.

Proportions, Linear Functions, and Models

15.   In Massachusetts there is a sales tax of 5%. The tax T, in dollars, is proportional to the price P of an object, also in dollars. The constant of proportionality is $k = 5\% = .05$. Write a formula that expresses the sales tax as a linear function of the price, and use your formula to compute the tax on a television set that costs \$289.00 and a toaster that costs \$37.50.

16.   Suppose $W = 213 - 17\,Z$. How does W change when Z changes from 3 to 7; from 3 to 3.4; from 3 to 3.02? Let ΔZ denote a change in Z and ΔW the change thereby produced in W. Is $\Delta W = m\,\Delta Z$ for some constant m? If so, what is m?

17.   a) In the following table, q is a linear function of p. Fill in the blanks in the table.

p	-3	0			7	13		π
q	7		4	1		0		

b) Find a formula to express Δq as a function of Δp, and another to express q as a function of p.

18.   **Thermometers**. There are two common scales in use to measure the temperature, called **Fahrenheit degrees** and the **Celsius degrees**. Let F and C, respectively, be the temperature on each of these scales. Each of these quantities is a linear function of the other; the relation between them in determined by the following table:

physical measurement	C	F
freezing point of water	0	32
boiling point of water	100	212

a) Which represents a larger change in temperature, a Celsius degree or a Fahrenheit degree?

b) How many Fahrenheit degrees does it take to make the temperature go up one Celsius degree? How many Celsius degrees does it take to make it go up one Fahrenheit degree?

c) What is the multiplier m in the equation $\Delta F = m \cdot \Delta C$? What is the multiplier μ in the equation $\Delta C = \mu \cdot \Delta F$? (The symbol μ is the Greek letter mu.) What is the relation between μ and m?

[Answer: $m = 9/5$; $\mu = 1/m$.]

d) Express F as a linear function of C. Graph this function. Put scales and labels on the axes. Indicate clearly the slope of the graph and its vertical intercept.

e) Express C as a linear function of F and graph this function. How are the graphs in parts (d) and (e) related? Give a clear and detailed explanation.

f) Is there any temperature that has the same reading on the two temperature scales? What is it? Does the temperature of the air ever reach this value? Where?

19. **The Greenhouse Effect.** The concentration of carbon dioxide (CO_2) in the atmosphere is increasing. The concentration is measured in parts per million (PPM). Records kept at the South Pole show an increase of .8 PPM per year during the 1960s.

a) At that rate, how many years does it take for the concentration to increase by 5 PPM; by 15 PPM?

b) At the beginning of 1960 the concentration was about 316 PPM. What would it be at the beginning of 1970; at the beginning of 1980?

c) Draw a graph that shows CO_2 concentration as a function of the time since 1960. Put scales on the axes and label everything clearly.

d) The *actual* CO_2 concentration at the South Pole was 324 PPM at the beginning of 1970 and 338 PPM at the beginning of 1980. Plot these values on your graph, and compare them to your calculated values.

[Answer: The two values for 1970 agree, but the actual value is 6 PPM larger than the calculated value in 1980.]

e) Using the actual concentrations in 1970 and 1980, calculate a new rate of increase in concentration. Using that rate, estimate what the increase in CO_2 concentration was between 1970 and 1990. Estimate the CO_2 concentration at the beginning of 1990.

f) Using the rate of .8 PPM per year that held during the 1960s, determine how many years before 1960 there would have been *no* carbon dioxide at all in the atmosphere.

20. **Thermal Expansion.** Measurements show that the length of a metal bar increases in proportion to the increase in temperature. An aluminum bar that is exactly 100 inches long when the temperature is 40°F becomes 100.0052 inches long when the temperature increases to 80°F.

a) How long is the bar when the temperature is 60°F? 100°F?

b) What is the multiplier that connects an increase in length ΔL to an increase in temperature ΔT?

[Answer: The constant is $k = .00013$ inches per °F.]

c) Express ΔL as a linear function of ΔT.

d) How long will the bar be when $T = 0$°F?

e) Express L as a linear function of T.

f) What temperature change would make $L = 100.01$ inches?

g) For a *steel* bar that is also 100 inches long when the temperature is 40° F, the relation between ΔL and ΔT is $\Delta L = .00067\,\Delta T$. Which expands more when the temperature is increased; aluminum or steel?

h) How long will this steel bar be when $T = 80$°F?

21. **Falling Bodies.** In the simplest model of the motion of a falling body, the velocity increases in proportion to the increase in the time that the body has been falling. If the velocity is given in feet per second, measurements show the constant of proportionality is approximately 32.

a) A ball is falling at a velocity of 40 feet/sec after 1 second. How fast is it falling after 3 seconds?

b) Express the change in the ball's velocity Δv as a linear function of the change in time Δt.

c) Express v as a linear function of t.

The model can be expanded to keep track of the *distance* that the body has fallen. If the distance d is measured in feet, the units of d' are feet per second; in fact, $d' = v$. So the model describing the motion of the body is given by the rate equations

$$d' = v \quad \text{feet per second;}$$
$$v' = 32 \quad \text{feet per second per second.}$$

d) At what rate is the distance increasing after 1 second?   After 2 seconds? After 3 seconds?

e) Is d a linear function of t? Explain your answer.

In many cases, the rate of change of a variable quantity is proportional to the quantity itself. Consider a human population as an example. If a city of 100,000 is increasing at the rate of 1500 persons per year, we would expect a similar city of 200,000 to be increasing at the rate of 3000 persons per year. That is, if P is the population at time t, then the **net growth rate** P' is proportional to P:

$$P' = k\,P.$$

In the case of the two cities, we have

$$P' = 1500 = k\,P = k \times 100000 \quad \text{so} \quad k = \frac{1500}{100000} = .015.$$

22.   In the equation $P' = k\,P$, above, explain why the units for k are

$$\frac{\text{persons per year}}{\text{person}}.$$

The number k is called the **per capita growth rate**. ("Per capita" means "per person"—"per *head*", literally.)

23.   **Poland and Afghanistan**. In 1985 the per capita growth rate in Poland was 9 persons per year per thousand persons. (That is, $k = 9/1000 = .009$.) In Afghanistan it was 21.6 persons per year per thousand.

a) Let P denote the population of Poland and A the population of Afghanistan. Write the equations that govern the growth rates of these populations.

b) In 1985 the population of Poland was estimated to be 37.5 million persons, that of Afghanistan 15 million. What are the net growth rates P' and A' (as distinct from the *per capita* growth rates)? Comment on the following assertion: When comparing two countries, the one with the larger per capita growth rate will have the larger net growth rate.

c) What was the average time interval between births in Poland in 1985? What was the time interval in Afghanistan?

24.   a) **Bacterial Growth.** A colony of bacteria on a culture medium grows at a rate proportional to the present size of the colony. When the colony weighed 32 grams it was growing at the rate of 0.79 grams per hour. Write an equation that links the growth rate to the size of the population.

[Answer: If P is the size of the colony, in grams, and P' is its growth rate in grams per hour, then $P' = .0247\,P$ grams per hour.]

b) What is ΔP if $\Delta t = 1$ minute? Estimate how long it would take to make $\Delta P = .5$ grams.

25.   **Radioactivity.** In radioactive decay, radium slowly changes into lead. If one sample of radium is twice the size of a second lump, then the larger sample will produce twice as much lead as the second in any given time. In other words, the rate of decay is proportional to the amount of radium present. Measurements show that 1 gram of radium decays at the rate of 1/2337 grams per year. Write an equation that links the decay rate to the size of the radium sample. How does your equation indicate that the process involves *decay* rather than *growth*?

26.   **Cooling.** Suppose a cup of hot coffee is brought into a room at 70°F. It will cool off, and it will cool off *faster* when the temperature difference between the coffee and the room is greater. The simplest assumption we can make is that the rate of cooling is proportional to this temperature difference (this is called Newton's law of cooling). Let C denote the temperature of the coffee, in °F, and C' the rate at which it is cooling, in °F per minute. The new element here is that C' is proportional, not to C, but to the *difference* between C and the room temperature of 70°F.

a) Write an equation that relates C' and C. It will contain a proportionality constant k. How did you indicate that the coffee is *cooling* and not *heating up*?

b) When the coffee is at 180°F it is cooling at the rate of 9°F per minute. What is k?

c) At what rate is the coffee cooling when its temperature is 120°F?

d) Estimate how long it takes the temperature to fall from 180°F to 120°F. Then make a better estimate, and explain why it is better.

§3. Using a Program

Computers

A computer changes the way we can use calculus as a tool, and it vastly enlarges the range of questions that we can tackle. No longer need we back away from a problem that involves a lot of computations. There are two aspects to the power of a computer. First, it is fast. It can do a million additions in the time it takes us to do one. Second, it can be programmed. By arranging computations into a loop—as we did on page 17—we can construct a program with only a few instructions that will carry out millions of repetitive calculations.

The purpose of this section is to give you practice using a computer program that estimates values of S, I, and R in the epidemic model. As you will see, it carries out the three rounds of calculations you have already done by hand. It also contains a loop that will allow you to do a hundred, or a million, rounds of calculations with no extra effort.

The Program SIR

The program on the following page calculates values of S, I, and R. It is a set of instructions—sometimes called **code**—that is designed to be read by you and by a computer. These instructions mirror the operations we performed by hand to generate the table on page 13. The code here is written as it would be in most programming languages. The line numbers, however, are not part of the program; they are there to help us refer to the lines. A computer reads the code one line at a time, starting at the top. Each line is a complete instruction which causes the computer to do something. The purpose of nearly every instruction in this program is to assign a numerical value to a symbol. Watch for this as we go down the lines of code.

Read a program line by line from the top

The first line, `t = 0`, is the instruction "Give t the value 0." The next four lines are similar. Notice, in the fifth line, how Δt is typed out as `deltat`. It is a common practice for the name of a variable to be several letters long. A few lines later S' is typed out as `Sprime`, for instance. The instruction on the sixth line is the first that does not assign a value to a symbol. Instead, it causes the computer to print the following on the computer monitor screen:

Lines 1–5

Line 6

$$0 \qquad 45400 \qquad 2100 \qquad 2500$$

Program: SIR

```
1     t = 0
2     S = 45400
3     I = 2100
4     R = 2500
5     deltat = 1
6     PRINT t, S, I, R
7     FOR k = 1 TO 3
8             Sprime = -.00001 * S * I
9             Iprime = .00001 * S * I - I / 14      } Step I
10            Rprime = I / 14
11            deltaS = Sprime * deltat
12            deltaI = Iprime * deltat
13            deltaR = Rprime * deltat
14            t = t + deltat
15            S = S + deltaS
16            I = I + deltaI
17            R = R + deltaR
18            PRINT t, S, I, R
19    NEXT k
```

Line 7   Skip over the line that says FOR k = 1 TO 3. It will be easier to understand after we've read the rest of the program.

Lines 8–10   Look at the first three indented lines. You should recognize them as coded versions of the rate equations

$$
\begin{aligned}
S' &= -.00001\,SI \\
I' &= .00001\,SI - I/14 \\
R' &= I/14
\end{aligned}
$$

for the measles epidemic. (The program uses * to denote multiplication.) They are instructions to assign numerical values to the symbols S', I', and R'. For instance, Sprime = -.00001 * S * I (line 8) says

Give S' the value $-.00001\,SI$;
use the current values of S and I to get $-.00001\,SI$.

Now the computer knows that the current values of S and I are 45400 and 2100, respectively. (Can you see why?) So it calculates the product

$-.00001 \times 45400 \times 2100 = -953.4$ and then gives S' the value -953.4. There is an extra step to calculate the product.

Lines 11–13

Notice that the first three indented lines are bracketed together and labelled "Step I," because they carry out Step I in the flow chart. The next three indented lines carry out Step II in the flow chart. They assign values to three more symbols—namely ΔS, ΔI, and ΔR—using the current values of S', I', R' and Δt.

Lines 14–17

The next four indented lines present a puzzle. They don't make sense if we read them as ordinary mathematics. In an expression like `t = t + deltat`, we would cancel the t's and conclude `deltat = 0`. The lines *do* make sense when we read them as computer instructions, however. As a computer instruction, `t = t + deltat` says

Make the new value of t equal to the current value of $t + \Delta t$.

Once again we have an instruction that assigns a numerical value to a symbol, but this time the symbol (t, in this case) already has a value before the instruction is carried out. The instruction gives it a *new* value. (Here the value of t is changed from 0 to 1.) Likewise, the instruction `S = S + deltaS` gives S a new value. What was the old value, and what is the new?

How a program computes new values

Compare the three lines of code that produce new values of S, I, and R with the original equations that we used to define Step III back on page 16:

`S = S + deltaS`	new S = current $S + \Delta S$
`I = I + deltaI`	new I = current $I + \Delta I$
`R = R + deltaR`	new R = current $R + \Delta R$

The words "new" and "current" aren't needed in the computer code because they are automatically understood to be there. Why? First of all, a symbol (like `S`) always has a *current* value, but an instruction can give it a *new* value. Second, a computer instruction of the form `A = B` is always understood to mean "new `A` = current `B`."

Notice that the instructions `A = B` and `B = A` mean different things. The second says "new B = current A." Thus, in `A = B`, A is altered to equal B, while in `B = A`, B is altered to to equal A. To emphasize that the symbol on the left is always the one affected, some programming languages use a modified equal sign, as in `A := B`. We sometimes read this as "A gets B".

The next line is another PRINT statement, exactly like the one on line 6. It causes the current values of t, S, I, and R to be printed on the computer monitor screen. But this time what appears is

$$1 \qquad 44446.6 \qquad 2903.4 \qquad 2650$$

The values were changed by the previous four instructions. It is important to remember that the computer carries out instructions in the order they are written. Had the second PRINT statement appeared right after line 13, say, the old values of t, S, I, and R would have appeared on the monitor screen a second time.

We will take the last line and line 7 together. They are the instructions for the loop. Consider the situation when we reach the last line. The variables t, S, I, and R now have their "day 1" values. To continue, we need an instruction that will get us back to line 8, because the instructions on lines 8–17 will convert the current (day 1) values of t, S, I, and R into their "day 2" values. That's what lines 7 and 19 do.

Here is the meaning of the instruction FOR k = 1 TO 3 on line 7:

Give k the value 1, and be prepared later to
give it the value 2 and then the value 3.

The variable k plays the role of a **counter**, telling us how many times we have gone around the loop. Notice that k did not appear in our hand calculations. However, when we said we had done three rounds of calculations, for example, we were really saying $k = 3$.

After the computer reads and executes line 7, it carries out all the instructions from lines 8 to 18, arriving finally at the last line. The computer then interprets the instruction NEXT k on the last line as follows:

Give k the next value that the FOR command allows, and
move back to the line immediately after the FOR command.

After the computer carries out this instruction, k has the value 2 and the computer is set to carry out the instruction on line 8. It then executes that instruction, and continues down the program, line by line, until it reaches line 19 once again. This sets the value of k to 3 and moves the computer back to line 8. Once again it continues down the program to line 19. This time there is no allowable value that k can be given, so the program stops.

The NEXT k command is different from all the others in the program. It is the only one that directs the computer to go to a different line. That action causes the program to **loop**. Because the loop involves all the indented instructions between the FOR statement and the NEXT statement, it is called a **FOR–NEXT loop**. This is just one kind of loop. Computer programs can contain other sorts that carry out different tasks. In the next chapter we will see how a DO–WHILE loop is used.

Exercises

The program SIR

The object of these exercises is to verify that the program SIR works the way the text says it does. Follow the instructions for running a program on the computer you are using.

1.   Run the program to confirm that it reproduces what you have already calculated by hand (table, page 13).

2.   On a copy of the program, mark the instructions that carry out the following tasks:

a)   give the input values of S, I, and R;

b)   say that the calculations take us 1 day into the future;

c)   carry out step II (see page 16);

d)   carry out step III;

e)   give us the output values of S, I, and R;

f)   take us once around the whole loop;

g)   say how many times we go around the loop.

3.   Delete all the lines of the program from line 7 onward (or else type in the first 6 lines). Will this program run? What will it do? Run it and report what you see. Is this what you expected?

4.   Starting with the original SIR program on page 50, delete lines 7 and 19. These are the ones that declare the FOR–NEXT loop. Will this program run? What will it do? Run it and report what you see. Is this what you expected?

5.   Using the 17-line program you constructed in the previous question, remove the PRINT statement from the last line and insert it between what appear as lines 13 and 14 on page 50. Will this program run? What will it do? Run it and report what you see. Is this what you expected?

6.   Starting with the original SIR program on page 50, change line 7 so it reads FOR k = 26 TO 28. Thus, the counter k takes the values 26, 27, and 28. Will this program run? What will it do? Run it and report what you see. Is this what you expected?

Programs to practice on

In this section there are a number of short programs for you to analyze and run.

Program 1	Program 2	Program 3
A = 2	A = 2	A = 2
B = 3	B = 3	B = 3
A = B	B = A	A = A + B
PRINT A, B	PRINT A, B	B = A + B
		PRINT A, B

7.   When Program 1 runs it will print the values of A and B that are current when the program stops. What values will it print? Type in this program and run it to verify your answers.

8.   What will Program 2 do when it runs? Type in this program and run it to verify your answers.

9.   After each line in Program 3 write the values that A and B have *after* that line has been carried out. What values of A and B will it print? Type in this program and run it to verify your answers.

The next three programs have an element not found in the program SIR. In each of them, there is a FOR–NEXT loop, and the counter k actually appears in the statements within the loop.

Program 4	Program 5	Program 6

```
FOR k = 1 TO 5     FOR k = 1 TO 5     x = 0
    A = k ^ 3          A = k ^ 3      FOR k = 1 TO 5
    PRINT A        NEXT k                 x = x + k
NEXT k             PRINT A                PRINT k, x
                                      NEXT k
```

10.   What output does Program 4 produce? Type in the code and run the program to confirm your answer.

11.   What is the difference between the code in Program 5 and the code in Program 4? What is the output of Program 5? Does it differ from the output of Program 4? If so, why?

12.   What output does Program 6 produce? Type in the code and run the program to confirm your answer.

[Answer: The output is

1	1
2	3
3	6
4	10
5	15

Notice that each x value is the sum of the first k whole numbers.]

Program 7	Program 8	Program 9

```
A = 0              A = 0              A = 0
B = 0              B = 1              B = 1
FOR k = 1 TO 5     FOR k = 1 TO 5     FOR k = 1 TO 5
    A = A + 1          A = A + B          A = A + B
    B = A + B          B = A + B          B = A + B
    PRINT A, B         PRINT A, B     NEXT k
NEXT k             NEXT k             PRINT A, B
```

13.   Program 7 prints five lines of output. What are they? Type in the program and run it to confirm your answers.

14.   What is the output of Program 8? Type in the program and run it to confirm your answers.

15.   Describe exactly how the *codes* for Programs 8 and 9 differ. How do the *outputs* differ?

Analyzing the measles epidemic

16.   Alter the program SIR to have it calculate estimates for S, I, and R over the first *six* days. Construct a table that shows those values.

17.   Alter the program to have it estimate the values of S, I, and R for the first *thirty* days.
a) What are the values of S, I, and R when $t = 30$?
[Answer: To the nearest tenth, $S = 133.1$, $I = 10329.7$, $R = 39537.2$.]
b) According to these figures, on what day does the infection peak?
[Answer: I is largest when $t = 13$.]

18.   By adding an appropriate PRINT statement after line 10 you can also get the program to print values for S', I', and R'. Do this, and check that you get the values shown in the table on page 13.

19.   According to these estimates, on what day do the largest number new infections occur? How many are there? Explain where you got your information.
[Answer: Our estimate for the largest number of new infections in one day is 4609.1; it occurs when $t = 8$. This is found by checking the rate of new infections, which is $-S'$.]

20.   On what day do you estimate that the largest number of recoveries occurs? Do you see a connection between this question and 17 b?

21.   On what day do you estimate the infected population grows most rapidly? Declines most rapidly? What value does I' have on those days?
[Answer: I is growing most rapidly when $t = 7$ and declining most rapidly when $t = 18$.]

22.   a) Alter the original SIR program so that it will go *backward* in time, with time steps of 1 day. Specify the changes you made in the

program. Use this altered program to obtain estimates for the values of S, I, and R yesterday. Compare your estimates with those in the text (page 13).

b) Estimate the values of S, I, and R *three* days before today.

23. According to the *S-I-R* model, when did the infection begin? That is, how many days before today was the estimated value of I approximately 1?

24. **There and back again.** Use the SIR program, modified as necessary, to carry out the calculations described in exercise 18 on page 23. Do your computer results agree with those you obtained earlier?

§4. Chapter Summary

The Main Ideas

- Natural processes like the spread of disease can often be described by **mathematical models**. Initially, this involves identifying **numerical quantities** and relations between them.

- A relation between quantities often takes the form of a **function**. A function can be described in many different ways; **graphs**, **tables**, and **formulas** are among the most common.

- **Linear functions** make up a special but important class. If y is a linear function of x, then $\boldsymbol{\Delta y = m \cdot \Delta x}$, for some constant m. The constant m is a **multiplier**, **slope**, and **rate of change**.

- If $y = f(x)$, then we can consider the **rate of change $\boldsymbol{y'}$** of y with respect to x. A mathematical model whose variables are connected by **rate equations** can be analyzed to predict how those variables will change.

- Predicted changes are **estimates** of the form $\boldsymbol{\Delta y = y' \cdot \Delta x}$.

- The computations that produce estimates from rate equations can be put into a **loop**, and they are readily carried out on a **computer**.

- A computer increases the **scope** and **complexity** of the problems we can consider.

Self-Testing

- You should be able to work with functions given in various forms, to find the output for any given input.

- You should be able to read a graph. You should also be able to construct the graph of a linear function directly, and the graph of a more complicated function using a computer graphing package.

- You should be able to determine the natural domain of a function given by a formula.

- You should be able to express proportional quantities by a linear function, and interpret the constant of proportionality as a multiplier.

- Given any two of these quantities for a linear function—multiplier, change in input, change in output—you should be able to determine the third.

- You should be able to model a situation in which one variable is proportional to its rate of change.

- Given the value of a quantity that depends on time and given its rate of change, you should be able to estimate values of the quantity at other times.

- For a set of quantities determined by rate equations and initial conditions, you should be able to estimate how the quantities change.

- Given a set of rate equations, you should be able to determine what happens when one of the quantities reaches a maximum or minimum, or remains unchanged over time.

- You should be able to understand how a computer program with a FOR–NEXT loop works.

Chapter Exercises

A Model of an Orchard

If an apple orchard occupies one acre of land, how many trees should it contain so as to produce the largest apple crop? This is an example of an **optimization** problem. The word *optimum* means "best possible"— especially, the best under a given set of conditions. These exercises seek an optimum by analyzing a simple mathematical model of the orchard. The model is the function that describes how the total yield depends on the number of trees.

An immediate impulse is just to plant a lot of trees, on the principle: more trees, more apples. But there is a catch: if there are too many trees in a single acre, they crowd together. Each tree then gets less sunlight and nutrients, so it produces fewer apples. For example, the relation between the *yield per tree*, Y, and the *number of trees*, N, may be like that shown in the graph drawn on the left, below.

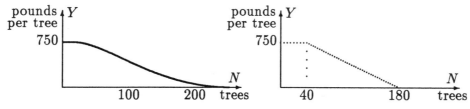

When there are only a few trees, they don't get each other's way, and they produce at the maximum level—say, 750 pounds per tree. Hence the graph starts off level. At some point, the trees become too crowded to produce *anything*! In between, the yield per tree drops off as shown by the curved middle part of the graph.

We want to choose N so that the *total yield*, T, will be as large as possible. We have $T(N) = Y(N) \cdot N$, but since we don't know $Y(N)$ very precisely, it is difficult to analyze $T(N)$. To help, let's replace $Y(N)$ by the approximation shown in the graph on the right. Now carry out an analysis using this graph to represent $Y(N)$.

1.   Find a formula for the straight segment of the new graph of $Y(N)$ on the interval $40 \leq N \leq 180$. What is the formula for $T(N)$ on the same interval?

2.   What are the formulas for $T(N)$ when $0 \leq N \leq 40$ and when $180 \leq N$? Graph T as a function of N. Describe the graph in words.

3.   What is the maximum possible total yield T? For which N is this maximum attained?

[Answer: The maximum yield is $75 \times 90^2/14$ pounds of apples, when there are 90 trees on the acre.]

4.   Suppose the endpoints of the sloping segment were P and Q, instead of 40 and 180, respectively. Now what is the formula for $T(N)$? (Note that P and Q are *parameters* here. Different values of P and Q will give different models for the behavior of the total output.) How many trees would then produce the maximum total output? Expect the maximum to depend on the parameters P and Q.

Rate Equations

5.   **Radioactivity.** From exercise 25 of §2 we know a sample of R grams of radium decays into lead at the rate

$$R' = \frac{-1}{2337} R \quad \text{grams per year.}$$

Using a step size of 10 years, estimate how much radium remains in a 0.072 gram sample after 40 years.   [Answr: $R \approx 0.0708$ grams.]

6.   **Poland and Afghanistan.** If P and A denote the populations of Poland and Afghanistan, respectively, then their net per capita growth rates imply the following equations:

$$P' = .009\,P \quad \text{persons per year;}$$
$$A' = .0216\,A \quad \text{persons per year.}$$

(See exercise 23 of §2.) In 1985, $P = 37.5$ million, $A = 15$ million. Using a step size of 1 year, estimate P and A in 1990.

7.   **Falling bodies.** If d and v denote the distance fallen (in feet) and the velocity (in feet per second) of a falling body, then the motion can be described by the following equations:

$$d' = v \quad \text{feet per second;}$$
$$v' = 32 \quad \text{feet per second per second.}$$

(See exercise 21 of §2.) Assume that when $t = 0$, $d = 0$ feet and $v = 10$ feet/sec. Using a step size of 1 second, estimate d and v after 3 seconds have passed.

Chapter 2

Successive Approximations

In this chapter we continue exploring the mathematical implications of the S-I-R model. In the last chapter we calculated future values of S, I, and R by assuming that the rates S', I', and R' stayed fixed for a whole day. Since the rates are *not* fixed—they change with S, I, and R—the values of S, I, and R we obtained have to be considered as estimates only. In this chapter we will see how to build a succession of better and better estimates that get us as close as we wish to the true values implied by the model.

This method of **successive approximation** is a basic tool of calculus. It is the one fundamentally new process you will encounter, the ingredient that sets calculus apart from the mathematics you have already studied. With it you will be able to solve a vast array of problems that other methods can't handle.

§1. Making Approximations

In chapter 1 we looked at the specific S-I-R model:

$$
\begin{aligned}
S' &= -.00001\, SI \\
I' &= .00001\, SI - I/14 \\
R' &= I/14
\end{aligned}
$$

with initial values at time $t = 0$:

$$
S = 45400 \quad I = 2100 \quad R = 2500\,.
$$

Rate equations
tell us where to
go next
We originally developed this model as a *description* of the relations among the different components of an epidemic. Almost immediately, though, we began using the rate equations in the model as a *recipe* for predicting what happens over the course of the epidemic: If we know at some time t the values of $S(t), I(t),$ and $R(t)$), then the equations tell us how to estimate values of the functions at other times. We used this approach in the last chapter to move backwards and forwards in time, calculating the values of S, I, and R as we went.

While we got numbers, there were some questions about how accurate these numbers were—that is, how exactly they represented the values implied by the model. In the process we called "there and back again" we used current values of S, I, and R to find the rates, used these rates to go forward one day, recalculate the rates, and come back to the present—and we got different values from the ones we started with! Resolving this discrepancy will be an important feature of the technique developed in this section.

The Longest March Begins with a Single Step

So far, in generating numbers from the S-I-R rate equations, we have assumed that the rates remained constant over an entire day, or longer. Since the rates aren't constant—they depend on the values of S, I, and R, which are always changing—the values we calculated for the variables at times other than the given initial time are, at best, estimates. These estimates, while incorrect, are not useless. Let's see how they behave in the "there and back again" process of chapter 1 as we recalculate the rates more and more frequently, producing a sequence of approximations to the values we are looking for.

There and Back Again Again

On page 14 in chapter 1 we used the rate equations to go forward a day and come back again. We started with the initial values

$$S(0) = 45400 \quad I(0) = 2100 \quad R(0) = 2500 \,,$$

calculated the rates, went forward a day to $t = 1$, recalculated the rates, and came back a day to $t = 0$. We ended up with the estimates

$$S(0) = 45737.1 \quad I(0) = 1820.3 \quad R(0) = 2442.6$$

—which are rather far from the values of $S(0)$, $I(0)$, and $R(0)$ we started with.

A clue to the resolution of this discrepancy appeared in problem 18, page 23. There you were asked to go forward two days and come back again in two different ways, using $\Delta t = 2$ (a total of 2 steps) in the first case and $\Delta t = 1$ (a total of 4 steps) in the second. Here are the resulting values calculated for $S(0)$ in each case and the discrepancy between this value and the original value $S(0) = 45400$:

step size	new $S(0)$	discrepancy
$\Delta t = 2$	46717.6	2217.6
$\Delta t = 1$	46021.3	1521.3

While the discrepancy is fairly large in either case, $\Delta t = 1$ clearly does better than $\Delta t = 2$. But if smaller is better, why stop at $\Delta t = 1$? What happens if we take even smaller time steps, get the corresponding new values of S, I, and R, and use these values to recalculate the rates each time?

Smaller steps generate a smaller discrepancy

Recall that the rate S' (or I' or R') is simply the multiplier which gives ΔS—the (estimated) change in S—for a given change Δt in t

$$\Delta S = S' \cdot \Delta t \,.$$

This relation holds for any value of Δt, integer or not. Once we have this value for ΔS, we can then calculate

$$\text{new (estimated) } S = \text{current (estimated) } S + \Delta S$$

in the usual way. Note that we have written "(estimated)" throughout to emphasize the fact that if S' is not constant over the entire time Δt, then the value we get for ΔS will typically be only an approximation to the real change in S.

Let's try going forward one day and coming back, using different values for Δt. As we reduce Δt the number of calculations will increase. The program SIR we used in the last chapter can still be used to do the tedious calculations. Thus if we decide to use 10 steps of size $\Delta t = .1$, we would just change two lines in that program:

```
deltat = .1
FOR k = 1 TO 10
```

If we now run SIR with these modifications we can verify the following sequence of values (The values have been rounded off, and the PRINT statement has been modified to show the new values of the rates at each step as well):

Estimated values of $S, I,$ and R
for step sizes $\Delta t = .1$

t	$S(t)$	$I(t)$	$R(t)$	$S'(t)$	$I'(t)$	$R'(t)$
0.0	45 400.0	2100.0	2500.0	−953.4	803.4	150.0
0.1	45 304.7	2180.3	2515.0	−987.8	832.1	155.7
0.2	45 205.9	2263.6	2530.6	−1023.3	861.6	161.7
0.3	45 103.6	2349.7	2546.7	−1059.8	892.0	167.8
⋮	⋮	⋮	⋮	⋮	⋮	⋮
1.0	44 278.7	3042.9	2678.4	−1347.7	1130.0	217.4

Having arrived at $t = 1$, we can now use SIR to turn around and go back to $t = 0$. Here's how:

- Change the initial line of the program to t = 1 to reflect our new starting time.

- Change the next three lines to use the values we just calculated for $S(1)$, $I(1)$, and $R(1)$ as our *starting* values in SIR.

The *sign* of deltat determines whether we move forward or backward in time

- Change the value of deltat to be -.1 (so each time the program executes the command t = t + deltat it reduces the value of t by .1).

With these changes SIR will yield the desired estimates for $S(0)$, $I(0)$, and $R(0)$, and we get

$$S(0) = 45433.5 \quad I(0) = 2072.3 \quad R(0) = 2494.3\,.$$

This is clearly a considerable improvement over the values obtained with $\Delta t = 1$.

With this promising result, the obvious thing to do is to try even smaller values of Δt, perhaps $\Delta t = .01$. We could continue using SIR, making the needed modifications each time. Instead, though, let's rewrite SIR slightly to make it better suited to our current needs. Look at the program SIRVALUE below and compare it with SIR.

Program: SIRVALUE

Program: SIR

```
tinitial = 0                                    t = 0
tfinal = 1                                      S = 45400
t = tinitial                                    I = 2100
S = 45400                                       R = 2500
I = 2100                                         deltat = .1
R = 2500                                         FOR k = 1 TO 10
numberofsteps = 10                                  Sprime = -.00001 * S * I
deltat = (tfinal - tinitial)/numberofsteps          Iprime = .00001 * S * I - I / 14
FOR k = 1 TO numberofsteps                          Rprime = I / 14
    Sprime = -.00001 * S * I                        deltaS = Sprime * deltat
    Iprime = .00001 * S * I - I / 14                deltaI = Iprime * deltat
    Rprime = I / 14                                 deltaR = Rprime * deltat
    deltaS = Sprime * deltat                        t = t + deltat
    deltaI = Iprime * deltat                        S = S + deltaS
    deltaR = Rprime * deltat                        I = I + deltaI
    t = t + deltat                                  R = R + deltaR
    S = S + deltaS                                  PRINT t, S, I, R
    I = I + deltaI                              NEXT k
    R = R + deltaR
NEXT k
PRINT t, S, I, R
```

You will see that the major change is to place the PRINT statement outside the loop, so only the final values of S, I, and R get printed. This speeds up the work, since otherwise, with $\Delta t = .001$, for instance, we would be asking the computer to print out 1000 lines—about 30 screens of text! Another change is that the value of deltat no longer needs to be specified—it is automatically determined by the values of tinitial, tfinal, and numberofsteps.

As written above, SIR and SIRVALUE both run for 10 steps of size 0.1. By changing the value of numberofsteps in the program we can quickly get estimates for $S(1)$, $I(1)$, and $R(1)$ for a wide range of values for Δt. Moreover, once we have these estimates we can use SIRVALUE again to go backwards in time to $t = 0$, by making changes similar to those we made in SIR earlier. First, we need to change the value of tinitial to 1 and the value of tfinal to 0. Notice that this automatically will make deltat a negative quantity, so that each time we run through the loop we step back in time. Second, we need to set the starting values of S, I, and R to the values we just obtained for

With a computer we can generate lots of data and look for patterns

$S(1)$, $I(1)$, and $R(1)$. With these changes, SIRVALUE will give us the corresponding estimated values for $S(0)$, $I(0)$, and $R(0)$.

If we use SIRVALUE with Δt ranging from 1 to .00001 (which means letting **numberofsteps** range from 1 to 100,000) we get the table below. This table lists the computed values of $S(1)$, $I(1)$, and $R(1)$ for each Δt, followed by the estimated value of $S(0)$ obtained by running SIRVALUE backward in time from these new values, and, finally, the discrepancy between this estimated value of $S(0)$ and the original value $S(0) = 45400$.

Estimated values of $S, I,$ and R when $t = 1$,
for step sizes $\Delta t = 10^{-N}$, $N = 0, \ldots 5$,
together with the corresponding backwards estimate for $S(0)$.

Δt	$S(1)$	$I(1)$	$R(1)$	new $S(0)$	discrepancy
1.0	44 446.6	2903.4	2650.0	45 737.0626	337.0626
0.1	44 278.6648	3042.9241	2678.4111	45 433.4741	33.4741
0.01	44 257.8301	3060.1948	2681.9751	45 403.3615	3.3615
0.001	44 255.6960	3061.9633	2682.3406	45 400.3363	.3363
0.0001	44 255.4821	3062.1406	2682.3773	45 400.0336	.0336
0.00001	44 255.4607	3062.1584	2682.3809	45 400.0034	.0034

Smaller steps generate a discrepancy which can be made as small as we like

There are several striking features of this table. The first is that if we go forward one day and come back again, we can get back as close as we want to our initial value of $S(0)$ *provided we recalculate the rates frequently enough*. After 200,000 rounds of calculations ($\Delta t = .00001$) we ended up only .0034 away from our starting value. In fact, there is a clear pattern to the values of the errors as we decrease the step size. In the exercises it is left for you to explore this pattern and show that similar results hold for I and for R.

A second feature is that as we read down the column under $S(1)$, we find each digit **stabilizes**—that is, after changing for a while, it eventually becomes fixed at a particular value. The initial digits 44 are the first to stabilize, and that happens by the time $\Delta t = 0.1$. Then the third digit 2 stabilizes, when $\Delta t = 0.01$. Roughly speaking, one more digit stabilizes at each successive level. The table is revealing to us, digit by digit, the true value of $S(1)$. By the fifth stage we learn that the integer part of $S(1)$ is 44255. By the sixth stage we can say that the true value of $S(1)$ is 44255.4 . . . .

When we write $S(1) = 44255.4\ldots$ we are expressing $S(1)$ to **one decimal place accuracy**. This says, first, that the decimal expansion of $S(1)$ begins with exactly the six digits shown and, second, that there are further digits after the 4 (represented by the three dots "..."). In this case, we can identify further digits simply by continuing the table. Since our step sizes have the form $\Delta t = 10^{-N}$, we just need to increase N. For example, to express $S(1)$ accurately to six decimal places, we need to stabilize the first eleven digits in our estimates of $S(1)$. The table suggests that Δt should probably be about 10^{-10}—i.e., $N = 10$.

Approximations lead to exact values

The true value of $S(1)$ emerges through a process that generates a sequence of successive approximations. We say $S(1) = 44255.4\ldots$ is the **limit** of this sequence as Δt is made smaller and smaller or, equivalently, as N is made larger and larger. We also say that the sequence of successive approximations **converges** to the limit $S(1)$. Here is a mathematical notation that expresses these statements more compactly:

$$S(1) \;=\; \lim_{\Delta t \to 0}\{\text{the estimate of } S(1)\} \qquad \text{or, equivalently,}$$
$$=\; \lim_{N \to \infty}\{\text{the estimate of } S(1) \text{ with } \Delta t = 10^{-N}\}.$$

The symbol ∞ stands for "infinity," and the expression $N \to \infty$ is often pronounced "as N goes to infinity." However, it is often more instructive to say "as N gets larger and larger, without bound."

You should check that similar patterns are occurring in the $I(1)$ and $R(1)$ columns as well.

> The limit concept lies at the heart of calculus. Later on we'll give a precise definition, but you should first see limits at work in a number of contexts and begin to develop some intuitions about what they are. This approach mirrors the historical development of calculus—mathematicians freely used limits for well over a century before a careful, rigorous definition was developed.

One Picture Is Worth a Hundred Tables

As we noted, the program SIRVALUE prints out only the final values of S, I, and R because it would typically take too much space to print out the intermediate values. However, if instead of printing these values we plot them graphically, we can convey all this intermediate information in a compact and comprehensible form.

Suppose, for instance, that we wanted to record the calculations leading up to $S(3)$ by plotting all the points. The graphs at the right plot all the pairs of values (t, S) that are calculated along the way for the cases $\Delta t = 1$ (4 points), $\Delta t = .1$ (31 points), and $\Delta t = .01$ (301 points).

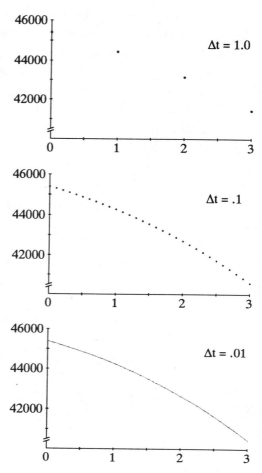

By the time we get to steps of size .01, the resulting graph begins to look like a continuous curve. This suggests that instead of simply plotting the points we might want to draw lines connecting the points as they're calculated.

We can easily modify SIRVALUE to do this—the only changes will be to replace the PRINT command with a command to draw a line and to move this command inside the loop (so that it is executed every time new values are computed). We will also need to add a line or two at the beginning to tell the computer to set up the screen to plot points. This usually involves opening a **window**—i.e., specifying the horizontal and vertical ranges the screen should depict. Since programming languages vary slightly in the way this is done, we use italicized text *"Set up GRAPHICS"* to make clear that this statement is **not** part of the program—you will have to express this in the form your programming language specifies. Similarly, the command

"Plot the line from (t, S)
   to (t + deltat, S + deltaS)"

will have to be stated in the correct format for your language. The computational core of SIRVALUE is unchanged. Here is what the new program looks like if we want to use $\Delta t = .1$ and connect the points with straight lines:

Program: SIRPLOT

```
Set up GRAPHICS
tinitial = 0
tfinal = 3
t = tinitial
S = 45400
I = 2100
R = 2500
numberofsteps = 30
deltat = (tfinal - tinitial)/numberofsteps
FOR k = 1 TO numberofsteps
        Sprime = -.00001 * S * I
        Iprime = .00001 * S * I - I / 14
        Rprime = I / 14
        deltaS = Sprime * deltat
        deltaI = Iprime * deltat
        deltaR = Rprime * deltat
        Plot the line from (t, S)
            to (t + deltat, S + deltaS)
        t = t + deltat
        S = S + deltaS
        I = I + deltaI
        R = R + deltaR
NEXT k
```

If we had wanted just to plot the points, we could have used a command of the form *Plot the point* (t, S) in place of the command to plot the line. We would also need to place that command before the loop so that the initial point corresponding to $t = 0$ gets plotted.

When we "connect the dots" like this we emphasize graphically the underlying assumption we have been making in all our estimates: that the function $S(t)$ is linear (i.e., it is changing at a constant rate) over each interval Δt. Let's see what the graphs look like when we do this for the three values of Δt we used above. To compare the results more readily we'll plot the graphs on the same set of axes. (We will look at a program for doing this in the next section.) We get the following picture:

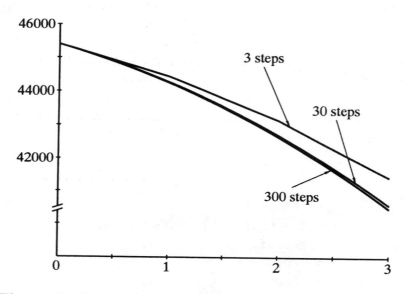

Graphs made up of line segments look like smooth curves if the segments are short enough

The graphs become indistinguishable from each other and increasingly look like smooth curves as the number of segments increases. If we plotted the 3000-step graph as well, it would be indistinguishable from the 300-step graph at this scale. If we now shift our focus from the end value $S(3)$ and look at all the intermediate values as well, we find that each graph gives an approximate value for $S(t)$ for *every* value of t between 0 and 3. We are seeing the entire function $S(t)$ over this interval.

Just as we wrote

$$S(3) = \lim_{N \to \infty} \{\text{the estimate of } S(3) \text{ with } \Delta t = 10^{-N}\}.$$

We can also write

$$\text{graph of } S(t) = \lim_{N \to \infty} \{\text{line-segment approximations with } \Delta t = 10^{-N}\}.$$

The way we see the graph of $S(t)$ emerging from successive approximations is our first example of a fundamental result. It has wide-ranging implications which will occupy much of our attention for the rest of the course.

Piecewise Linear Functions

Let's examine the implications of this approach more closely by considering the "one-step" ($\Delta t = 3$) approximation to $S(t)$ and the "three-step" ($\Delta t = 1$) approximation over the time interval $0 \leq t \leq 3$. In the first case we are making the simplifying assumption that S decreases at the rate $S' = -953.4$ persons per day for the entire three days. In the second case we use three shorter steps of length $\Delta t = 1$, with the slopes of the corresponding segments given by the table on page 13 in Chapter 1, summarized below (note that since $\Delta t = 1$ we have $\Delta S = S'$):

t	S	$\Delta S = S'$
0	45 400.0	-953.4
1	44 446.6	$-1\,290.5$
2	43 156.1	$-1\,720.4$
3	41 435.7	

Here are the corresponding graphs we get:

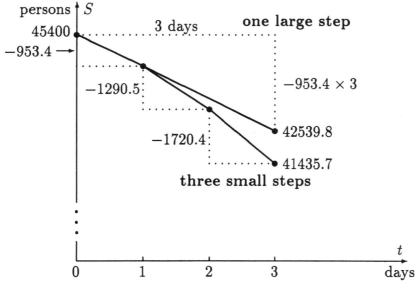

Two approximations to S during the first three days

The "one-step" estimate. Assuming that S decreases at the rate $S' = -953.4$ persons per day for the entire three days is equivalent to assuming that S follows the upper graph—a straight line with slope

−953.4 persons/day. In other words, the one-step approach approximates S by a **linear function** of t. If we use the notation $S_1(t)$ to denote this (one-step) linear approximation we have

One-step estimate: $S(t) \approx S_1(t) = 45400 - 953.4\,t$.

Because the one-step estimate is actually a function we can find the value of $S_1(t)$ for *all* t in the interval $0 \leq t \leq 3$, not just $t = 3$, and thereby get corresponding estimates for $S(t)$ as well.
For example,

$$
\begin{aligned}
S_1(2) &= 45400 - 953.4 \times 2 = 43493.2 \\
S_1(1.7) &= 45400 - 953.4 \times 1.7 = 43779.22 \, .
\end{aligned}
$$

The "three-step" estimate. With three smaller steps of size $\Delta t = 1$, we get a function whose graph is composed of three line segments, each with a slope equal to the rate of new infections at the beginning of the corresponding day. Let's call this function $S_3(t)$. The three-step estimate $S_3(t)$ is hence not a linear function, strictly speaking. However, since its graph is made up of several straight pieces, it is called a **piecewise linear function.**

With a bit of effort we could write down the three linear formulas that apply to each segment of $S_3(t)$, but it is easier just to define $S_3(t)$ by its graph. Since the slopes of the three segments of $S_3(t)$ are progressively more negative, the piecewise linear graph gets progressively steeper as t increases. This explains why the value $S_3(3)$ is *lower* than the value of $S_1(3)$.

By the time we are dealing with the 300-step function $S_{300}(t)$ we can't even tell by its graph that it is piecewise linear unless we zoom in very close. In principle, though, we could still write down a simple linear formula for each of its segments (see the exercises).

An appraisal. The graph of $S_1(t)$ gives us a rough idea of what is happening to the true function $S(t)$ during the first three days. It starts off at the same rate as S', but subsequently the rates move apart. The value of S_1' never changes, while S' changes with the (ever-changing) values of S and I.

The graph of $S_3(t)$ is a distinct improvement because it changes its direction twice, modifying its slope at the beginning of each day to come back into agreement with the rate equation. But since the three-step graph is still piecewise linear, it continues to suffer from the same shortcoming as the one-step: once we restrict our attention to a single straight segment (for example, where $1 \leq t \leq 2$), then the three-step graph *also* has a constant slope, while S' is always changing. Nevertheless, $S_3(t)$ does satisfy the rate equation in our original model three times—at the beginning of each segment—and isn't too far off at other times. When we get to $S_{300}(t)$ we have a function which satisfies the rate equation at 300 times and is very close in between.

Each of these graphs gives us an idea of the behavior of the true function $S(t)$ during the time interval $0 \leq t \leq 3$. None is strictly correct, but none is hopelessly wrong, either. All are **approximations** to the truth. Moreover, $S_3(t)$ is a **better** approximation than $S_1(t)$— because it reflects at least some of the variability in S'—and $S_{300}(t)$ is better still. Thus, even before we have a clear picture of the shape of the true function $S(t)$, we would expect it to be closer to $S_{300}(t)$ than to $S_3(t)$. As we saw above, when we take piecewise linear approximations with smaller and smaller step sizes, it is reasonable to think that they will approach the true function S in the limit. Expressing this in the notation we have used before,

the function $S(t) = \lim_{N \to \infty} \{$the chain of linear functions with $\Delta t = 10^{-N}\}$.

Approximate versus Exact

You may find it unsettling that our efforts give us only a sequence of approximations to $S(3)$, and not the exact value, or only a sequence of piecewise-linear approximations to $S(t)$, not the "real" function itself. In what sense can we say we "know" the number $S(3)$ or the function $S(t)$? The answer is: in the same sense that we "know" a number like $\sqrt{2}$ or π. There are two distinct aspects to the way we know a number. On the one hand, we can **characterize** a number precisely and completely:

What does it mean to "know" a number like π?

π: the ratio of the circumference of a circle to its diameter;
$\sqrt{2}$:   the positive number whose square is 2;

On the other hand, when we try to **construct** the decimal expansion of a number, we usually get only approximate and incomplete results. For example, when we do calculations by hand we might use the rough estimates $\sqrt{2} \approx 1.414$ and $\pi \approx 3.1416$. With a desk-top computer we might have $\sqrt{2} \approx 1.414\,213\,562\,373\,095$ and $\pi \approx 3.141\,592\,653\,589\,793$, but these are still approximations, and we are really saying

$$\sqrt{2} \;=\; 1.414\,213\,562\,373\,09\ldots$$
$$\pi \;=\; 3.141\,592\,653\,589\,79\ldots\ .$$

The *complete* decimal expansions for $\sqrt{2}$ and π are unknown! The exact values exist as limits of approximations that involve successively longer strings of digits, but we never see the limits—only approximations. In the final section of this chapter we will see ways of generating these approximations for $\sqrt{2}$ and for π.

What we say about π and $\sqrt{2}$ is true for $S(3)$ in exactly the same way. We can *characterize* it quite precisely, and we can *construct* approximations to its numerical value to any desired degree off accuracy. Here, for example, is a characterization of $S(3)$:

> The S-I-R problem for which $a = .00001$ and $b = 1/14$ and for which $S = 45400$, $I = 2100$, $R = 2500$ when $t = 0$ determines three functions $S(t)$, $I(t)$, and $R(t)$. The number $S(3)$ is the value that the function $S(t)$ has when $t = 3$.

You should try to extend this argument to describe the sense in which we "know" the function $S(t)$ by knowing its piecewise-linear approximations. Try to convince yourself that this is operationally no different from the way we "know" functions like $f(x) = \sqrt{x}$. In each instance we can **characterize** the function completely, but we can only **construct** an approximation to most values of the function or to its graph.

All this discussion of approximations may strike you as an unfortunate departure from the accuracy and precision you may have been led to expect in mathematics up until now. In fact, it is precisely this ability to make quick and accurate approximations to problems that is one of the most powerful features of mathematics. This is what goes on everytime you use your calculator to evaluate $\log 3$ or $\sin 37$. Your calculator doesn't really know what these numbers are—but it does know how to approximate them quickly to 12 decimal places. Similar kinds of approximations are also at the heart of how bridges are built and spaceships are sent to the moon.

A Caution: The fact that computers and calculators are really only dealing with approximations when we think they are being ex-

Being able to approximate a number to 12 decimal places is usually as good as knowing its value precisely

However ...

act occasionally leads to problems, the most common of which involves
roundoff errors. You can probably generate a relatively harmless
manifestation of this on your computer with the SIRVALUE program.
Modify the `PRINT` line so it prints out the final value of t to 10 or 12
digits, and try running it with a high value for `numberofsteps`, say 1
million or 10 million. You would expect the final value of t to be exactly
1 in every case, since you are adding `deltat = 1/numberofsteps` to itself `numberofsteps` times. The catch is that the computer doesn't store
the exact value `1/numberofsteps` unless `numberofsteps` is a power of
2. In all other cases it will only be using an approximation, and if
you add up enough quantities that are slightly off, their cumulative
error will begin to show. We will encounter a somewhat less benign
manifestation of roundoff error in the next chapter.

Exercises

There and back again

1. a) Look at the table on page 66. What is your best guess of the
exact value of $I(1)$? (Use the "..." notation introduced on page 67.)
b) What is the exact value of $R(1)$?

2. We noted that the discrepancy (the difference between the new
estimate for $S(0)$ and the original value) seemed to decrease as Δt
decreased.
a) What is your best estimate (using only the information in the table)
for the value of Δt needed to produce a discrepancy of .001 ?
b) More generally, express as precisely as you can the apparent relation
between the size of the discrepancy and the size of Δt.

3. a) Suppose you wanted to try going three days forward and then
coming back, using $\Delta t = .01$. What changes would you have to make
in SIRVALUE to do this?
b) Make a table similar to the one on page 65 for going three days
forward and coming back for $\Delta t = 1, .1, .01,$ and.001.
c) In this new table how does the size of the discrepancy for a given
value of Δt compare with the value in the original table?
d) What value of Δt do you think you would need to determine the
integer parts of $S(3)$ and $R(3)$ exactly?

Piecewise linear functions

4. What are the formulas for the three linear segments of the three-step approximation $S_3(t)$ discussed in the text? (Use the $y = y_0 + m \cdot \Delta x$ form.)

5. Using this three-step approximation, what is $S_3(1.7)$? What is $S_3(2.5)$?

[Answer: $S_3(2.5) = 42295.9$]

6. How would you modify SIRVALUE to get $S_{3000}(3)$? Do it; what do you get?

7. What additional changes would you make to get the values of t, S, and S' at the beginning of the 193rd segment of $S_{300}(t)$? [HINT: You only need to alter the FOR k = 1 TO numberofsteps line and the PRINT line.]

8. Suppose we wanted to determine the value of $S_{300}(2.84135)$.

a) In which of the 300 segments of the graph of $S_{300}(t)$ would we look to find this information?

b) What are the values of t, S, and S' at the beginning of this segment?

c) What is the equation of this segment?

d) What is $S_{300}(2.84135)$?

9. How would you modify SIRVALUE to calculate estimates for S, I, and R when $t = -6$, using $\Delta t = .05$? Do it; what do you get?

10. We want to use SIRPLOT to look at the graph of $S(t)$ over the first 20 days, using $\Delta t = .01$.

a) What changes would we have to make in the program?

b) Sketch the graph you get when you make these changes.

c) If you wanted to plot the graph of $I(t)$ over this same time interval, what additional modifications to SIRPLOT would be needed? Make them, and sketch the result. When does the infection appear to hit its peak?

d) Modify SIRPLOT to sketch on the same graph all three functions over the first 70 days. Sketch the result.

The DO-WHILE loop

A difficulty in giving a precise answer to the last question was that we had to get all the values for 20 days, then go back to estimate by eye when the peak occurred. It would be helpful if we could write a program that ran until it reached the point we were looking for, and then stopped. To do this, we need a different kind of loop—a *conditional* loop that keeps looping only while some specified condition is true. A DO-WHILE loop is one useful way to do this. Here's how the modified SIRPLOT program would look:

```
Set up GRAPHICS
tinitial = 0
t = tinitial
S = 45400
I = 2100
R = 2500
Iprime = .00001 * S * I - I / 14
deltat = .01
DO WHILE Iprime > 0
        Sprime = -.00001 * S * I
        Iprime = .00001 * S * I - I / 14
        Rprime = I / 14
        deltaS = Sprime * deltat
        deltaI = Iprime * deltat
        deltaR = Rprime * deltat
        Plot the line from (t, S)
           to (t + deltat, S + deltaS)
        t = t + deltat
        S = S + deltaS
        I = I + deltaI
        R = R + deltaR
LOOP
PRINT t
```

The changes we have made are:

- Since we don't know what the final time will be, eliminate the `tfinal = 20` and the `numberofsteps = 2000` commands.

- Since the condition in our loop is keyed to the value of I', we have to calculate the initial value of I' before the loop starts.

- Instead of `deltat = (tfinal - tinitial)/numberofsteps` use the statement `deltat = .01`.

- The key change is to replace the `FOR k = 1 TO numberofsteps` line by the line `DO WHILE Iprime > 0`.

- To denote the end of the loop we replace the `NEXT k` command with the command `LOOP`.

- After the `LOOP` command add the line `PRINT t`.

The net effect of all this is that the program will continue working as before, calculating values and plotting points (`t, S`), but *only as long as the condition in the* `DO WHILE` *statement is true.* The condition we used here was that I' had to be positive—this is the condition that ensures that values of I are still getting bigger. While this condition is true, we can always get a larger value for I by going forward another increment Δt. As soon as the condition is false—i.e., as soon as I' is negative—the values for I will be decreasing, which means we have passed the peak and so want to stop.

11.   Make these modifications; what value for t do you get?

12.   You could modify the `PRINT t` command to also print out other quantities.

a) What is the value of I at its peak?

b) What is the value of S when I is at its peak? Does this agree with what we predicted in chapter 1?

13.   Suppose you change the initial value of S to be 5400 and run the previous program. Now what happens? Why?

14.   Suppose we wanted to know how long the infection lasts. We could use DO-WHILE and keep stepping forward using, say $\Delta t = .1$, so long as $I \geq 1$. As soon as I was less than 1 we would want to stop and see what the value of t was.

a) What modifications would you make in SIRVALUE to get this information?

[Answer: (partial) Replace the `FOR k = 1 TO numberofsteps` line with `DO WHILE I >= 1`.]

b) Run your modified program. What value do you get for t?

[Answer: $t = 160.2$]

c) Run the program using $\Delta t = .01$. Now what is your estimate for the duration of the infection? What can you say about the actual time required for I to drop below 1?

15. a) If we think the epidemic started with a single individual, we can go backwards in time until I is no longer greater than 1, and see what the corresponding time is. Do this for $\Delta t = -1, -.1$, and $-.01$. What is your best estimate for the time that the infection arrived?

b) How many Recovered were there at the start of the epidemic?

§2.   The Mathematical Implications—Euler's Method

Approximate Solutions

In the last section we approximated the function $S(t)$ by piecewise-linear approximations using steps of size $\Delta t = 1, .1$, and $.01$. This process can clearly be extended to produce approximations with an *arbitrary* number of steps. For any given step size Δt, the result is a piecewise linear graph whose segments are Δt days wide. This graph then provides us with an estimate for $S(t)$ for every value of t in the interval $0 \leq t \leq 3$. We call the process for constructing the approximations **Euler's method**, after the Swiss mathematician Leonhard Euler (1707–1783). Euler was interested in the general problem of finding the functions determined by a set of rate equations, and in 1768 he proposed this method to approximate them. The method can indeed be used to get approximate solutions for an enormous range of rate equations, and for this reason we will make it a basic tool.

To begin to get a sense of the general utility of Euler's method, let's use it in a new setting. Here is a simple problem that involves just a single variable y that depends on t.

$$\text{rate equation: } \quad y' = .1\, y \left(1 - \frac{y}{1000}\right),$$
$$\text{initial condition: } \quad y = 100 \text{ when } t = 0.$$

To solve the problem we must find the function $y(t)$ determined by this rate equation and initial condition. We'll work without a context, in order to emphasize the purely mathematical nature of Euler's method. However, this rate equation is one member of a family called **logistic equations** that are frequently used in population models. We will explore this context in the exercises.

Euler's method requires that we specify the step size and an interval over which we want to approximate the solution. Let's choose $\Delta t = .5$ and construct the function that approximates $y(t)$ on the interval $0 \leq t \leq 75$. We can then make the suitable modifications in SIRPLOT or SIRVALUE to get the approximation we want. Suppose we want to view the approximation graphically. Here's what the modified SIRPLOT would look like:

Program: modified SIRPLOT

```
Set up GRAPHICS

tinitial = 0
tfinal = 75
t = tinitial
y = 100
numberofsteps = 150
deltat = (tfinal - tinitial)/numberofsteps
FOR k = 1 TO numberofsteps
      yprime = .1 * y * (1 - y / 1000)
      deltay = yprime * deltat
      Plot the line from (t, y)
         to (t + deltat, y + deltay)
      t = t + deltat
      y = y + deltay
NEXT k
```

As before, words in italics, like "*Plot the line from*" need to be translated into the specific formulation required by the computer language you are using.

When we run this program, we get the following graph (axes and scales have been added):

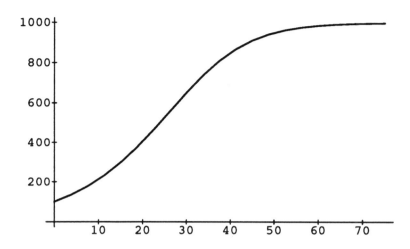

As before, what we see here is only an approximation to the true solution $y(t)$. How can we get some idea of how good this approximation is?

Exact Solutions

Euler's method produces, for any chosen step size, an *approximate* solution to a rate equation problem. In the last section we saw that we can improve the accuracy of the approximation by making the steps smaller and using more of them.

For example, we have already found an approximate solution to the problem

$$y' = .1y \left(1 - \frac{y}{1000}\right); \qquad y(0) = 100$$

on the interval $0 \leq t \leq 75$, using 150 steps of size $\Delta t = 1/2$. Consider a **sequence** of Euler approximations to this problem that are obtained by doubling the number of steps from one stage to the next. To be systematic, let the first approximation have 1 step, the next 2, the next 4, and so on. The number of steps thus has the form 2^{j-1}, where $j = 1, 2, 3, \ldots$. If we use $y_j(t)$ to denote the approximating function with 2^{j-1} steps, then we have an unending list:

$$y_1(t) \ : \ \text{Euler's approximation with 1 step}$$
$$y_2(t) \ : \ \text{Euler's approximation with 2 steps}$$
$$y_3(t) \ : \ \text{Euler's approximation with 4 steps}$$
$$\vdots \qquad\qquad \vdots$$
$$y_j(t) \ : \ \text{Euler's approximation with } 2^{j-1} \text{ steps}$$
$$\vdots \qquad\qquad \vdots$$

Here are the graphs of $y_j(t)$ for $j = 1, 2, \ldots, 7$ plus the graph of $y_{11}(t)$:

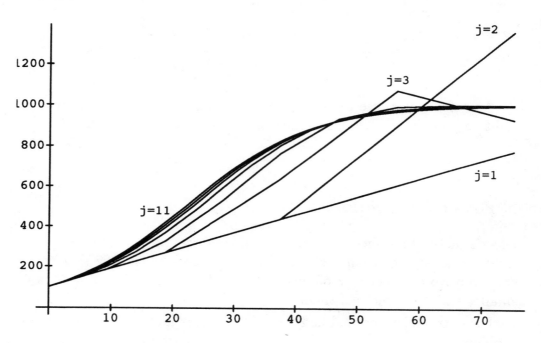

These functions form a sequence of **successive approximations** to the true solution $y(t)$, which is obtained by taking the **limit**, as we did in the last section:

$$y(t) = \lim_{j \to \infty} y_j(t).$$

Functions and graphs can be limits, too

Earlier we noticed how the digits in the estimates for $S(3)$ stabilized. If we plot the approximations $y_j(t)$ together we'll find that they stabilize, too. Each graph in the sequence is different from the preceding one, but the differences diminish the larger j becomes. Eventually, when j is large enough, the graph of y_{j+1} does not differ noticeably from the graph of y_j. That is, the position of the graph **stabilizes** in the coordinate plane. In this example, at the scale in the graph above, this happens around $j = 11$. If we had drawn the graph of y_{15} or y_{20}, it would not have been distinguishable from the graph of y_{11}.

The program SEQUENCE shown below plots the first 14 approximations to $y(t)$. It demonstrates how the graphs of $y_j(t)$ stabilize to define $y(t)$ as their limit.

Program: SEQUENCE

A sequence of graphs for $y' = .1y(1 - y/1000)$; $y(0) = 100$

Set up GRAPHICS
```
FOR j = 1 TO 14
    tinitial = 0
    tfinal = 75
    t = tinitial
    y = 100
    numberofsteps = 2 ^ (j - 1)
    deltat = (tfinal - tinitial) / numberofsteps
    FOR k = 1 TO numberofsteps
          yprime = .1 * y * (1 - y / 1000
          deltay = yprime * deltat
```
          Plot the line from `(t, y)`
              to `(t + deltat, y + deltay)`
          Color the line with color j
```
          t = t + deltat
          y = y + deltay
    NEXT k
NEXT j
```

(bracket spanning the inner portion) **Program: modified SIRPLOT**

Notice that SEQUENCE contains the program SIRPLOT embedded in a loop that executes SIRPLOT 14 times. In this way SEQUENCE plots 14 different graphs. The only new element that has been added to SIRPLOT is "*Color the line with color* j". When you express this in your programming language it instructs the computer to draw the j-th graph using color number j in the computer's "palette." In the exercises you are asked to use the program SEQUENCE to explore the solutions to a number of rate equation problems.

Approximate solutions versus exact

By constructing successive approximations to the solution of a rate equation problem, using a sequence of step sizes `deltat` $= \Delta t$ that shrink to 0, we obtain the *exact* solution in the limit.

In practice, though, all we can ever get are particular approximations. However, we can control the level of precision in our approximations by adjusting the step size. If we are dealing with a model of some real process, then this is typically all we need. For example, when it comes to interpreting the *S-I-R* model, we might be satisfied to predict that there will be about 40500 susceptibles remaining in the population after three days. The table on page 50 indicates we would get that level of precision using a step size of about $\Delta t = 10^{-2}$. Greater precision than this may be pointless, because the modelling process—which converts reality to mathematics—is itself only an approximation.

The question we asked in the last section—In what sense do we know a number?—applies equally to the functions we obtain using Euler's method. That is, even if we can characterize a function quite precisely as the solution of a particular rate equation, we may be able to evaluate it only approximately.

A Caution

We have now seen how to take a set of rate equations and find approximations to the solution of these equations to any degree of accuracy desired. It is important to remember that all these mathematical manipulations are only drawing inferences about the model. We are essentially saying that *if* the original equations capture the internal dynamics of the situation being modelled, *then* here is what we would expect to see. It is still essential at some point to go back to the reality being modeled and check these predictions to see whether our original assumptions were in fact reasonable, or need to be modified. As Alfred North Whitehead has said:

> There is no more common error than to assume that, because prolonged and accurate mathematical calculations have been made, the application of the result to some fact of nature is absolutely certain.

Exercises

Approximate solutions

1. Modify SIRVALUE and SIRPLOT to analyze the population of Poland (see exercise 23 of chapter 1, page 47). We assume the population $P(t)$ satifies the conditions

$$P' = .009\,P \quad \text{and} \quad P(0) = 37,500,000,$$

where t is years since 1985. We want to know P 100 years into the future; you can assume that P does not exceed 100,000,000.

a) Estimate the population in 2085.

b) Sketch the graph that describes this population growth.

The Logistic Equation

Suppose we were studying a population of rabbits. If we turn 100 rabbits loose in a field and let $y(t)$ be the nuber of rabbits at time t, we would like to know how this function behaves. The next several exercises are designed to explore the behavior of the rate equation

$$y' = .1y\left(1 - \frac{y}{1000}\right); \qquad y(0) = 100$$

and see why it might be a reasonable model for this system.

2. By modifying SIRVALUE, obtain a sequence of estimates for $y(37)$ that allow you to specify the exact value of $y(37)$ to two decimal places accuracy.

3. a) Referring to the graph of $y(t)$ obtained in the text, what can you say about the behavior of y as t gets large?

b) Suppose we had started with $y(0) = 1000$. How would the population have changed over time? Why?

c) Suppose we had started with $y(0) = 1500$. How would the population have changed over time? Why?

d) Suppose we had started with $y(0) = 0$. How would the population have changed over time? Why?

e) The number 1000 in the denominator of the rate equation is called the **carrying capacity** of the system. Can you give a physical interpretation for this number?

4.　Obtain graphical solutions for the rate equation for different values of the carrying capacity. What seems to be happening as the carrying capacity is increased? (Don't restrict yourself to $t = 37$ here.)

5.　Keep everything in the original problem unchanged except for the constant .1 out front. Obtain graphical solutions with the value of this constant = .05, .2, .3, and .6. How does the behavior of the solution change as this constant changes?

6.　Returning to the original logistic equation, modify SIRVALUE or DO-WHILE to find the value for t such that $y(t) = 900$.

7.　Suppose we wanted to fit a logistic rate equation to a population, starting with $y(0) = 100$. Suppose further that we were comfortable with the 1000 in the denominator of the equation, but weren't sure about the .1 out front. If we knew that $y(20) = 900$, what would be a better value for this constant?

Using SEQUENCE

8.　Each Euler approximation is made up of a certain number of straight line segments. What instruction in the program SEQUENCE determines the number of segments in a particular approximation? The first graph drawn has only a single segment. How many does the fifth have? How many does the fourteenth have?

9.　What is the slope of the first graph? What are the slopes of the two parts of the second graph?

10.　Modify SEQUENCE to construct a sequence of Euler approximations to the function $y(t)$ that satisfies the conditions

$$y' = .2\,y(5 - y) \quad \text{and} \quad y(0) = 1.$$

Construct approximations on the interval $0 \le t \le 10$. [You need to change the final t value in the program, and you also need to ensure that the graphs will fit on your screen.]

a) What is $y(10)$? [If you add the line PRINT j, y just before the line NEXT j, a sequence of 14 estimates for $y(10)$ will appear on the screen with the graphs.]

b) Make a rough sketch of the graph that is the limit of these approximations. The right half of the limit graph has a distinctive feature; what is it?

[Answer: The graph levels off as t increases; the right half becomes a straight line with slope 0.]

c) Without doing any calculations, can you estimate the value of $y(50)$? How did you arrive at this value?

d) Change the **initial condition** from $y(0) = 1$ to $y(0) = 9$. Construct the sequence of Euler approximations and make a rough sketch of the limit graph. What is $y(10)$ now?

11.   Modify SEQUENCE to construct a sequence of Euler approximations for population of Poland (from exercise 1, above). Sketch the limit graph $P(t)$, and mark the values of $P(0)$ and $P(100)$ at the two ends.

12.   Construct a sequence of Euler approximations to the function $y(t)$ that satisfies the conditions

$$y' = 2t \quad \text{and} \quad y(0) = 0$$

over the interval $0 \le t \le 2$. Note that this time the rate y' is given in terms of t, not y. Euler's method works equally well. Estimate $y(2)$. How accurate is your estimate?

13.   Construct a sequence of Euler approximations to the function $y(t)$ that satisfies the conditions

$$y' = \frac{4}{1 + t^2} \quad \text{and} \quad y(0) = 0$$

over the interval $0 \le t \le 1$. Estimate $y(1)$. How accurate is your estimate? [Note: the exact value of $y(1)$ is π, which your estimates may have led you to expect. By using special methods we shall develop much later we can prove that $y(1) = \pi$.]

§3.   Approximate Solutions

Our efforts to find the functions that were determined by the rate equations for the S-I-R model have brought to light several important issues:

- We often have to deal with a question that does not have a simple, straightforward answer; perhaps we are trying to determine a quantity (like the square root of 2, or $S(3)$ in the S-I-R model), to find some function (like $S(t)$), or to understand a process (like an epidemic, or buying and selling in a market). An **approximation** can get us started.

- In many instances, we can make repeated improvements in the approximation. If these **successive approximations** get arbitrarily close to the unknown, and they do it quickly enough, that may answer the question for all practical purposes. In many cases, there is no alternative.

- The information that successive approximations give us is conveyed in the form of a **limit**.

- The method of successive approximations can be used to evaluate many kinds of mathematical objects, including numbers, graphs, and functions.

- **Limit processes** give us a valuable tool to probe difficult questions. They lie at the heart of calculus.

Even the process of building a mathematical model for a physical system can be seen as an instance of successive approximations. We typically start with a simple model (such as the S-I-R model) and then add more and more features to it (e.g., in the case of the S-I-R model we might divide the population into different subgroups, have the parameters in the model depend on the season of the year, make immunity of limited duration, etc.). Is it always possible, at least in theory, to get a sequence of approximating mathematical models that approaches reality in the limit?

In the following chapters we will apply the process of successive approximation to many different kinds of problems. For example, in chapter 3 the problem will be to get a better understanding of the notion of a rate of change of one quantity with respect to another. Then, in chapter 4, we will return to the task of solving rate equations using Euler's method. Chapter 6 introduces the integral, defining it through a sequence of successive approximations. As you study each chapter, pause to identify the places where the method of successive approximations is being used. This can give you insight into the special role that calculus plays within the broader subject of mathematics.

To illustrate the general utility of the method, we end this chapter by returning to the problem raised in section 1 of constructing the values of $\sqrt{2}$ and π to an arbitrary number of decimal places.

Calculating π—The Length of a Curve

Humans were grappling with the problem of calculating π at least 3000 years ago. In his work *Measurement of the Circle*, Archimedes (287–212 B.C.) used the method of successive approximations to calculate $\pi = 3.14\ldots$. He did this by starting with a circle of radius 1, constructing an inscribed and a circumscribed hexagon, and calculating the lengths of their perimeters. The perimeter of the circumscribed hexagon was clearly an overestimate for π, while the perimeter of the inscribed hexagon was an underestimate. He then improved these estimates by going from hexagons to inscribed and circumscribed 12-sided polygons and again calculating the perimeters. He repeated this process of doubling the number of sides until he had inscribed and circumscribed polygons with 96 sides. These left him with his final estimate

$$3.1409\ldots = 3\frac{284\frac{1}{4}}{2017\frac{1}{4}} < \pi < 3\frac{667\frac{1}{2}}{4673\frac{1}{2}} = 3.1428\ldots$$

In grade school we learned a nice simple formula for the length of a circle, but that was about it. We were never taught formulas for the lengths of other simple curves like elliptic or parabolic arcs, for a very good reason—there are no such formulas. There are various physical approaches we might take. For example, we could get a rough approximation by laying a piece of string along the curve, then picking up the string and measuring it with a ruler. Instead of a physical solution, we can use the essence of Archimedes' insight of approximating a circle by an inscribed "polygon"—what we have earlier called a piecewise linear graph—to determine the length of any curve. The basic idea is reminiscent of the way we made successive approximations to the functions $S(t)$, $I(t)$, and $R(t)$ in the first section of this chapter. Here is how we will approach the problem:

- approximate the curve by a chain of straight line segments;

- measure the lengths of the segments;

- use the sum of the lengths as an approximation to the true length of the curve.

Repeat this process over and over, each time using a chain that has shorter segments (and therefore more of them) than the last one. The length of the curve emerges as the limit of the sums of the lengths of the successive chains.

Distance Formula If we are given two points $P_1(x_1, y_1)$ and $P_2(x_2, y_2)$ in the plane, then the distance between them is just

$$\begin{aligned} d &= \sqrt{\Delta x^2 + \Delta y^2} \\ &= \sqrt{(x_2 - x_1)^2 + (y_2 - y_1)^2} \end{aligned}$$

That this follows directly from the Pythagorean theorem can be seen from the picture below:

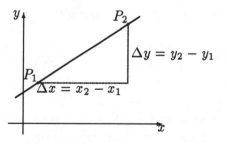

We'll demonstrate how this process works on a parabola. Specifically, consider the graph of $y = x^2$ on the interval $0 \le x \le 1$. At the right we have sketched the graph and our initial approximation. It is a chain with two segments whose end points are equally spaced along the x-axis.

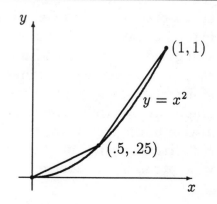

We can use the distance formula to find the lengths of the two segments.

$$\text{first segment}: \quad \sqrt{(.5-0)^2 + (.25-0)^2} = .559016994$$
$$\text{second segment}: \quad \sqrt{(1-.5)^2 + (1-.25)^2} = .901387819$$

Their total length is the sum

$$.559016994 + .901387819 = 1.460404813.$$

The following program prints out the lengths of the two **segments** and their **total** length.

Program: LENGTH
Estimating the length of $y = x^2$ over $0 \le x \le 1$

```
DEF fnf (x) = x ^ 2
xinitial = 0
xfinal = 1
n = 2
deltax = (xfinal - xinitial) / n
total = 0
FOR k = 1 TO n
      xl = xinitial + (k - 1) * deltax
      xr = xinitial + k * deltax
      yl = fnf(xl)
      yr = fnf(xr)
      segment = SQR((xr - xl) ^ 2 + (yr - yl) ^ 2)
      total = total + segment
      PRINT k, segment
NEXT k
PRINT n, total
```

Finding Roots with a Computer

When we casually turn to our calculator and ask it for the value of $\sqrt{2}$, what does it really do? Like us, the calculator can only add, subtract, multiply, and divide. Anything else we ask it to do must be reducible to these operations. In particular, the calculator doesn't really "know"

Calculators and computers really work by making approximations the value of $\sqrt{2}$. What it does know is how to approximate $\sqrt{2}$ to, say, 12 significant figures using only elementary arithmetic. In this section we will look at two ways we might do this. Apart from the fact that both approaches use successive approximations, they are remarkably different in flavor. One works graphically, using a computer graphing package, and the other is a numerical algorithm that is about 4000 years old.

A geometric approach

Exercise 9 on page 40 considered the problem of finding the roots of $f(x) = 1 - 2x^2$. A bit of algebra confirms that $\sqrt{2}/2$ is a root—i.e., $f(\sqrt{2}/2) = 0$. The question is: what is the *numerical value* of $\sqrt{2}/2$?

We'll answer this question by constructing a sequence of approximations that add digits, one at a time, to an estimate for $\sqrt{2}/2$. Since the root lies at the point where the graph of f crosses the x-axis, we just magnify the graph at this point over and over again, "trapping" the point between x values that can be made arbitrarily close together.

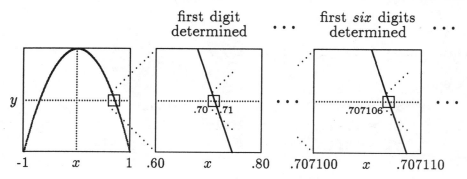

The graph of $y = 1 - 2x^2$ under successive magnifications

If we make each stage a ten-fold magnification over the previous one, then, as we zoom in on the next smaller interval that contains the root, one more digit in our estimate will be stabilized. The first six stages are described in the table below. They tell us $\sqrt{2}/2 = .707106\ldots$ to six decimal places accuracy.

Since this method of finding roots requires only that we be able to plot successive magnifications of the graph of f on a computer screen, the method can be applied to any function that can be entered into a computer.

The positive root of $1 - 2x^2$

when the root lies between:		the decimal expansion
lower value	upper value	of the root begins with
.70	.80	.7
.700	.710	.70
.7070	.7080	.707
.70710	.70720	.7071
.707100	.707110	.70710
.7071060	.7071070	.707106
⋮	⋮	⋮

An algebraic approach – the Babylonian algorithm

About 4000 years ago Babylonian builders had a method for construct-ing the square root of a number from a sequence of successive approxi-mations. To demonstrate the method, we'll construct $\sqrt{5}$. We want to find x so that

$$x^2 = 5 \qquad \text{or} \qquad x = \frac{5}{x}.$$

The second expression may seem a peculiar way to characterize x, but it is at least equivalent to the first. The advantage of the second ex-pression is that it gives us *two* numbers to consider—x and $5/x$.

For example, suppose we guess that $x = 2$. Of course this is incor-rect, because $2^2 = 4$, not 5. The two numbers we get from the second expression are 2 and $5/2 = 2.5$. One is smaller than $\sqrt{5}$, the other is larger (because $2.5^2 = 6.25$). Perhaps their *average* is a better estimate. The average is 2.25, and $2.25^2 = 5.0625$.

Although 2.25 is not $\sqrt{5}$, it is better estimate than either 2.5 or 2. If we change x to 2.25, then

$$x = 2.25, \quad \frac{5}{x} = 2.222, \quad \text{and their average is } 2.236.$$

Is 2.236 a better estimate than 2.5 or 2.222? Indeed it is: $2.236^2 = 4.999696$. If we now change x to 2.236, a remarkable thing happens:

$$x = 2.236, \quad \frac{5}{x} = 2.236, \quad \text{and their average is } 2.236!$$

In other words, if we want accuracy to three decimal places, we have already found $\sqrt{5}$. All the digits have stabilized.

Suppose we want *greater* accuracy? Our routine readily obliges. If we set $x = 2.236000$, then

$$x = 2.236000\,, \quad \frac{5}{x} = 2.236136\,, \quad \text{and their average is } 2.236068\,.$$

In fact, $\sqrt{5} = 2.236068\ldots$ is accurate to six decimal places.

Here is a summary of the argument we have just developed, expressed in terms of $\sqrt{a}$ for an arbitrary positive number a.

> **If x is an estimate for $\sqrt{a}$,**
> **then the average of x and a/x**
> **is a better estimate.**

Once we choose an *initial* estimate, this argument constructs a sequence of successive approximations to $\sqrt{a}$. The process of constructing the sequence is called the **Babylonian algorithm** for square roots.

A procedure that tells us how to carry out a sequence of steps, one at a time, to reach a specific goal is called an **algorithm**. Many algebraic processes are algorithms. The word is a Latinization of the name of the Arab astronomer Muhammad al-Khwārizmī (c. 780A.D. – c. 850A.D.). The title of his seminal book *Hisāb al-jabr wal-muqā bala* (830A.D.)— usually referred to simply as *al-Jabr*—has a Latin form that is even more familiar to mathematics students.

The Babylonian algorithm takes the current estimate x for $\sqrt{a}$ that we have at each stage and says "replace x by the average of x and a/x." This kind of instruction is ideally suited to a computer, because A = B in a computer program means "replace the current value of A by the current value of B." In the program BABYLON printed below, the algorithm is realized by a FOR - NEXT loop with a single line that reads "x = (x + a / x) / 2".

The three-step procedure that we used in chapter 1 to obtain values of S, I, and R in the epidemic problem is also an algorithm, and for that reason it was a straightforward matter to express it as the computer program SIRVALUE.

Program: BABYLON
An algorithm to find $\sqrt{a}$

```
a = 5
x = 2
n = 6
FOR k = 1 TO n
    x = (x + a / x) / 2
    PRINT x
NEXT k
```

Output:

2.25
2.236111
2.236068
2.236068
2.236068
2.236068

Exercises

1.   By using a computer to graph $y = x^2 - 2^x$, find the solutions of the equation $x^2 = 2^x$ to four decimal place accuracy.

[Answer: There are three roots, and the smallest is $-.7666\ldots$ to four decimal place accuracy.]

The program LENGTH

2.   Run the program LENGTH to verify that it gives the lengths of the individual segments and their total length.

3.   What line in the program gives the instruction to work with the function $f(x) = x^2$? What line indicates the number of segments to be measured?

4.   Each segment has a left and a right endpoint. What lines in the program designate the x- and y-coordinates of the left endpoint; the right endpoint?

5.   Where in the program is the length of the k-th segment calculated? The segment is treated as the hypoteneuse of a triangle whose length is measured by the Pythagorean theorem. How is the base of that triangle denoted in the program? How is the altitude of that triangle denoted?

6.   Modify the program so that it uses 20 segments to estimate the length of the parabola. What is the estimated value?

[Answer: 1.478 756 512.]

7.   Modify the program so that it estimates the length of the parabola using 200, 2,000, 20,000, 200,000, and 2,000,000 segments. Compare your results with those tabulated below. [The work is speeded up considerably if you delete the PRINT statement that appears inside the loop.]

Number of line segments	Sum of their lengths
2	1.460 404 813
20	1.478 756 512
200	1.478 940 994
2 000	1.478 942 839
20 000	1.478 942 857
200 000	1.478 942 857
2 000 000	1.478 942 857

8.   What is the length of the parabola $y = x^2$ over the interval $0 \leq x \leq 1$, correct to 8 decimal places? What is the length, correct to 12 decimal places?

9.   Starting at the origin, and moving along the parabola $y = x^2$, where are you when you've gone a total distance of 10?

10.   Modify the program to find the length of the curve $y = x^3$ over the interval $0 \leq x \leq 1$. Find a value that is correct to 8 decimal places.

11.   **Back to the circle.** Consider the unit circle centered at the origin. Pythagoras' Theorem shows that a point (x, y) is on the circle if and only if $x^2 + y^2 = 1$. If we solve this for y in terms of x, we get $y = \pm\sqrt{1 - x^2}$, where the plus sign gives us the upper half of the circle and the minus sign gives the lower half. This suggests that we look at the function $g(x) = \sqrt{1 - x^2}$. The arclength of $g(x)$ over the interval $-1 \leq x \leq 1$ should then be exactly π.

a) Divide the interval into 100 pieces—what is the corresponding length?

b) How many pieces do you have to divide the interval into to get an accuracy equal to that of Archimedes?

c) Find the length of the curve $y = g(x)$ over the interval $-1 \leq x \leq 1$, correct to eight decimal places accuracy.

12. This question concerns the function $h(x) = \dfrac{4}{1 + x^2}$.

a) Sketch the graph of $y = h(x)$ over the interval $-2 \le x \le 2$.

b) Find the length of the curve $y = h(x)$ over the interval $-2 \le x \le 2$.

13. Find the length of the curve $y = \sin x$ over the interval $0 \le x \le \pi$.

The program BABYLON

14. Run the program BABYLON on a computer to verify the tabulated estimates for $\sqrt{5}$.

15. Run BABYLON with at least 14-digit precision. What is the estimated value of $\sqrt{5}$ in this circumstance? How many steps were needed to get this value? Use the square of this estimate as a measure of its accuracy. What is the square?

[Answer: The estimate $2.236\,067\,977\,499\,79$ is obtained at the fourth step. Its square is $5.000\,000\,000\,000\,001$.]

16. Use the Babylonian algorithm to find $\sqrt{80}$.

a) First use 2 as your initial estimate. How many steps are needed for the calculations to stabilize—that is, to reach a value that doesn't change from one step to the next?

b) Since $9^2 = 81$, a good first estimate for $\sqrt{80}$ is 9. How many steps are needed this time for the calculations to stabilize? Are the final values in (a) and (b) the same?

17. Use the Babylonian algorithm to find $\sqrt{250}$ and $\sqrt{1990}$. If you use 2 as the inital estimate in each case, how many steps are needed for the calculations to stabilize? If you use the integer nearest to the final answer as your initial estimate, then how many steps are needed? Square your answers to measure their accuracy.

18. The Babylonian algorithm is considered to be **very fast**, in the sense that each stage roughly doubles the number of digits that stabilize. Does your work on the preceding exercises confirm this observation? By comparison, is the routine that got the estimates for $S(3)$ (computed with the program SIRVALUE) faster or slower than the Babylonian algorithm?

§4.   Chapter Summary

The Main Ideas

- The exact numerical value of a quantity may not be known; the value is often given by an **approximation**.

- A numerical quantity is often given as the **limit** of a sequence of **successive approximations**.

- When a particular digit in a sequence of successive approximations **stabilizes**, that digit is assumed to appear in the limit.

- **Euler's method** is a procedure to **approximate** a function defined by a set of rate equations and initial conditions. The approximation is a piecewise linear function.

- The exact function defined by a set of rate equations and initial conditions can be expressed as the **limit** of a sequence of **successive Euler approximations** with smaller and smaller step sizes.

Self-Testing

- You should be able to use a program to **construct a sequence** of estimates for S, I, and R, given a specific S-I-R model with initial conditions.

- You should be able to **modify** the SIRVALUE and SIRPLOT programs to construct a sequence of estimates for the values of functions defined by other rate equations and initial conditions.

- You should be able to use programs that construct a sequence of **Euler approximations** for the function defined by a rate equation with an initial condition. The programs should provide both tabular and graphical output.

- You should be able to estimate the values of the roots of an equation $f(x) = 0$ using a **computer graphing package**.

- You should be able to find a square root using the **Babylonian algorithm**.

• You should be able to find the length of any piece of any curve.

Chapter Exercises

1. a) We have considered the logistic equation

$$y' = .1y \left(1 - \frac{y}{1000}\right); \qquad y(0) = 100$$

This specifies y' as a function of y. Sketch the graph of this function. That is, plot y values on the horizontal axis; for each y, plot the value of y' given by the logistic equation vertically. What shape does the graph have?

b) For what value of y does this graph take on its largest value? Where does this y value appear in the graph of $y(t)$ versus t?

c) For what values of y does the graph of y' versus y cross the y-axis? Where do these y values appear in the graph of $y(t)$ versus t?

Grids on Graphs

When writing a graphing program it is often useful to have the computer draw a grid on the screen. This makes it easier to estimate numerical values, for instance. We can use a simple FOR–NEXT loop inside a program to do this. For instance, suppose we had written a program (SIRPLOT or SEQUENCE, for instance) with the graphics already in place. Suppose the screen window covered values from 0 to 100 horizontally, and 0 to 50,000 vertically. If we wanted to draw 21 vertical lines (including both ends) spaced 5 units apart and 11 horizontal lines spaced 5,000 units apart, the following two loops inserted in the program would work:

```
FOR k = 0 TO 20
      Plot the line from (5 * k, 0)
         to (5 * k, 50000)
NEXT k
FOR k = 0 TO 10
      Plot the line from (0, 5000 * k)
         to (100, 5000 * k)
NEXT k
```

You should make sure you see how these loops work and that you can modify them as needed.

2.   If the screen window runs from −20 to 120 horizontally, and 250 to 750 vertically, how would you modify the loops above to create a vertical grid spaced 10 units apart and a horizontal grid spaced 25 units apart? Answer:

```
FOR k = 0 TO 15
        Plot the line from (-20 + 10 * k, 250)
            to (-20 + 10 * k, 750)
NEXT k
FOR k = 0 TO 20
        Plot the line from (-20, 250 + 25 * k)
            to (120, 250 + 25 * k)
NEXT k
```

3.   Go back to our basic $S - I - R$ model. Modify SIRPLOT to calculate and plot on the same graph the values of $S(t)$, $I(t)$, and $R(t)$ for t going from 0 to 120, using a stepsize of $\Delta t = .1$. Include a grid with a horizontal spacing of 5 days, and a vertical spacing of 2000 people. You should get something that looks like

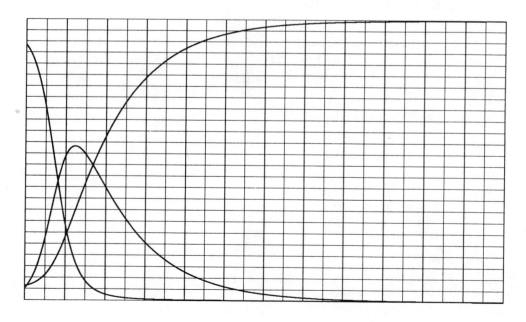

Chapter 3

The Derivative

In developing the *S-I-R* model in chapter 1 we took the idea of the rate of change of a population as intuitively clear. The rate at which one quantity changes with respect to another is a central concept of calculus and leads to a broad range of insights. The chief purpose of this chapter is to develop a fuller understanding—both analytic and geometric—of the connection between a function and its rate of change. To do this we will introduce the concept of the **derivative** of a function.

§1.   Rates of Change

The Changing Time of Sunrise

The sun rises at different times, depending on the date and location. At 40° N latitude (New York, Beijing, and Madrid are about at this latitude) in the year 1990, for instance, the sun rose at

The time of sunrise is a function

$$
\begin{array}{ll}
7{:}16 & \text{on January 23} \\
5{:}58 & \text{on March 24} \\
4{:}52 & \text{on July 25.}
\end{array}
$$

Clearly the time of sunrise is a function of the date. If we represent the time of sunrise by T (in hours and minutes) and the date by d (the day of the year), we can express this functional relation in the form $T = T(d)$. For example, from the table above we find $T(23) = 7{:}16$. It is not obvious from the table, but it is also true that the the rate at which the time of sunrise is changing is different at different times of the year—T' varies as d varies. We can see how the rate varies by looking at some further data for sunrise at the same latitude, taken from *The Nautical Almanac for the Year 1990*:

date	time	date	time	date	time
January 20	7:18	March 21	6:02	July 22	4:49
23	7:16	24	5:58	25	4:52
26	7:14	27	5:53	28	4:54

Calculate the rate using earlier and later dates

Let's use this table to estimate the rate at which the time of sunrise is changing on January 23. We'll use the times three days earlier and three days later, and compare them. On January 26 the sun rose 4 minutes earlier than on January 20. This is a change of -4 minutes in 6 days, so the rate of change is

$$\frac{-4 \text{ minutes}}{6 \text{ days}} \approx -.67 \text{ minutes per day.}$$

We say this is the **rate** at which sunrise is changing on January 23, and we write

$$T'(23) \approx -.67 \; \frac{\text{minutes}}{\text{day}}.$$

The rate is negative because the time of sunrise is decreasing—the sun is rising earlier each day.

Similarly, we find that around March 24 the time of sunrise is changing approximately $-9/6 = -1.5$ minutes per day, and around July 25 the rate is $5/6 \approx +.8$ minutes per day. The last value is positive, since the time of sunrise is increasing—the sun is rising later each day in July. Since March 24 is the 83rd day of the year and July 25 is the 206th, using our notation for rate of change we can write

$$T'(83) \approx -1.5 \; \frac{\text{minutes}}{\text{day}}; \qquad T'(206) \approx .8 \; \frac{\text{minutes}}{\text{day}}.$$

Notice that, in each case, we have calculated the rate on a given day by using times *shortly before* and *shortly after* that day. We will continue this pattern wherever possible. In particular, you should follow it when you do the exercises about a falling object, at the end of the section.

Once we have the rates, we can estimate the time of sunrise for dates not given in the table. For instance, January 28 is five days after

January 23, so the total change in the time of sunrise from January 23 to January 28 should be approximately

$$\Delta T \approx -.67 \frac{\text{minutes}}{\text{day}} \times 5 \text{ days} = -3.35 \text{ minutes}.$$

In whole numbers, then, the sun rose 3 minutes earlier on January 28 than on January 23. Since sunrise was at 7:16 on the 23rd, it was at 7:13 on the 28th.

By letting the change in the number of days be negative, we can use this same reasoning to tell us the time of sunrise on days shortly *before* the given dates. For example, March 18 is −6 days away from March 24, so the change in the time of sunrise should be

$$\Delta T \approx -1.5 \frac{\text{minutes}}{\text{day}} \times -6 \text{ days} = +9 \text{ minutes}.$$

Therefore, we can estimate that sunrise occurred at 5:58 + 0:09 = 6:07 on March 18.

Changing Rates

Suppose instead of using the tabulated values for March we tried to use our January data to *predict* the time of sunrise in March. Now March 24 is 60 days after January 23, so the change in the time of sunrise should be approximately

$$\Delta T \approx -.67 \frac{\text{minutes}}{\text{day}} \times 60 \text{ days} = -40.2 \text{ minutes},$$

and we would conclude that sunrise on March 24 should be at about 7:16 − 0:40 = 6:37, which is more than half an hour later than the actual time! This is a problem we met often in estimating future values in the *S-I-R* model. When we use the formula above to estimate ΔT, we implicitly assume that the time of sunrise changes at the fixed rate of −.67 minutes per day over the entire 60-day time-span. But this turns out not to be true: the rate actually varies, and the variation is too great for us to get a useful estimate. Only with a much smaller time-span does the rate not vary too much.

Predictions over long time spans are less reliable

Here is the same lesson in another context. Suppose you are travelling in a car along a busy road at rush hour and notice that you are going 50 miles per hour. You would be fairly confident that in the next 30 seconds (1/120 of an hour) you will travel about

$$\Delta \text{ position} \approx 50 \, \frac{\text{miles}}{\text{hour}} \times \frac{1}{120} \text{ hour} = \frac{5}{12} \text{ mile} = 2200 \text{ feet.}$$

The actual value ought to be within 50 feet of this, making the estimate accurate to within about 2% or 3%. On the other hand, if you wanted to estimate how far you would go in the next 30 minutes, your speed would probably fluctuate too much for the calculation

$$\Delta \text{ position} \approx 50 \, \frac{\text{miles}}{\text{hour}} \times \frac{1}{2} \text{ hours} = 25 \text{ miles}$$

to have the same level of reliability.

Other Rates, Other Units

In the *S-I-R* model the rates we analyzed were **population growth rates**. They told us how the three populations changed over time, in units of persons per day. If we were studying the growth of a colony of mold, measuring its size by its weight (in grams), we could describe *its* population growth rate in units of grams per hour. In discussing the motion of an automobile, the rate we consider is the **velocity** (in miles per hour), which tells us how the distance from some starting point changes over time. We also pay attention to the rate at which *velocity* changes over time. This is called **acceleration**, and can be measured in miles per hour per hour.

Examples of
rates

While many rates do involve changes with respect to time, other rates do not. Two examples are the survival rate for a disease (survivors per thousand infected persons) and the dose rate for a medicine (milligrams per pound of body weight). Other common rates are the annual birth rate and the annual death rate, which might have values like 19.3 live births per 1,000 population and 12.4 deaths per 1,000 population. Any quantity expressed as a percentage, such as an interest rate or an unemployment rate, is a rate of a similar sort. An unemployment rate of 5%, for instance, means 5 unemployed workers per 100 workers. There are many other examples of rates in the economic world that make use of a variety of units—exchange rates (e.g., francs per dollar), marginal return (e.g., dollars of profit per dollar of change in price).

Sometimes we even want to know the rate of change of one rate with respect to a second. For example, automobile fuel economy (in miles per gallon—the first rate) changes with speed (in miles per hour—the second rate), and we can measure the rate of change of fuel economy with speed. Take a car that goes 22 miles per gallon of fuel at 50 miles per hour, but only 19 miles per gallon at 60 miles per hour. Then its fuel economy is changing approximately at the rate

The rate of change of a rate

$$\frac{\Delta \text{ fuel economy}}{\Delta \text{ speed}} = \frac{19 - 22 \text{ miles per gallon}}{60 - 50 \text{ miles per hour}}$$

$$= -.3 \text{ miles per gallon } per \text{ mile per hour.}$$

Exercises

A falling object. These questions deal with an object that is held motionless 10,000 feet above the surface of the ocean and then dropped. Start a clock ticking the moment it is dropped, and let D be the number of feet it has fallen after the clock has run t seconds. The following table shows some of the values of t and D.

time (seconds)	distance (feet)
0	0.00
1	15.07
2	56.90
3	121.03
4	203.76
5	302.00
6	413.16
7	535.10

1. What units do you use to measure velocity—that is, the rate of change of distance with respect to time—in this problem?

2. a) Make a careful graph that shows these eight data points. Put *time* on the horizontal axis. Label the axes and indicate the units you are using on each.

b) The slope of any line in the graph has the units of a *velocity*. Explain why.

3.   Make three estimates of the velocity of the falling object at the 2 second mark using the distances fallen between these times:
   i) from 1 second to 2 seconds;
   ii) from 2 seconds to 3 seconds;
   iii) from 1 second to 3 seconds.

[Answer: The three velocity estimates are 41.83, 64.13, and 52.98.]

4.   a) Each of the estimates in the previous question corresponds to the slope of a particular line you can draw in your graph. Draw those lines and label each with the corresponding velocity.

b) Which of the three estimates in the previous question do you think is best? Explain your choice.

5.   Using your best method, estimate the velocity of the falling object after 4 seconds have passed.

6.   Is the object speeding up or slowing down as it falls? How can you tell?

7.   Approximate the velocity of the falling object after 7 seconds have passed. Use your answer to estimate the number of feet the object has fallen after 8 seconds have passed. Do you think your estimate is too high or too low? Why?

8.   **Poland and Afghanistan**. The population of Afghanistan was estimated to be 15 million at the beginning of 1985. During the year, for every 1000 persons, there were 48.9 births and 27.3 deaths. This gives a *net* per capita birth rate of 21.6 persons per thousand persons per year.

a) What is the growth rate of the population, in persons per year? Make a clear distinction between the growth rate and the net per capita birth rate. Estimate the population size at the beginning of 1986.

b) The population of Poland was estimated to be 37.5 million in 1985, and the net per capita birth rate was 9.0 persons per thousand persons per year. What is the growth rate, in persons per year? Estimate the population size in 1986.

c) In which country is the population growing more rapidly, Poland or Afghanistan? Explain your view.

§2. Microscopes and Local Linearity

The Graph of Data

This section is about seeing rates geometrically. We
know from chapter 1 that
we can visualize the rate of
change of a *linear* function
as the slope of its graph.
Can we say the same thing
about the sunrise function?
The graph of this function
appears at the right; it plots
the time of sunrise (over the

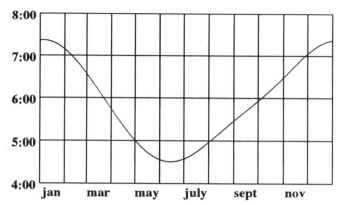

course of a year at 40° N latitude), as a function of the date. The graph
is curved, so the sunrise function is not linear. There is no immediately
obvious connection between rate and slope. In fact, it isn't even clear
what we might mean by the *slope* of this graph! We can make it clear
by using a **microscope**.

Imagine we have a microscope that allows us to "zoom in" on the
graph near each of the the three dates we considered in §1. If we put
each magnified image in a window, then we get the following:

Zoom in on the
graph with a
microscope

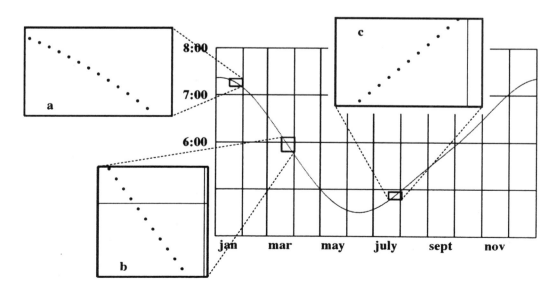

Notice how different the graph looks under the microscope. First of all, it now shows up clearly as a collection of separate points—one for each day of the year. Second, the points in a particular window lie on a line that is essentially straight. The straight lines in the three windows have very different slopes, but that is only to be expected.

The graph looks straight under a microscope

What is the connection between these slopes and the rates of change we calculated in the last section? To decide, we should calculate the slope in each window. This involves choosing a pair of points (d_1, T_1) and (d_2, T_2) on the graph and calculating the ratio

$$\frac{\Delta T}{\Delta d} = \frac{T_2 - T_1}{d_2 - d_1}.$$

In window **a** we'll take the two points that lie three days on either side of the central date, January 23. These points have coordinates $(20, 7{:}18)$ and $(26, 7{:}14)$ (table, page 102). The slope is thus

$$\frac{\Delta T}{\Delta d} = \frac{7{:}14 - 7{:}18}{26 - 20} = \frac{-4 \text{ minutes}}{6 \text{ days}} = -.67 \frac{\text{minutes}}{\text{day}}.$$

Slope and rate calculations are the same

If we use the same approach in the other two windows we find that the line in window **b** has slope -1.5 min/day, while the line in window **c** has slope $+.8$ min/day. These are exactly the same calculations we did in §1 to determine the rate of change of the time of sunrise around January 23, March 24, and July 25, and they produce the same values we obtained there:

$$T'(23) \approx -.67 \frac{\text{min}}{\text{day}} \qquad T'(83) \approx -1.5 \frac{\text{min}}{\text{day}} \qquad T'(206) \approx .8 \frac{\text{min}}{\text{day}}.$$

This is a crucial observation which we use repeatedly in other contexts; let's pause and state it in general terms:

> **The rate of change of a function at a point is equal to the slope of its graph at that point, if the graph looks straight when we view it under a microscope.**

The Graph of a Formula

Rates and slopes are really the same thing—that's what we learn by using a microscope to view the graph of the sunrise function. But

the sunrise graph consists of a finite number of disconnected points—
a very common situation when we deal with data. In such cases it
doesn't make sense to magnify the graph too much. For instance, we
would get no useful information from a window that was narrower than
the space between the data points. There is no such limitation if we
use a microscope to look at the graph of a function given by a formula,
though. We can zoom in as close as we wish and still see a continuous
curve or line. By using a high-power microscope, we can learn even
more about rates and slopes.

High-power magnification is possible with a formula

Consider this rather complicated-looking function:

$$f(x) = \frac{2 + x^3 \cos x + 1.5^x}{2 + x^2}.$$

Let's find $f'(27)$, the rate of change of f when $x = 27$. We need to zoom
in on the graph of f at the point $(27, f(27)) = (27, 69.859043)$. We do
this in stages, producing a succession of windows that run clockwise
from the upper left. Notice how the graph gets straighter with each
magnification.

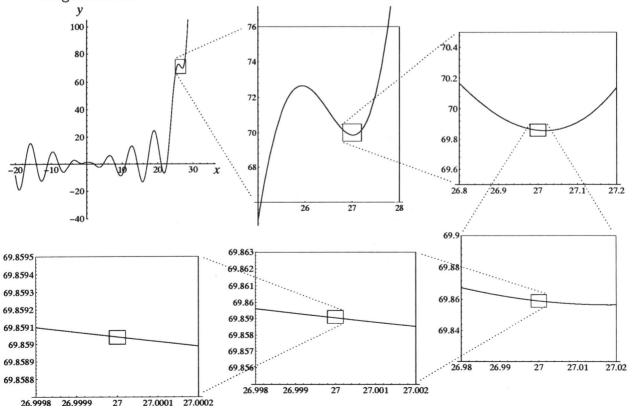

The field of view
of a window We will need a way to describe the part of the graph that we see in
a window. Let's call it the **field of view**. The field of view of each
window is only one-tenth as wide as the previous one, and the field
of view of the last window is only one-millionth of the first! The last
window shows what we would see if we looked at the original graph
with a million-power microscope.

The microscope used to study functions is real, but it is different from the one a biologist uses to
study micro-organisms. Our microscope is a computer graphing program that can "zoom in" on
any point on a graph. The computer screen is the window you look through, and you determine
the field of view when you set the size of the interval over which the graph is plotted.

Here is our point of departure: **the rate $f'(27)$ is the slope of
the graph of f when we magnify the graph enough to make it
look straight.** But how much is enough? Which window should we
use? The following table gives the slope $\Delta y/\Delta x$ of the line that appears
in each of the last four windows in the sequence. For Δx we take the
difference between the x-coordinates of the points at the ends of the
line, and for Δy we take the difference between the y-coordinates. In
particular, the width of the field of view in each case is Δx.

Δx	Δy	$\Delta y/\Delta x$
.08	$-2.115\,566\,15 \times 10^{-2}$	$-.264\,445\,769$
.008	$-2.178\,202\,41 \times 10^{-3}$	$-.272\,275\,301$
.0008	$-2.178\,828\,82 \times 10^{-4}$	$-.272\,353\,602$
.00008	$-2.178\,834\,89 \times 10^{-5}$	$-.272\,354\,362$

As you can see, it *does* matter how much we magnify. The slopes
$\Delta y/\Delta x$ in the table are not quite the same, so we don't yet have a
definite value for $f'(27)$. The table gives us an idea how we *can* get a
definite value, though. Notice that the slopes get more and more alike,
the more we magnify. In fact, under successive magnifications the first
five digits of $\Delta y/\Delta x$ have **stabilized**. We saw in chapter 2 how to
think about a sequence of numbers whose digits stabilize one by one.
The slope is a
limit We should treat the values of $\Delta y/\Delta x$ as **successive approximations**
to the slope of the graph. The exact value of the slope is then the
limit of these approximations as the width of the field of view shrinks
to zero:

$$f'(27) = \text{the slope of the graph} = \lim_{\Delta x \to 0} \frac{\Delta y}{\Delta x}.$$

In the limit process we take $\Delta x \to 0$ because Δx is the width of the field of view. Since five digits of $\Delta y / \Delta x$ have stabilized, we can write

$$f'(27) = -.27235\ldots.$$

To find $f'(x)$ at some other point x, proceed the same way. Magnify the graph at that point repeatedly, until the value of the slope stabilizes. The method is very powerful. In the exercises you will have an opportunity to use it with other functions.

By using a microscope of arbitrarily high power, we have obtained further insights about rates and slopes. In fact, with these insights we can now state definitively what we mean by the slope of a curved graph and the rate of change of a function.

Definition. The **slope** of a graph at a point is the *limit* of the slopes seen in a microscope at that point, as the field of view shrinks to zero.

Definition. The **rate of change** of a function at a point is the slope of its graph at that point. Thus the rate of change is also a limit.

To calculate the value of the slope of the graph of $f(x)$ when $x = a$, we have to carry out a limit process. We can break down the process into these four steps:

1. Magnify the graph at the point $(a, f(a))$ until it appears straight.
2. Calculate the slope of the magnified segment.
3. Repeat steps 1 and 2 with successively higher magnifications.
4. Take the limit of the succession of slopes produced in step 3.

Local Linearity

The crucial property of a microscope is that it allows us to look at a graph **locally**, that is, in a small neighborhood of a particular point. The two functions we have been studying in this section have curved graphs—like most functions. But *locally*, their graphs are straight—or nearly so. This is a remarkable property, and we give it a name. We

A microscope gives a local view

say these functions are **locally linear**. In other words, a locally linear function looks like a linear function, locally.

The graph of a linear function has a definite slope at every point, and so does a *locally* linear function. For a linear function, the slope is easy to calculate, and it has the same value at every point. For a locally linear function, the slope is harder to calculate; it involves a limit process. The slope also varies from point to point.

All the standard function are locally linear at almost all points
How common is local linearity? All the standard functions you already deal with are locally linear almost everywhere. To see why we use the qualifying phrase "almost everywhere," look at what happens when we view the graph of $y = f(x) = x^{2/3}$ with a microscope. At any point other than the origin, the graph is locally linear. For instance, if we zoom in on the point $(1,1)$, here's what we see:

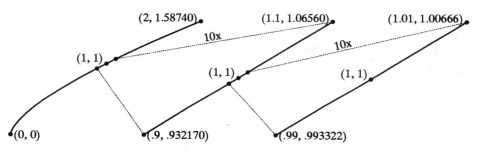

As the field of view shrinks, the graph looks more and more like a straight line. Using the highest magnification given, we estimate the slope of the graph—and hence the rate of change of the function—to be

$$f'(1) \approx \frac{\Delta y}{\Delta x} = \frac{1.006656 - .993322}{1.01 - .99} = \frac{.013334}{.02} = .6667.$$

Similarly, if we zoom in on the point $(.001, .01)$ we get:

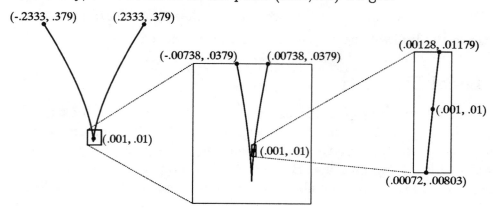

In the last window the graph looks like a line of slope

$$f'(.001) \approx \frac{.0118 - .0082}{.00128181 - .00074254} = \frac{.0036}{.00053937} = 6.674.$$

At the origin, though, something quite different happens:

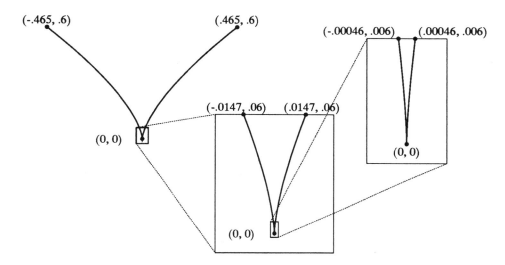

The graph simply looks more and more sharply pointed the closer we zoom in to the origin—it never looks like a straight line. However, the origin turns out to be the only point where the graph does not eventually look like a straight line.

In spite of these examples, it is important to realize that local linearity is a very special property. There are some functions that fail to be locally linear anywhere! Such functions are called **fractals**. No matter how much you magnify the graph of a fractal at any point, it continues to look non-linear—bent and "pointy" in various ways. In recent years fractals have been used in problems where the more common (locally linear) functions are inadequate. For instance, they describe irregular shapes like coastlines and clouds, and they model the way molecules are knocked about in a fluid (this is called *Brownian motion*). However, calculus does not deal with such functions. On the contrary,

Fractals are locally non-linear objects

> **Calculus studies functions that are locally linear almost everywhere.**

Exercises

Using a microscope

1.   Use a computer microscope to do the following. (A suggestion: first look at each graph over a fairly large interval.)

a) With a window of size $\Delta x = .002$, estimate the rate $f'(1)$ where $f(x) = x^4 - 8x$.

b) With a window of size $\Delta x = .0002$, estimate the rate $g'(0)$ where $g(x) = 10^x$.

c) With a window of size $\Delta t = .05$, estimate the slope of the graph of $y = t + 2^{-t}$ at $t = 7$.

d) With a window of size $\Delta z = .0004$, estimate the slope of the graph of $w = \sin z$ at $z = 0$.

2.   Use a computer microscope to determine the following values, correct to 1 decimal place. Obtain estimates using a *sequence* of windows, and shrink the field of view until the first 2 decimal places stabilize. Show *all* the estimates you constructed in each sequence.

a) $f'(1)$ where $f(x) = x^4 - 8x$.

b) $h'(0)$ where $h(s) = 3^s$.

c)   The slope of the graph of $w = \sin z$ at $z = \pi/4$.

d)   The slope of the graph of $y = t + 2^{-t}$ at $t = 7$.

e)   The slope of the graph of $y = x^{2/3}$ at $x = -5$.

3.   For each of the following functions, magnify its graph at the indicated point until the graph appears straight. Determine the equation of that straight line. Then verify that your equation is correct by plotting it as a second function in the same window you are viewing the given function. (The two graphs should "share phosphor"!)

a) $f(x) = \sin x$ at $x = 0$;

b) $\varphi(t) = t + 2^{-t}$ at $t = 7$;

c) $H(x) = x^{2/3}$ at $x = -5$.

4.   Consider the function that we investigated in the text:

$$f(x) = \frac{2 + x^3 \cos x + 1.5^x}{2 + x^2}.$$

a) Determine $f(0)$.

b) Make a sketch of the graph of f on the interval $-1 \leq x \leq 1$. Use the same scale on the horizontal and vertical axes so your graph shows slopes accurately.

c) Sketch what happens when you magnify the previous graph so the field of view is only $-.001 \leq x \leq .001$.

d) Estimate the slope of the line you drew in the part (c).

e) Estimate $f'(0)$. How many decimal places of accuracy does your estimate have?

f) What is the equation of the line in part (c)?

5.   A function that occurs in several different contexts in physical problems is
$$g(x) = \frac{\sin x}{x}.$$
Use a graphing program to answer the following questions.

a) Estimate the rate of change of g at the following points to two decimal place accuracy:

$$g'(1), \qquad g'(2.79), \qquad g'(\pi), \qquad g'(3.1).$$

b) Find three values of x where $g'(x) = 0$.

c) In the interval from 0 to 2π, where is g decreasing the most rapidly? At what rate is it decreasing there?

d) Find a value of x for which $g'(x) = -0.25$.

e) Although $g(0)$ is not defined, the function $g(x)$ seems to behave nicely in a neighborhood of 0. What seems to be true about $g(x)$ and $g'(x)$ when x is near 0?

f) According to your graphs, what value does $g(x)$ approach as $x \to 0$? What value does $g'(x)$ approach as $x \to 0$?

Rates from graphs; graphs from rates

6.   a) Sketch the graph of a function f that has $f(1) = 1$ and $f'(1) = 2$.

b) Sketch the graph of a function f that has $f(1) = 1$, $f'(1) = 2$, and $f(1.1) = -5$.

7.   A and B start off at the same time, run to a point 50 feet away, and return, all in 10 seconds. A graph of distance from the starting point as a function of time for each runner appears below. It tells where each runner is during this time interval.

a) Who is in the lead during the race?

b) At what time(s) is A farthest ahead of B? At what time(s) is B farthest ahead of A?

c) Estimate how fast A and B are going after one second.

d) Estimate the velocities of A and B during each of the ten seconds. Be sure to assign *negative* velocities to times when the distance to the starting point is *shrinking*. Use these estimates to sketch graphs of the velocities of A and B versus time. (Although the velocity of B changes rapidly around $t = 5$, assume that the graph of B's distance *is* locally linear at $t = 5$.)

e) Use your graphs in (d) to answer the following questions. When is A going faster than B? When is B going faster than A? Around what time is A running at -5 feet/second (i.e., running 5 feet/second *toward* the starting point)?

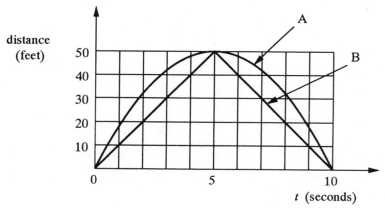

8.   For each of the following functions draw a graph which reflects the given information. Restate the given information in the language and notation of rate of change, paying particular attention to the units in which any rate of change is expressed.

a) A woman's height h (in inches) depends on her age t (in years). Babies grow very rapidly for the first two years, then more slowly until the adolescent growth spurt; much later, many women actually become shorter because of loss of cartilage and bone mass in the spinal column.

b) The number R of rabbits in a meadow varies with time t (in years). In the early years food is abundant and the rabbit population grows rapidly. However, as the population of rabbits approaches the "carrying capacity" of the meadow environment, the growth rate slows, and the population never exceeds the carrying capacity. Each year, during the harsh conditions of winter, the population dies back slightly, although it never gets quite as low as its value the previous year.

c) In a fixed population of couples who use a contraceptive, the average number N of children per couple depends on the effectiveness E (in percent) of the contraceptive. If the couples are using a contraceptive of low effectiveness, a small increase in effectiveness has a small effect on the value of N. As we look at contraceptives of greater and greater effectiveness, small additional increases in effectiveness have larger and larger effects on N.

9. If we graph the distance travelled by a parachutist in freefall as a function of the length of time spent falling, we would get a picture something like the following:

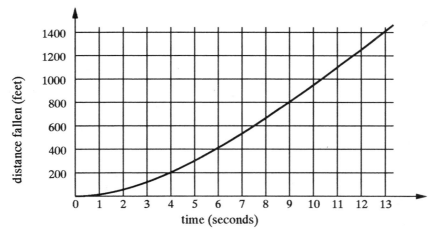

a) Use this graph to make estimates of the parachutist's velocity at the end of each second.

b) Describe what happens to the velocity as time passes.

c) How far do you think the parachutist would have fallen by the end of 15 seconds?

10. True or false. If you think a statement is true, give your reason; if you think a statement is false, give a counterexample—i.e., an example

that shows why it must be false.

a) If $g'(t)$ is positive for all t, we can conclude that $g(214)$ is positive.

b) If $g'(t)$ is positive for all t, we can conclude that $g(214) > g(17)$.

c) Bill and Samantha are driving separate cars in the same direction along the same road. At the start Samantha is 1 mile in front of Bill. If their speeds are the same at every moment thereafter, at the end of 20 minutes Samantha will be 1 mile in front of Bill.

d) Bill and Samantha are driving separate cars in the same direction along the same road. They start from the same point at 10 am and arrive at the same destination at 2 pm the same afternoon. At some time during the four hours their speeds must have been exactly the same.

When local linearity fails

11.   The absolute value function $f(x) = |x|$ is *not* locally linear at $x = 0$. Explore this fact by zooming in on the graph at $(0, 0)$. Describe what you see in successively smaller windows. Is there any change?

12.   Find three points where the function $f(x) = |\cos x|$ fails to be locally linear. Sketch the graph of f to demonstrate what is happening.

13.   Zoom in on the graph of $y = x^{4/5}$ at $(0, 0)$. In order to get an accurate picture, be sure that you use the same scales on the horizontal and vertical axes. Sketch what you see happening in successive windows. Is the function $x^{4/5}$ locally linear at $x = 0$?

14.   Is the function $x^{4/5}$ locally linear at $x = 1$? Explain your answer.

15.   This question concerns the function $K(x) = x^{10/9}$.

a) Sketch the graph of K on the interval $-1 \leq x \leq 1$. Compare K to the absolute value function $|x|$. Are they similar or dissimilar? In what ways? Would you say K is locally linear at the origin, or not?

b) Magnify the graph of K at the origin repeatedly, until the field of view is no bigger than $\Delta x \leq 10^{-10}$. As you magnify, be sure the scales on the horizontal and vertical axes remain the same, so you get a true picture of the slopes. Sketch what you see in the final window.

c) After using the microscope do you change your opinion about the local linearity of K at the origin? Explain your response.

§3. The Derivative

Definition

One of our main goals in this chapter is to make precise the notion of the rate of change of a function. In fact, we have already done that in the last section. We defined the rate of change of a function at a point to be the slope of its graph at that point; we defined the slope, in turn, by a four-step limit process. Thus, the precise definition of a rate of change involves a limit, and it involves geometric visualization—we think of a rate as a slope. We introduce a new word—*derivative*—to embrace both of these concepts as we now understand them.

> **Definition.** The **derivative** of the function $f(x)$ at $x = a$ is its rate of change at $x = a$, which is the same as the slope of its graph at $(a, f(a))$. The derivative of f at a is denoted $\boldsymbol{f'(a)}$.

Later in this section we will extend our interpretation of the derivative to include the idea of a *multiplier*, as well as a rate and a slope. Besides providing us with a single word to describe rates, slopes, and multipliers, the term "derivative" also reminds us that the quantity $f'(a)$ is *derived* from information about the function f in a particular way. It is worth repeating here the four steps by which we derive $f'(a)$:

1. Magnify the graph at the point $(a, f(a))$ until it appears straight.

2. Calculate the slope of the magnified segment.

3. Repeat steps 1 and 2 with successively higher magnifications.

4. Take the limit of the succession of slopes produced in step 3.

We can express this limit in analytic form in the following way:

$$f'(a) = \lim_{\Delta x \to 0} \frac{\Delta y}{\Delta x} = \lim_{h \to 0} \frac{f(a + h) - f(a - h)}{2h}.$$

The **difference quotient**

$$\frac{\Delta y}{\Delta x} = \frac{f(a + h) - f(a - h)}{2h}$$

is the usual way we estimate the slope of the magnified graph of f at the point $(a, f(a))$. As the figure on the top of the next page shows, the calculation involves two points equally spaced on either side of $(a, f(a))$.

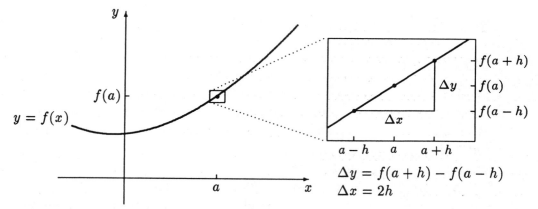

Estimating the value of the derivative $f'(a)$

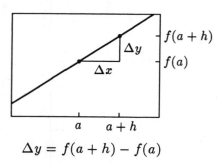

$\Delta y = f(a + h) - f(a)$

Choosing points in a window. To estimate the slope in the window above, we chose two particular points, $(a-h, f(a-h))$ and $(a+h, f(a+h))$. However, *any* two points in the window would give us a valid estimate. Our choice depends on the situation. For example, if we are working with formulas, we want simple expressions. In that case we would probably replace $(a - h, f(a - h))$ by $(a, f(a))$. We do that in the window on the left. The resulting slope is

$$\frac{\Delta y}{\Delta x} = \frac{f(a + h) - f(a)}{h}.$$

While the limiting value of $\Delta y/\Delta x$ doesn't depend on the choices you make, the *estimates* you produce with a fixed Δx can be closer to or farther from the true value. The exercises will explore this.

Data versus formulas. The derivative is a limit. To find that limit we have to be able to zoom in arbitrarily close, to make Δx arbitrarily small. For functions given by data, that is usually impossible; we can't use any Δx smaller than the spacing between the data points. Thus, a data function of this sort does not have a derivative, strictly speaking. However, by zooming in as much as the data allows, we get the most precise description possible for the rate of change of the function. In these circumstances it makes a difference which points we choose in a window to calculate $\Delta y/\Delta x$. In the exercises you will have a chance to see how the precision of your estimate depends on which points you choose to calculate the slope.

A data function might not have a derivative

For a function given by a formula, it is possible to find the value of the derivative exactly. In fact, the derivative of a function given by a formula is itself given by a formula. Later in this chapter we will describe some general rules which will allow us to produce the formula for the derivative without going through the successive approximation process each time. In chapter 5 we will discuss these rules more fully.

There are rules for finding derivatives

Practical considerations. The derivative is a limit, and there are always practical considerations to raise when we discuss limits. As we saw in chapter 2, we cannot expect to construct the entire decimal expansion of a limit. In most cases all we can get are a specified number of digits. For example, in §2 we found that

$$\text{if } f(x) = \frac{2 + x^3 \cos x + 1.5^x}{2 + x^2}, \text{ then } f'(27) = -.27235\ldots.$$

The same digits without the "..." give us **approximations.** Thus we can write $f'(27) \approx -.27235$; we also have $f'(27) \approx -.2723$ and $f'(27) \approx -.272$. Which approximation is the right one to use depends on the context. For example, if f appears in a problem in which all the other quantities are known only to one or two decimal places, we probably don't need a very precise value for $f'(27)$. In that case we don't have to carry the sequence of slopes $\Delta y / \Delta x$ very far. For instance, to write $f'(27) \approx -.272$, we only need to reduce Δx to the point where the slopes $\Delta y / \Delta x$ have values that all begin $-.2723\ldots.$ By the table on page 110, $\Delta x = .0008$ is sufficient.

Language and Notation

• If f has a derivative at a, we also say **f is differentiable at a.** If f is differentiable at every point a in its domain, we say **f is differentiable.**

> The German philosopher Gottfried Wilhelm Leibniz (1646–1716) developed calculus about the same time Newton did. While Newton dealt with derivatives in more or less the way we do, Leibniz introduced a related idea which he called a *differential*. Thus, Leibniz would say that a function that had a derivative was *differentiable*.

• Do *locally linear* and *differentiable* mean the same thing? The awkward case is a function whose graph is vertical at a point (for example, $y = \sqrt[3]{x}$ at the origin). On the one hand, it makes sense to say that the function is locally linear at such a point, because the graph looks straight under a microscope. On the other hand, the derivative itself is undefined, because the line is vertical. So the function is locally linear, but not differentiable, at that point.

There is another way to view the matter. We can say, instead, that a vertical line *does* have a slope, and its value is *infinity* (∞). From this point of view, if the graph of f is vertical at $x = a$, then $f'(a) = \infty$. In other words, f *does* have a derivative at $x = a$; its value just happens to be ∞.

Which view is "right"? Neither; we can choose either. Our choice is a matter of convention. (In some countries cars travel on the left; in others, on the right. That's a convention, too.) However, we will follow the second alternative. One advantage is that we will be able to use the derivative to indicate where the graph of a function is vertical. Another is that *locally linear* and *differentiable* then mean exactly the same thing.

Convention: if
the graph of f is
vertical at a,
write $f'(a) = \infty$

• Suppose $y = f(x)$ and the quantities x and y appear in a context in which they have units. Then the derivative of $f'(x)$ *also* has units, because it is the rate of change of y with respect to x. The units for the derivative must be

$$\text{units for } f' = \frac{\text{units for output } y}{\text{units for input } x}.$$

We have already seen several examples—persons per day, miles per hour, milligrams per pound, dollars of profit per dollar change in price—and we will see many more.

• There are several notations for the derivative. You should be aware of them because they are all in common use and because they reflect different ways of viewing the derivative. We have been writing the derivative of $y = f(x)$ as $f'(x)$. Leibniz wrote it as dy/dx. This notation has several advantages. It resembles the quotient $\Delta y/\Delta x$ that we use to approximate the derivative. Also, because dy/dx looks like a rate, it helps remind us that a derivative is a rate. Later on, when we consider the chain rule to find derivatives, you'll see that it can be stated very vividly using Leibniz's notation.

Leibniz's
notation

Calculus was originally thought of as a method of calculation. Besides using the derivative dy/dx, Leibniz did calculations with the individual symbols dx and dy, which he called *differentials*.

The other notation still encountered is due to Newton. It occurs primarily in physics and is used to denote rates with respect to time. If a quantity y is changing over time, then the Newton notation expresses the derivative of y as $\dot{y}$ (that's the variable y with a dot over it).

Newton's
notation

The Microscope Equation

A Context: Driving Time

If you make a 400 mile trip at an average speed of 50 miles per hour, then the trip takes 8 hours. Suppose you increase the average speed by 2 miles per hour. How much time does that cut off the trip?

One way to approach this question is to start with the basic formula

$$\text{speed} \times \text{time} = \text{distance}.$$

The distance is known to be 400 miles, and we really want to understand how *time* T depends on *speed* s. We get T as a function of s by rewriting the last equation:

Travel time depends on speed

$$T \text{ hours} = \frac{400 \text{ miles}}{s \text{ miles per hour}}.$$

To answer the question, just set $s = 52$ miles per hour in this equation. Then $T = 7.6923$ hours, or about 7 hours, 42 minutes. Thus, compared to the original 8 hours, the higher speed cuts 18 minutes off your driving time.

What happens to the driving time if you increase your speed by 4 miles per hour, or 5, instead of 2? What happens if you go slower, say 2 or 3 miles per hour slower? We could make a fresh start with each of these questions and answer them, one by one, the same way we did the first. But taking the questions one at a time misses the point. What we really want to know is the general pattern:

> If I'm travelling at 50 miles per hour, how much does any given increase in speed decrease my travel time?

How does travel time respond to changes in speed?

We already know how T and s are related: $T = 400/s$ hours. This question, however, is about the connection between a *change* in speed of

$$\Delta s = s - 50 \text{ miles per hour}$$

and a *change* in arrival time of

$$\Delta T = T - 8 \text{ hours}.$$

To answer it we should change our point of view slightly. It is not the relation between s and T, but between Δs and ΔT, that we want to understand.

Since we are considering speeds s that are only slightly above or below 50 miles per hour, Δs will be small. Consequently, the arrival time T will be only slightly different from 8 hours, so ΔT will also be small. Thus we want to study small changes in the function $T = 400/s$ near $(s, T) = (50, 8)$. The natural tool to use is a microscope.

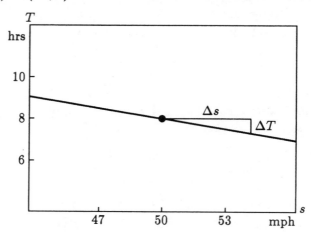

How travel time changes with speed around 50 miles per hour

In the microscope window above we see the graph of $T = 400/s$, magnified at the point $(s, T) = (50, 8)$. The field of view was chosen so that s can take values about 6 mph above or below 50 mph. The graph looks straight, and its slope is $T'(50)$. In the exercises you are asked to determine the value of $T'(50)$; you should find $T'(50) \approx -.16$. (Later on, when we have rules for finding the derivative of a formula, you will see that $T'(50) = -.16$ *exactly*.) Since the quotient $\Delta T/\Delta s$ is also an estimate for the slope of the line in the window, we can write

The slope of the graph in the microscope window

$$\frac{\Delta T}{\Delta s} \approx -.16 \text{ hours } per \text{ mile per hour.}$$

If we multiply both sides of this approximate equation by Δs miles per hours, we get

$$\Delta T \approx -.16\, \Delta s \text{ hours.}$$

How travel time changes with speed

This equation answers our question about the general pattern relating changes in travel time to changes in speed. It says that the changes are *proportional*. For every mile per hour increase in speed, travel time decreases by about .16 hours, or about $9\frac{1}{2}$ minutes. Thus, if the speed is 1 mph over 50 mph, travel time is cut by about $9\frac{1}{2}$ minutes. If we double the increase in speed, that doubles the savings in time: if the speed is 2 mph over 50 mph, travel time is cut by about 19 minutes. Compare this with a value of about 18 minutes that we got with the exact equation $T = 400/s$.

Notice that we are using Δs and ΔT in a slightly more restricted way than we have previously. Up to now, Δs measured the horizontal distance between *any* two points on a graph. Now, however, Δs just measures the horizontal distance from the fixed point $(s, T) = (50, 8)$ (marked with a large dot) that sits at the center of the window. Likewise, ΔT just measures the vertical distance from this point. The central point therefore plays the role of an origin, and Δs and ΔT are the *coordinates* of a point measured from that origin. To underscore the fact that Δs and ΔT are really coordinates, we have added a Δs-axis and a ΔT-axis in the window below. Notice that these coordinate axes have their own labels and scales.

Δs and ΔT now have a special meaning

Every point in the window can therefore be described in two different coordinate systems. The two different sets of coordinates of the point labelled P, for instance, are $(s, T) = (53, 7.52)$ and $(\Delta s, \Delta T) = (3, -.48)$. The first pair says "When your speed is 53 mph, the trip will take 7.52 hrs." The second pair says "When you increase your speed by 3 mph, you will decrease travel time by .48 hrs." Each statement can be translated into the other, but each statement has its own point of reference.

Δs and ΔT are coordinates in the window

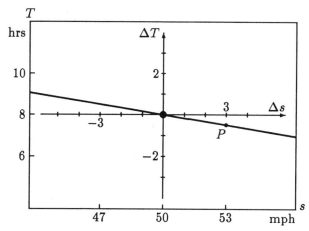

The microscope equation: $\Delta T \approx -.16\,\Delta s$ hours

We call $\Delta T \approx -.16\,\Delta s$ the **microscope equation** because it tells us how the microscope coordinates Δs and ΔT are related.

In fact, we now have two different ways to describe how travel time is related to speed. They can be compared in the following table.

	GLOBAL	LOCAL
coordinates:	$s,\ T$	$\Delta s,\ \Delta T$
equation:	$T = 400/s$	$\Delta T \approx -.16\,\Delta s$
properties:	exact non-linear	approximate linear

Global vs. local descriptions

We say the microscope equation is *local* because it is intended to deal only with speeds near 50 miles per hour. There is a different microscope equation for speeds near 40 miles per hour, for instance. By contrast, the original equation is *global*, because it works for all speeds. While the global equation is exact it is also non-linear; this can make it more difficult to compute. The microscope equation is approximate but linear; it is easy to compute. It is also easy to put into words:

At 50 miles per hour, the travel time of a 400 mile journey decreases $9\frac{1}{2}$ minutes for each mile per hour increase in speed.

The connection between the global equation and the microscope equation is shown in the following illustration.

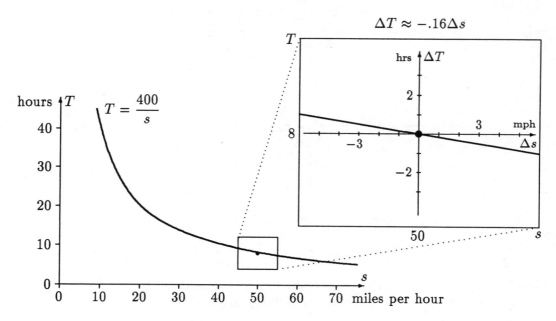

Local Linearity and Multipliers

The reasoning that led us to a microscope equation for travel time can be applied to any locally linear function. If $y = f(x)$ is locally linear, then at $x = a$ we can write

$$\boxed{\text{The microscope equation:} \quad \Delta y \approx f'(a) \cdot \Delta x}$$

We know an equation of the form $\Delta y = m \cdot \Delta x$ tells us that y is a linear function of x, in which m plays the role of slope, rate, and multiplier. The microscope equation therefore tells us that **y is a linear function of x when x is near a**—at least approximately. In this almost-linear relation, the derivative $f'(a)$ plays the role of slope, rate, and multiplier.

The microscope equation is just the idea of local linearity expressed analytically rather than geometrically—that is, by a formula rather than by a picture. Here is a chart that shows how the two descriptions of local linearity fit together.

The microscope equation is the analytic form of local linearity

$y = f(x)$ is locally linear at $x = a$:

GEOMETRICALLY	ANALYTICALLY
When magnified at $(a, f(a))$,	When x is near a,
the graph of f is almost straight,	y is almost a linear function of x,
and the slope of the line is $f'(a)$.	and the multiplier is $f'(a)$.

microscope window

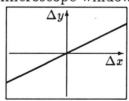

microscope equation

$$\boxed{\Delta y \approx f'(a) \cdot \Delta x}$$

Of course, the graph in a microscope window is not *quite* straight. The analytic counterpart of this statement is that the microscope equation is not *quite* exact—the two sides of the equation are only approximately equal. We write "$\approx$" instead of "$=$". However, we can make the graph look even straighter by increasing the magnification—or, what is the same thing, by decreasing the field of view. Analytically, this increases the exactness of the microscope equation.

In the microscope equation $\Delta y \approx f'(a) \cdot \Delta x$, **the derivative is the multiplier that tells how y responds to changes in x.** In particular, a small increase in x produces a change in y that depends on the sign and magnitude of $f'(a)$ in the following way:

- $f'(a)$ is large and positive $\Rightarrow$ large increase in y,

- $f'(a)$ is small and positive $\Rightarrow$ small increase in y,

- $f'(a)$ is large and negative $\Rightarrow$ large decrease in y,

- $f'(a)$ is small and negative $\Rightarrow$ small decrease in y.

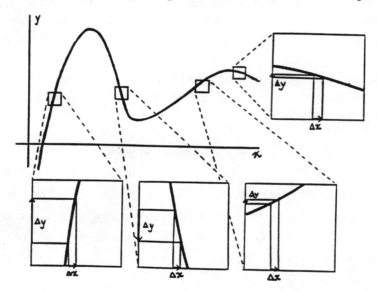

For example, suppose we are told the value of the derivative is 2. Then any small change in x induces a change in y approximately twice as large. If, instead, the derivative is $-1/5$, then a small change in x produces a change in y only one fifth as large, and in the opposite direction. That is, if x increases, then y decreases, and vice-versa.

The microscope equation is the recipe for building solutions to rate equations

The microscope equation should look familiar to you. It has been with us from the beginning of the course. Our "recipe" $\Delta S = S' \cdot \Delta t$ for predicting future values of S in the S-I-R model is just the microscope equation for the function $S(t)$. (Although we wrote it with an "$=$" instead of an "$\approx$" in the first chapter, we noted that ΔS would provide us only an *estimate* for the new value of S.)

The derivative is one of the fundamental concepts of the calculus, and one of its most important roles is in the microscope equation. Besides giving us a tool for building solutions to rate equations, the microscope equation helps us do estimation and error analysis, the subject of the next section.

We conclude with a summary that compares linear and locally linear functions. Note there are two differences, but only two: 1) the equation for local linearity is only an approximation; 2) it holds only locally—i.e., near a given point.

If $y = f(x)$ is linear,	If $y = f(x)$ is locally linear,
then $\Delta y = m \cdot \Delta x$;	then $\Delta y \approx f'(a) \cdot \Delta x$;
the constant m is	the derivative $f'(a)$ is
rate, slope, and multiplier	rate, slope, and multiplier
for all x.	for x near a.

Exercises

Computing Derivatives

1. Sketch graphs of the following functions and use these graphs to determine which function has a derivative that is always positive (except at $x = 0$, where neither the function nor its derivative is defined).

$$y = \frac{1}{x} \quad y = \frac{-1}{x} \quad y = \frac{1}{x^2} \quad y = \frac{-1}{x^2}$$

What feature of the graph told you whether the derivative was positive?

2. For each of the functions f below, approximate its derivative at the given value $x = a$ in two different ways. First, use a computer microscope (i.e., a graphing program) to view the graph of f near $x = a$. Zoom in until the graph looks straight and find its slope. Second, use a calculator to find the value of the quotient

$$\frac{f(a + h) - f(a - h)}{2h}$$

for $h = .1, .01, .001, \ldots, .000001$. Based on these values of the quotients, give your best estimate for $f'(a)$, and say how many decimal places of accuracy it has.

a) $f(x) = 1/x$ at $x = 2$.   c) $f(x) = x^3$ at $x = 200$.

b) $f(x) = \sin(7x)$ at $x = 3$.   d) $f(x) = 2^x$ at $x = 5$.

3.   In a later section we will establish that the derivative of $f(x) = x^3$ at $x = 1$ is exactly 3: $f'(1) = 3$. This question concerns the freedom we have to choose points in a window to estimate $f'(1)$ (see page 120). Its purpose is to compare two quotients, to see which gets closer to the exact value of $f'(1)$ for a fixed "field of view" Δx. The two quotients are

$$Q_1 = \frac{\Delta y}{\Delta x} = \frac{f(a+h) - f(a-h)}{2h} \quad \text{and} \quad Q_2 = \frac{\Delta y}{\Delta x} = \frac{f(a+h) - f(a)}{h}.$$

In this problem $a = 1$.

a) Construct a table that shows the values of Q_1 and Q_2 for each $h = 1/2^k$, where $k = 0, 1, 2, \ldots, 8$. If you wish, you can use this program to compute the values:

```
a = 1
FOR k = 0 TO 8
       h = 1 / 2 ^ k
       q1 = ((a + h) ^ 3 - (a - h) ^ 3) / (2 * h)
       q2 = ((a + h) ^ 3 - a ^ 3) / h
       PRINT h, q1, q2
NEXT k
```

b) How many digits of Q_1 stabilize in this table? How many digits of Q_2?

c) Which is a better estimator—Q_1 or Q_2? To indicate how much better, give the value of h for which the *better* estimator provides an estimate that is as close as the best estimate provided by the *poorer* estimator.

[Answer: Q_1 is much better than Q_2. The best Q_2 estimate is 3.012, when $h = 1/2^8 = .004$, but Q_1 gives 3.004 when $h = 1/2^4 = .063$, an h-value that is 16 times larger.]

4.   Repeat all the steps of the last question for the function $f(x) = \sqrt{x}$ at $x = 9$. The exact value of $f'(9)$ is $1/6$.

Comment: Note that, in §1, we estimated the rate of change of the sunrise function using an expression like Q_1 rather than one like Q_2. The previous exercises should persuade you this was deliberate. We were trying to get the most representative estimates, given the fact that we could not reduce the size of Δx arbitrarily.

5. Use one of the methods of problem 2 to estimate the value of the derivative of each of the following functions at $x = 0$:

$$y = 2^x, \quad y = 3^x, \quad y = 10^x, \quad \text{and} \quad y = (1/2)^x.$$

These are called **exponential** functions, because the input variable x appears in the exponent. How many decimal places accuracy do your approximations to the derivatives have?

6. In this problem we look again at the exponential function $f(x) = 2^x$ from the previous problem.

a) Use the rules for exponents to put the quotient

$$\frac{f(a + h) - f(a)}{h}$$

in the simplest form you can.

b) We know that

$$f'(0) = \lim_{h \to 0} \frac{f(h) - f(0)}{h}.$$

Use this fact, along with the algebraic result of part (a), to explain why $f'(a) = f'(0) \cdot 2^a$.

7. Apply all the steps of the previous question to the exponential function $f(x) = b^x$ with an arbitrary base b. Show that $f'(x) = f'(0) \cdot b^x$.

8. a) For which values of x is the absolute value function $y = |x|$ differentiable?

b) At each point where $y = |x|$ is differentiable, find the value of the derivative.

The microscope equation

9.   Write the microscope equation for each of the following functions at the indicated point. (To find the necessary derivative, consult problem 2.)

a) $f(x) = 1/x$ at $x = 2$.

b) $f(x) = \sin(7x)$ at $x = 3$.

c) $f(x) = x^3$ at $x = 200$.

d) $f(x) = 2^x$ at $x = 5$.

10.   This question uses the microscope equation for $f(x) = 1/x$ at $x = 2$ that you constructed in the previous question.

a) Draw the graph of what you would see in the microscope if the field of view is .2 units wide.

b) If we take $x = 2.05$, what is Δx in the microscope equation? What estimate does the microscope equation give for Δy? What estimate does the microscope equation then give for $f(2.05) = 1/2.05$? Calculate the true value of $1/2.05$ and compare the two values; how far is the microscope estimate from the true value?

[Answer: The true value is .4878048... and the microscope estimate is .4875. The difference is about .0003.]

c) What estimate does the microscope equation give for $1/2.02$? How far is this from the true value?

d) What estimate does the microscope equation give for $1/1.995$? How far is this from the true value?

11.   This question concerns the travel time function $T = 400/s$ hours, discussed in the text.

a) How many hours does a 400-mile trip take at an average speed of 40 miles per hour?

b) Find the microscope equation for T when $s = 40$ miles per hour.

c) At what rate does the travel time decrease as speed increases around 40 mph—in hours *per* mile per hour?

d) According to the microscope equation, how much travel time is saved by increasing the speed from 40 to 45 mph?

[Answer: There is a savings of 1.25 hours.]

e) According to the microscope equation based at 50 mph (as done in the text), how much time is *lost* by decreasing the speed from 50 to 45 mph?

f) The last two parts both predict the travel time when the speed is 45 mph. Do they give the same result?

[Answer: The total travel times are not equal. The 40 mph microscope equation predicts the trip takes 8.75 hours at 45 mph, while the 50 mph microscope equation predicts it takes 8.8 hours. (The exact formula predicts 8.89 hours!)]

12.   a) Suppose $y = f(x)$ is a function for which $f(5) = 12$ and $f'(5) = .4$. Write the microscope equation for f at $x = 5$.

b) Draw the graph of what you would see in the microscope. Do you need a formula for f itself, in order to do this?

c) If $x = 5.3$, what is Δx in the microscope equation? What estimate does the microscope equation give for Δy? What estimate does the microscope equation then give for $f(5.3)$?

d) What estimates does the microscope equation give for the following: $f(5.23)$, $f(4.9)$, $f(4.82)$, $f(9)$? Do you consider these estimates to be equally reliable?

13.   a) Suppose $z = g(t)$ is a function for which $g(-4) = 7$ and $g'(-4) = 3.5$. Write the microscope equation for g at $t = -4$.

b) Draw the graph of what you see in the microscope.

c) Estimate $g(-4.2)$ and $g(-3.75)$.

d) For what value of t near -4 would you estimate that $g(t) = 6$? For what value of t would you estimate $g(t) = 8.5$?

14.   If $f(a) = b$, $f'(a) = -3$ and if k is small, which of the following is the best estimate for $f(a + k)$?

$$a + 3k, \ b + 3k, \ a + 3b, \ b - 3k, \ a - 3k, \ 3a - b, \ a^2 - 3b, \ f'(a + k)$$

15.   If f is differentiable at a, which of the following, for small values of h, are reasonable estimates of $f'(a)$?

$$\frac{f(a + h) - f(a - h)}{h} \qquad \frac{f(a + h) - f(a - h)}{2h}$$

$$\frac{f(a + h) - f(h)}{h} \qquad \frac{f(a + 2h) - f(a - h)}{3h}$$

16. Suppose a person has travelled D feet in t seconds. Then $D'(t)$ is the person's velocity at time t; $D'(t)$ has units of feet per second.

a) Suppose $D(5) = 30$ feet and $D'(5) = 5$ feet/second. Estimate the following:

$$D(5.1) \qquad D(5.8) \qquad D(4.7)$$

b) If $D(2.8) = 22$ feet, while $D(3.1) = 26$ feet, what would you estimate $D'(3)$ to be?

17. Fill in the blanks.

a) If $f(3) = 2$ and $f'(3) = 4$, a reasonable estimate of $f(3.2)$ is ____.

b) If $g(7) = 6$ and $g'(7) = .3$, a reasonable estimate of $g(6.6)$ is ____.

c) If $h(1.6) = 1$, $h'(1.6) = -5$, a reasonable estimate of $h(\_\_\_)$ is 0.

d) If $F(2) = 0$, $F'(2) = .4$, a reasonable estimate of $F(\_\_\_)$ is .15.

e) If $G(0) = 2$ and $G'(0) = \_\_\_$, a reasonable estimate of $G(.4)$ is 1.6.

f) If $H(3) = -3$ and $H'(3) = \_\_\_$, a reasonable estimate of $H(2.9)$ is -1.

18. In manufacturing processes the profit is usually a function of the number of units being produced, among other things. Suppose we are studying some small industrial company that produces n units in a week and makes a corresponding weekly profit of P. Assume $P = P(n)$.

a) If $P(1000) = \$500$ and $P'(1000) = \$2/\text{unit}$, then

$$P(1002) \approx \rule{1.5cm}{0.4pt} \qquad P(995) \approx \rule{1.5cm}{0.4pt} \qquad P(\rule{1.5cm}{0.4pt}) \approx \$512$$

b) If $P(2000) = \$3000$ and $P'(2000) = -\$5/\text{unit}$, then

$$P(2010) \approx \rule{1.5cm}{0.4pt} \qquad P(1992) \approx \rule{1.5cm}{0.4pt} \qquad P(\rule{1.5cm}{0.4pt}) \approx \$3100$$

c) If $P(1234) = \$625$ and $P(1238) = \$634$, then what is an estimate for $P'(1236)$?

§4. Estimation and Error Analysis

Making Estimates

The Expanding House

In the book *The Secret House – 24 hours in the strange and unexpected world in which we spend our nights and days* (Simon and Schuster, 1986), David Bodanis describes many remarkable events that occur at the microscopic level in an ordinary house. At one point he explains how sunlight heats up the structure, stretching it imperceptibly in every direction through the day until it has become several cubic inches larger than it was the night before. Is it plausible that a house can become several cubic inches larger as it expands in the heat of the day? In particular, how much longer, wider, and taller would it have to become if it were to grow in volume by, let us say, 3 cubic inches?

How much does a house expand in the heat?

For simplicity, assume the house is a cube 200 inches on a side. (This is about 17 feet, so the house is the size of a small, two-story cottage.) If s is the length of a side of *any* cube, in inches, then its volume is

$$V = s^3 \text{ cubic inches.}$$

Our question is about how V changes with s when s is about 200 inches. In particular, we want to know which Δs would yield a ΔV of 3 cubic inches. This is a natural question for the microscope equation

$$\Delta V \approx V'(200) \cdot \Delta s.$$

According to exercise 2c in the previous section, we can estimate the value of $V'(200)$ to be about 120,000, and the appropriate units for V' are cubic inches per inch. Thus

$$\Delta V \approx 120000 \, \Delta s$$
$$3 \text{ cubic inches} \approx 120000 \, \frac{\text{cubic inches}}{\text{inches}} \times \Delta s \text{ inches,}$$

so $\Delta s \approx 3/120000 = .000025$ inches—many times thinner than a human hair!

This value is much too small. To get a more realistic value, let's suppose the house is made of wood and the temperature increases about 30°F from night to day. Then measurements show that a 200-inch length of wood will actually become about $\Delta s = .01$ inches longer. Consequently the volume will actually expand by about

$$\Delta V \approx 120000 \frac{\text{cubic inches}}{\text{inches}} \times .01 \text{ inches} = 1200 \text{ cubic inches}.$$

This increase is 400 times as much as Bodanis claimed; it is about the size of a small computer monitor. So even as he opens our eyes to the effects of thermal expansion, Bodanis dramatically understates his point.

Estimates versus Exact Values

Don't lose sight of the fact that the values we derived for the expanding house are *estimates*. In some cases we can get the exact values. Why don't we, whenever we can?

For example, we can calculate exactly how much the volume increases when we add $\Delta s = .01$ inches to $s = 200$ inches. The increase is from $V = 200^3 = 8,000,000$ cubic inches to

$$V = (200.01)^3 = 8001200.060001 \text{ cubic inches}.$$

Thus, the *exact* value of ΔV is 1200.060001 cubic inches. The estimate is off by only about .06 cubic inches. This isn't very much, and it is even less significant when you think of it as a percentage of the volume (namely 1200 cubic inches) being calculated. The percentage is

$$\frac{.06 \text{ cubic inches}}{1200 \text{ cubic inches}} = .00005 = .005\%.$$

That is, the difference is only 1/200 of 1% of the calculated volume.

To get the exact value we had to cube two numbers and take their difference. To get the estimate we only had to do a single multiplica-

Exact values can be harder to calculate than estimates.

tion. Estimates made with the microscope equation are always easy to calculate—they involve only linear functions. Exact values are usually harder to calculate. As you can see in the example, the extra effort may not gain us extra information. That's one reason why we don't always calculate exact values when we can.

Here's another reason. Go back to the question: How large must Δs be if $\Delta V = 3$ cubic inches? To get the exact answer, we must solve for Δs in the equation

$$
\begin{aligned}
3 \;=\; \Delta V &= (200 + \Delta s)^3 - 200^3 \\
&= 200^3 + 3(200)^2 \Delta s + 3(200)(\Delta s)^2 + (\Delta s)^3 - 200^3.
\end{aligned}
$$

Simplifying, we get

$$
3 = 120000\,\Delta s + 600(\Delta s)^2 + (\Delta s)^3.
$$

This is a cubic equation for Δs; it *can* be solved, but the steps are complicated. Compare this with solving the microscope equation:

$$
3 = 120000\,\Delta s.
$$

Thus, another reason we don't calculate exact values at every opportunity is that the calculations can be daunting. The microscope estimates are always straightforward.

Propagation of Error

From Measurements to Calculations

We can view all the estimates we made for the expanding house from another perspective—the lack of precision in measurement. To begin with, just think of the house as a cubical box that measures 200 inches on a side. Then the volume must be 8,000,000 cubic inches. But measurements are never exact, and any uncertainty in measuring the length of the side will lead to an uncertainty in calculating the volume. Let's say your measurement of length is accurate to within .5 inch. In other words, you believe the true length lies between 199.5 inches and 200.5 inches, but you are uncertain precisely where it lies within that interval. How uncertain does that make your calculation of the volume?

There is a direct approach to this question: we can simply say that the volume must lie between $199.5^3 = 7{,}940{,}149.875$ cubic inches and $200.5^3 = 8{,}060{,}150.125$ cubic inches. In a sense, these values are almost too precise. They don't reveal a general pattern. We would like to know how an uncertainty—or error—in measuring the length of the side of a cube propagates to an error in calculating its volume.

How uncertain is the calculated value of the volume?

Let's take another approach. If we measure s as 200 inches, and the true value differs from this by Δs inches, then Δs is the error in measurement. That produces an error ΔV in the calculated value of V. The microscope equation for the expanding house (page 135) tells us how ΔV depends on Δs when $s = 200$:

$$\Delta V \approx 120000\,\Delta s.$$

Since we now interpret Δs and ΔV as errors, the microscope equation becomes the **error propagation** equation:

$$\text{error in } V \text{ (cu. in.)} \approx 120000 \left(\frac{\text{cu. in.}}{\text{inch}}\right) \times \text{error in } s \text{ (inches).}$$

The microscope equation describes how errors propagate

Thus, for example, an error of 1/2 inch in measuring s propagates to an error of about 60,000 cubic inches in calculating V. This is about 35 cubic feet, the size of a large refrigerator! Putting it another way:

if $s = 200 \pm.5$ inches, then $V \approx 8,000,000 \pm 60,000$ cubic inches.

If we keep in mind the error propagation equation $\Delta V \approx 120000\,\Delta s$, we can quickly answer other questions about measuring the same cube. For instance, suppose we wanted to determine the volume of the cube to within 5,000 cubic inches. How accurately would we have to measure the side? Thus we are given $\Delta V = 5000$, and we conclude $\Delta s \approx 5000/120000 \approx .04$ inches. This is just a little more than 1/32 inch.

Relative Error

Suppose we have a second cube whose side is twice as large ($s = 400$ inches), and once again we measure its length with an error of 1/2 inch. Then the error in the calculated value of the volume is

$$\Delta V \approx V'(400) \cdot \Delta s = 480000 \times .5 = 240,000 \text{ cubic inches.}$$

Bigger numbers have bigger errors

(In the exercises you will be asked to show $V'(400) = 480,000$.) The error in our calculation for the bigger cube is four times what it was for the smaller cube, even though the length was measured to the same accuracy in both cases! There is no mistake here. In fact, the volume of the second cube is eight times the volume of the first, so the numbers we are dealing with in the second case are roughly eight times as large. We should not be surprised if the error is larger, too.

In general, we must expect that the size of an error will depend on the size of the numbers we are working with. We expect big numbers to have big errors and small numbers to have small errors. In a sense, though, an error of 1 inch in a measurement of 50 inches is no worse than an error of 1/10-th of an inch in a measurement of 5 inches: both errors are 1/50-th the size of the quantity being measured.

> A watchmaker who measures the tiny objects that go into a watch only as accurately as a carpenter measures lumber would never make a watch that worked; likewise, a carpenter who takes the pains to measure things as accurately as a watchmaker does would take forever to build a house. The scale of allowable errors is dictated by the scale of the objects they work on.

The errors Δx we have been considering are called **absolute errors**; their values depend on the size of the quantities x we are working with. To reduce the effect of differences due to the size of x, we can look instead at the error as a *fraction* of the number being measured or calculated. This fraction $\Delta x/x$ is called **relative error**. Consider two measurements: one is 50 inches with an error of ± 1 inch; the other is 2 inches with an error of $\pm .1$ inch. The *absolute* error in the second measurement is much smaller than in the first, but the *relative* error is $2\frac{1}{2}$ times larger. (The first relative error is .02 inch per inch, the second is .05 inch per inch.) To judge how good or bad a measurement really is, we usually take the relative error instead of the absolute error.

Absolute and relative error

Let's compare the propagation of relative and absolute errors. For example, the absolute error in calculating the volume of a cube whose side measures s is

$$\Delta V \approx V'(s) \cdot \Delta s.$$

The absolute errors are proportional, but the multiplier $V'(s)$ depends on the size of s. (We saw above that the multiplier is 120,000 cubic inches per inch when $s = 200$ inches, but it grows to 480,000 cubic inches per inch when $s = 400$ inches.)

In §5, which deals with formulas for derivatives, we will see that $V'(s) = 3s^2$. If we substitute $3s^2$ for $V'(s)$ in the propagation equation for absolute error, we get

$$\Delta V \approx 3s^2 \cdot \Delta s.$$

To see how relative error propagates, we should divide this equation by $V = s^3$:

$$\frac{\Delta V}{V} \approx \frac{3s^2 \cdot \Delta s}{s^3} = 3\frac{\Delta s}{s}$$

The relative errors are proportional, but the multiplier is always 3; it doesn't depend on the size of the cube, as it did for absolute errors.

Return to the case where $\Delta s = .5$ inch and $s = 200$ inches. Since Δs and s have the same units, the relative error $\Delta s/s$ is "dimensionless"— it has no units. We can, however, describe $\Delta s/s$ as a *percentage*:

Percentage error
is relative error

$\Delta s/s = .5/200 = .25\%$, or $1/4$ of 1%. For this reason, relative error is sometimes called **percentage error**. It tells us the error in measuring a quantity as a *percentage* of the value of that quantity. Since the percentage error in volume is

$$\frac{\Delta V}{V} = \frac{60{,}000 \text{ cu. in.}}{8{,}000{,}000 \text{ cu. in.}} = .0075 = .75\%$$

we see that the percentage error in volume is 3 times the percentage error in length—*and this is independent of the length and volumes involved*. This is what the propagation equation for relative error says: A 1% error in measuring s, whether $s = .0002$ inches or $s = 2000$ inches, will produce a 3% error in the calculated value of the volume.

Exercises

Estimation

1. a) Suppose you are going on a 110 mile trip. Then the time T it takes to make the trip is a function of how fast you drive:

$$T(v) = \frac{110 \text{ miles}}{v \text{ miles per hour}} = 110\,v^{-1} \text{ hours} .$$

If you drive at $v = 55$ miles per hour, T will be 2 hours. Calculate $T'(55)$ and write an English sentence interpreting this number.

b) More generally, if you and a friend are driving separate cars on a 110 mile trip, and if her speed is 1% greater than yours, then her travel time is less. How much less, as a percentage of yours?

2. a) Suppose you have 600 square feet of plywood which you are going to use to construct a cubical box. Assuming there is no waste, what will its volume be?

b) Find a general formula which expresses the volume V of the box as a function of the area A of plywood available.

c) What is $V'(600)$?

d) Use this multiplier to estimate the additional amount of plywood you would need to increase the volume of the box by 10 cubic feet.

e) In the original problem, if you had to allow for wasting 10 square feet of plywood in the construction process, by how much would this decrease the volume of the box?

f) In the original problem, if you had to allow for wasting 2% of the plywood in the construction process, by what percentage would this decrease the volume of the box?

3. a) Let $R(s) = 1/s$. You can use the fact that $R'(s) = -1/s^2$, to be established in §5.

b) Since $R(100) = $ _____ and $R'(100) = $ _____, we can make the following approximations:

$$1/97 \approx \text{\_\_\_\_\_} \qquad 1/104 \approx \text{\_\_\_\_\_} \qquad R(\text{\_\_\_\_\_}) \approx .0106\,.$$

4. Using the fact that the derivative of $f(x) = \sqrt{x}$ is $f'(x) = 1/(2\sqrt{x})$, you can estimate the square roots of numbers that are close to perfect squares.

a) For instance $f(4) = $ _____ and $f'(4) = $ _____, so $\sqrt{4.3} \approx$ _____.
b) Use the values of $f(4)$ and $f'(4)$ to approximate $\sqrt{5}$ and $\sqrt{3.6}$.
c) Use the values of $f(100)$ and $f'(100)$ to approximate $\sqrt{101}$ and $\sqrt{99.73}$.

Error analysis

5. a) If you measure the side of a square to be 12.3 inches, with an uncertainty of $\pm.05$ inch, what is your relative error?

b) What is the area of the square? Write an error propagation equation that will tell you how uncertain you should be about this value.

c) What is the relative error in your calculation of the area?

d) If you wanted to calculate the area with an error of less than 1 square inch, how accurately would you have to measure the length of the side? If you wanted the error to be less than .1 square inch, how accurately would you have to measure the side?

6. a) Suppose the side of a square measures x meters, with a possible error of Δx meters. Write the equation that describes how the error in length propagates to an error in the area. (The derivative of $f(x) = x^2$ is $f'(x) = 2x$; see §5.)

b) Write an equation that describes how the *relative* error in length propagates to a *relative* error in area.

7.   You are trying to measure the height of a building by dropping a stone off the top and seeing how long it takes to hit the ground, knowing that the distance d (in feet) an object falls is related to the time of fall, t (in seconds), by the formula $d = 16t^2$. You find that the time of fall is 2.5 seconds, and you estimate that you are accurate to within a quarter of a second. What do you calculate the height of the building to be, and how much uncertainty do you consider your calculation to have?

8.   You see a flash of lightning in the distance and note that the sound of thunder arrives 5 seconds later. You know that at 20°C sound travels at 343.4 m/sec. This gives you an estimate of

$$5 \text{ sec} \times 343.4 \ \frac{\text{meters}}{\text{sec}} = 1717 \text{ meters}$$

for the distance between you and the spot where the lightning struck. You also know that the velocity v of sound varies as the square root of the temperature T measured in degrees Kelvin (the Kelvin temperature = Celsius temperature + 273), so

$$v(T) = k\sqrt{T}$$

for some constant k.

a)   Use the information given here to determine the value of k.

b)   If your estimate of the temperature is off by 5 degrees, how far off is your estimate of the distance to the lightning strike? How significant is this source of error likely to be in comparison with the imprecision with which you measured the 5 second time lapse? (Suppose your uncertainty about the time is .25 seconds.) Give a clear analysis justifying your answer.

9.   We can measure the distance to the moon by bouncing a laser beam off a reflector placed on the moon's surface and seeing how long it takes the beam to make the round trip. If the moon is roughly 400,000 km away, and if light travels at 300,000 km/sec, how accurately do we have to be able to measure the length of the time interval to be able to determine the distance to the moon to the nearest .1 meter?

§5. A Global View

Derivative as Function

Up to now we have looked upon the derivative as a *number*. It gives us information about a function at a *point*—the rate at which the function is changing, the slope of its graph, the value of the multiplier in the microscope equation. But the numerical value of the derivative varies from point to point, and these values can also be considered as the values of a new function—the derivative function—with its own graph. Viewed this way the derivative is a *global* object.

> The derivative is a function in its own right

The connection between a function and its derivative can be seen very clearly if we look at their graphs. To illustrate, we'll use the function $I(t)$ that describes how the size of an infected population varies over time, from the *S-I-R* problem we analyzed in chapter 1. The graph of I appears below, and directly beneath it is the graph of I', the derivative of I. The graphs are lined-up vertically: the values of $I(a)$ and $I'(a)$ are recorded on the same vertical line that passes through the point $t = a$ on the t-axis.

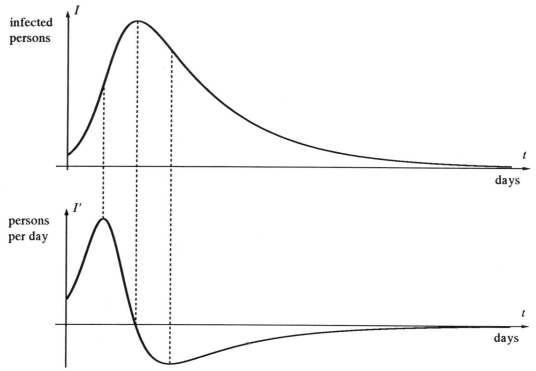

The height of I'
is the slope of I

To understand the connection between the graphs, keep in mind that the derivative represents a slope. Thus, at any point t, the *height* of the lower graph (I') tells us the *slope* of the upper graph (I). At the points where I is increasing, I' is positive—that is, I' lies *above* its t-axis. At the point where I is increasing most rapidly, I' reaches its highest value. In other words, where the graph of I is steepest, the graph of I' is highest. At the point where I is decreasing most rapidly, I' has its lowest value.

Next, consider what happens when I itself reaches its maximum value. Since I is about to switch from increasing to decreasing, the derivative must be about to switch from positive to negative. Thus, at the moment when I is largest, I' must be zero. Note that the highest point on the graph of I lines up with the point where I' crosses the t-axis. Furthermore, if we zoomed in on the graph of I at its highest point, we would find a horizontal line—in other words, one whose slope is zero.

All functions and their derivatives are related the same way that I and I' are. In the following table we list the various features of the graph of a function; alongside each is the corresponding feature of the graph of the derivative.

function	derivative
increasing	positive
decreasing	negative
horizontal	zero
steep (rising or falling)	large (positive or negative)
gradual (rising or falling)	small (positive or negative)
straight	horizontal

By using this table, you should be able to make a rough sketch of the graph of the derivative, when you are given the graph of a function. You can also read the table from right to left, to see how the graph of a function is influenced by the graph of its derivative.

For instance, suppose the graph of the function $L(x)$ is

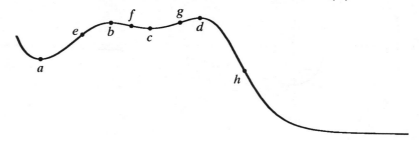

Then we know that its derivative L' must be 0 at points a, b, c, and d; that the derivative must be positive between a and b and between c and d, negative otherwise; that the derivative takes on relatively large values at e and g (positive) and at f and h (negative); that the derivative must approach 0 at the right endpoint and be large and negative at the left endpoint. Putting all this together we conclude that the graph of the derivative L' must look something like following:

Finding a derivative "by eye"

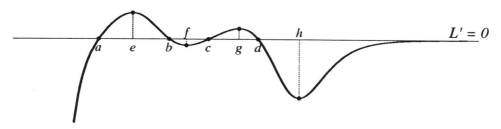

Conversely, suppose all we are told about a certain function G is that the graph of its derivative G' looks like this:

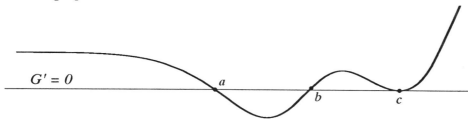

Then we can infer that the function G itself is decreasing between a and b and is increasing everywhere else; that the graph of G is horizontal at a, b, and c; that both ends of the graph of G slope upward from left to right—the left end more or less straight, the right getting steeper and steeper.

Formulas for Derivatives

Basic Functions

If a function is given by a formula, then its derivative also has a formula. The formula is produced by a definite process, called **differentiation**. In this section we look at some of the basic aspects, and in the next we will take up the chain rule, which is the key to the whole process. Then in chapter 5 we will review differentiation systematically.

Most formulas are constructed by combining only a few basic functions in various ways. For instance, the formula

$$3\,x^7 - \frac{\sin x}{8\sqrt{x}},$$

uses the basic functions x^7, $\sin x$, and $\sqrt{x}$. In fact, since $\sqrt{x} = x^{1/2}$, we can think of x^7 and $\sqrt{x}$ as two different instances of a single basic "power of x"—which we can write as x^p.

The following table lists some of the more common basic functions with their derivatives. The number c is an arbitrary constant, and so is the power p. The last function in the table is the exponential function with base b. Its derivative involves a parameter k_b that varies with the base b. For instance, exercise 6 in §3 established that, when $b = 2$, then $k_2 \approx .69$. Exercise 7 established, for any base b, that k_b is the value of the derivative of b^x when $x = 0$. We will have more to say about the parameter k_b in the next chapter.

function	derivative
c	0
x^p	px^{p-1}
$\sin x$	$\cos x$
$\cos x$	$-\sin x$
$\tan x$	$sec^2 x$
b^x	$k_b \cdot b^x$

For example,

- the derivative of $1/x = x^{-1}$ is $-x^{-2} = -1/x^2$;

- the derivative of $\sqrt{w} = w^{1/2}$ is $\frac{1}{2}w^{-1/2} = \frac{1}{2\sqrt{w}}$;

- the derivative of x^π is $\pi x^{\pi-1}$; and

- the derivative of π^x is $k_\pi \cdot \pi^x \approx 1.4\,\pi^x$.

Compare the last two functions. The first, x^π, is a **power function**—it is a power of the input x. The second, π^x, is an **exponential function**—the input x appears in the exponent. When you differentiate a power function, the exponent drops by 1; when you differentiate an exponential function, the exponent doesn't change.

Basic Rules

Since basic functions are combined in various ways to make formulas, we need to know how to differentiate *combinations*. For example, suppose we add the basic functions $g(x)$ and $h(x)$, to get $f(x) = g(x) + h(x)$.

Then f is differentiable, and $f'(x) = g'(x) + h'(x)$. Actually, this is true for *all* differentiable functions g and h, not just basic functions. It says: "The rate at which f changes is the sum of the separate rates at which g and h change." Here are some examples that illustrate the point.

The addition rule

$$\text{If} \quad f(x) = \tan x + x^{-6}, \quad \text{then} \quad f'(x) = \frac{1}{\cos^2 x} - 6x^{-7}.$$

$$\text{If} \quad f(w) = 2^w + \sqrt{w}, \quad \text{then} \quad f'(w) = k_2 \, 2^w + \frac{1}{2\sqrt{w}} \ (\text{and } k_2 \approx .69).$$

Likewise, if we multiply any differentiable function g by a constant c, then the product $f(x) = cg(x)$ is also differentiable and $f'(x) = cg'(x)$. This says: "If f is c times as large as g, then f changes at c times the rate of g." Thus the derivative of $5 \sin x$ is $5 \cos x$. Likewise, the derivative of $(5x)^2$ is $50\,x$. (This took an extra calculation.) However, the rule does *not* tell us how to find the derivative of $\sin(5x)$, because $\sin(5x) \neq 5\sin(x)$. We will need the chain rule to work this one out.

The constant multiple rule

The rules about sums and constant multiples of functions are just the first of several basic rules for differentiating combinations of functions. We will describe how to handle products and quotients of functions in chapter 5. For the moment we summarize in the following table the rules we have already covered.

function	derivative
$f(x) + g(x)$	$f'(x) + g'(x)$
$c \cdot f(x)$	$c \cdot f'(x)$

With just the few facts already laid out we can differentiate a variety of functions given by formulas. In particular, we can differentiate any **polynomial** function:

$$P(x) = a_n x^n + a_{n-1} x^{n-1} + \cdots + a_2 x^2 + a_1 x + a_0.$$

Here $a_n, a_{n-1}, \ldots, a_2, a_1, a_0$ are various constants, and n is a positive integer, called the **degree** of the polynomial. A polynomial is a sum of constant multiples of integer powers of the input variable. A polynomial of degree 1 is just a linear function.

All the rules presented up to this point are illustrated in the following examples; note that the first three involve polynomials.

function	derivative
$7x + 2$	7
$5x^4 - 2x^3$	$20x^3 - 6x^2$
$5x^4 - 2x^3 + 17$	$20x^3 - 6x^2$
$3u^{15} + .5u^8 - \pi u^3 + u - \sqrt{2}$	$45u^{14} + 4u^7 - 3\pi u^2 + 1$
$6 \cdot 10^z + 17/z^5$	$6 \cdot k_{10}\, 10^z - 85/z^6$
$3 \sin t - 2t^3$	$3 \cos t - 6t^2$
$\pi \cos x - \sqrt{3} \tan x + \pi^2$	$-\pi \sin x - \dfrac{\sqrt{3}}{\cos^2 x}$

The first two functions have the same derivative because they differ only by a constant, and the derivative of a constant is zero. The constant k_{10} that appears in the fourth example is approximately 2.30.

Exercises

Sketching the graph of the derivative

1.   Sketch the graphs of two different functions that have the same derivative. (For example, can you find two *linear* functions that have the same derivative?)

2.   Here are the graphs of four related functions: s, its derivative s', another function $c(t) = s(2t)$, and *its* derivative $c'(t)$. The graphs are out of order. Label them with the correct names s, s', c, and c'.

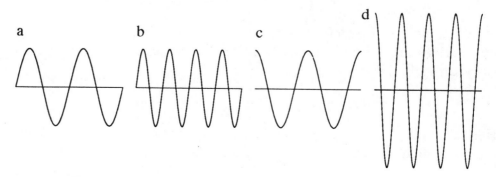

3.   For each of the functions graphed below, sketch the graph of its derivative.

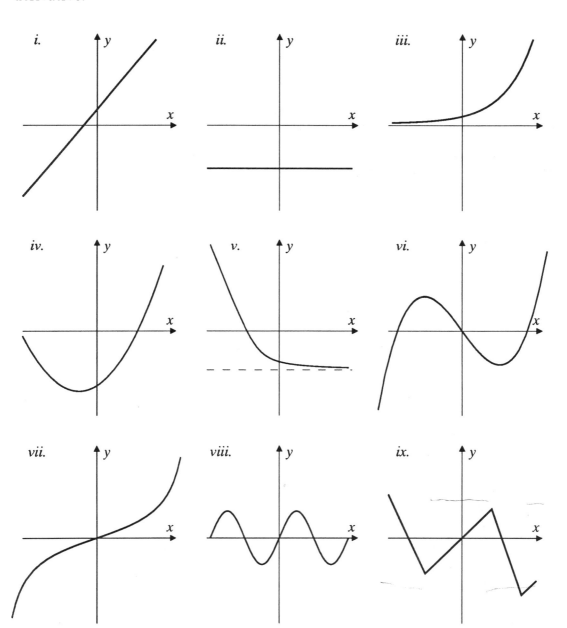

4. a) Suppose a function $y = g(x)$ satisfies $g(0) = 0$ and $0 \le g'(x) \le 1$ for all values of x in the interval $0 \le x \le 3$. Explain carefully why the graph of g must lie entirely in the triangular region shaded below:

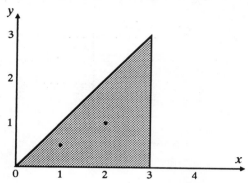

b) Suppose you learn that $g(1) = .5$ and $g(2) = 1$. Draw the smallest shaded region in which you can guarantee that the graph of g must lie.

5. Suppose h is differentiable over the interval $0 \le x \le 3$. Suppose $h(0) = 0$, and that

$$
\begin{array}{ccccccccc}
.5 & \le & h'(x) & \le & 1 & \quad \text{for} \quad & 0 & \le x \le & 1 \\
0 & \le & h'(x) & \le & .5 & \quad \text{for} \quad & 1 & \le x \le & 2 \\
-1 & \le & h'(x) & \le & 0 & \quad \text{for} \quad & 2 & \le x \le & 3
\end{array}
$$

Draw the smallest shaded region in the x, y-plane in which you can guarantee that the graph of $y = h(x)$ must lie.

Differentiation

6. Find formulas for the derivatives of the following functions; that is, *differentiate* them.

a) $3x^7 - .3x^4 + \pi x^3 - 17$

b) $\sqrt{3}\,\sqrt{x} + \dfrac{7}{x^5}$

c) $2w^8 - \sin w + \dfrac{1}{3w^2}$

d) $4 \cos u - 3 \tan u + \sqrt[3]{u}$

e) $\sqrt[4]{16} - \sqrt[4]{s}$

f) $\sqrt{7} \cdot 2^z + (1/2)^z$

g) $-\dfrac{a}{2}t^2 + v_0 t + d_0$ (a, v_0, and d_0 are constants)

7.   Use a computer graphing utility for this exercise. Graph on the same screen the following three functions:

   1. the function f given below, on the indicated interval;

   2. the function $g(x) = (f(x + .01) - f(x - .01))/.02$ that estimates the slope of the graph of f at x;

   3. the function $h(x) = f'(x)$, where you use the differentiation rules to find f'.

a) $f(x) = x^4$ on $-1 \le x \le 1$.

b) $f(x) = x^{-1}$ on $1 \le x \le 8$.

c) $f(x) = \sqrt{x}$ on $.25 \le x \le 9$.

d) $f(x) = \sin x$ on $0 \le x \le 2\pi$.

The graphs of g and h should coincide—or "share phosphor"—in each case. Do they?

8.   In each case below, find a function $f(x)$ whose derivative $f'(x)$ is

a) $12\, x^{11}$

b) $5x^7$

c) $\cos x + \sin x$

d) $ax^2 + bx + c$

e) 0

f) $\dfrac{5}{\sqrt{x}}$

9.   What is the slope of the graph of $y = x - \sqrt{x}$ at $x = 4$? At $x = 100$? At $x = 10000$?

10.   a) For which values of x is the function $x - x^3$ increasing?

b) Where is the graph of $y = x - x^3$ rising most steeply?

c) At what points is the graph of $y = x - x^3$ horizontal?

d) Make a sketch of the graph of $y = x - x^3$ that reflects all these results.

11.   a) Sketch the graph of the function $y = 2x + \dfrac{5}{x}$ on the interval $.2 \le x \le 4$.

b) Where is the lowest point on that graph? Give the value of the x-coordinate *exactly*. [Answer: $x = \sqrt{2.5}$.]

12.   What is the slope of the graph of $y = \sin x + \cos x$ at $x = \pi/4$?

13.   a) Write the microscope equation for $y = \sin x$ at $x = 0$.

b) Using the microscope equation, estimate the following values: $\sin .3$, $\sin .007$, $\sin(-.02)$. Check these values with a calculator.

[Answer: $\sin .3 \approx .3$, according to the microscope equation, and $\sin .3 = .2955202$, according to a calculator. Be sure you are entering x in radians! The error in the approximation is less than $.005$.]

14.   a) Write the microscope equation for $y = \tan x$ at $x = 0$.

b) Estimate the following values: $\tan .007$, $\tan .3$, $\tan(-.02)$. Check these values with a calculator.

15.   a) Write the microscope equation for $y = \sqrt{x}$ at $x = 3600$.

b) Use the microscope equation to estimate $\sqrt{3628}$ and $\sqrt{3592}$. How far are these estimates from the values given by a calculator?

[Answer: $\sqrt{3592} \approx 59.933333$ according to the microscope equation, while $\sqrt{3592} = 59.933296$ according to a calculator. The difference is less than $.00005$.]

16.   If the radius of a spherical balloon is r inches, its volume is $\frac{4}{3}\pi r^3$ cubic inches.

a) At what rate does the volume increase, in cubic inches per inch, when the radius is 4 inches?

b) Write the microscope equation for the volume when $r = 4$ inches.

c) When the radius is 4 inches, approximately how much does it increase if the volume is increased by 50 cubic inches?

d) Suppose someone is inflating the balloon at the rate of 10 cubic inches of air per second. If the radius is 4 inches, at what rate is it increasing, in inches per second?

17.   A ball is held motionless and then dropped from the top of a 200 foot tall building. After t seconds have passed, the distance from the ground to the ball is $d = f(t) = -16t^2 + 200$ feet.

a) Find a formula for the velocity $v = f'(t)$ of the ball after t seconds. Check that your formula agrees with the given information that the initial velocity of the ball is 0 feet/second.

b) Draw graphs of both the velocity and the distance as functions of time. What time interval makes physical sense in this situation? (For example, does $t < 0$ make sense? Does the distance formula make sense after the ball hits the ground?)

c) At what time does the ball hit the ground? What is its velocity then?

18. A second ball is tossed straight up from the top of the same building with a velocity of 10 feet per second. After t seconds have passed, the distance from the ground to the ball is $d = f(t) = -16t^2 + 10t + 200$ feet.

a) Find a formula for the velocity of the second ball. Does the formula agree with given information that the initial velocity is $+10$ feet per second? Compare the velocity formulas for the two balls; how are they similar, and how are they different?

b) Draw graphs of both the velocity and the distance as functions of time. What time interval makes physical sense in this situation?

c) Use your graph to answer the following questions. During what period of time is the ball rising? During what period of time is it falling? When does it reach the highest point of its flight?

d) How high does the ball rise?

19. What is the velocity formula for a third ball that is thrown *downward* from the top of the building with a velocity of 40 feet per second? Check that your formula gives the correct initial velocity.

a) What is the distance formula for the third ball? Check that it satisfies the initial condition (namely, that the ball starts at the top of the building).

b) When does this ball hit the ground? How fast is it going then?

20. A steel ball is rolling along a 20-inch long straight track so that its distance from the midpoint of the track (which is 10 inches from either end) is $d = 3\sin t$ inches after t seconds have passed. (Think of the track as aligned from left to right. Positive distances mean the ball is to the right of the center; negative distances mean it is to the left.)

a) Find a formula for the velocity of the ball after t seconds. What is happening when the velocity is positive; when it is negative; when it equals zero? Write a sentence or two describing the motion of the ball.

b) How far from the midpoint of the track does the ball get? How can you tell?

c) How fast is the ball going when it is at the midpoint of the track? Does it ever go faster than this? How can you tell?

21. A forester who wants to know the height of a tree walks 100 feet from its base, sights to the top of the tree, and finds the resulting angle to be 57 degrees.

a) What height does this give for the tree?

b) If the measurement of the angle is certain only to 5 degrees, what can you say about the uncertainty of the height found in part (a)? (Note: you need to express angles in *radians* to use calculus: π radians = 180 degrees.)

22.   a) In the preceding problem, what percentage error in the height of the tree is produced by a 1 degree error in measuring the angle?

b) What would the percentage error have been if the angle had been 75 degrees instead of 57 degrees? 40 degrees?

c) If you can measure angles to within 1 degree accuracy and you want to measure the height of a tree that's roughly 150 feet tall by means of the technique in the preceding problem, how far away from the tree should you stand to get your best estimate of the tree's height? How accurate would your answer be?

§6. The Chain Rule

Combining Rates of Change

Let's return to the expanding house that we studied in §4. When the temperature T increased, every side s of the house got longer; when s got longer, the volume V got larger. We already discussed how V responds to changes in s, but that's only part of the story. What we'd really like to know is this: exactly how does the volume V respond to changes in temperature T? We can work this out in stages: first we see how V responds to changes in s, and then how s responds to changes in T.

Stage 1. Recall that our "house" is a cube that measures 200 inches on a side. The microscope equation from §4 describes how V responds to changes in s:

$$\Delta V \approx 120000 \ \frac{\text{cubic inches of volume}}{\text{inch of length}} \cdot \Delta s \text{ inches.}$$

How volume responds to changes in length

Stage 2. Physical experiments with wood show that a 200 inch length of wood increases about .0004 inches in length per degree Fahrenheit. This is a *rate*, and we can build a second microscope equation with it:

$$\Delta s \approx .0004 \ \frac{\text{inches of length}}{\text{degree F}} \cdot \Delta T \text{ degrees F,}$$

where ΔT measures the change in temperature, in degrees Fahrenheit.

How length responds to changes in temperature

We can combine the two stages because Δs appears in both. Replace Δs in the first equation by the right-hand side of the second equation. The result is

$$\Delta V \approx 120000 \ \frac{\text{cubic inches}}{\text{inch}} \times .0004 \ \frac{\text{inches}}{\text{degree F}} \cdot \Delta T \text{ degrees F.}$$

We can condense this to

$$\Delta V \approx 48 \ \frac{\text{cubic inches}}{\text{degree F}} \cdot \Delta T \text{ degrees F.}$$

How volume responds to changes in temperature

This is a *third* microscope equation, and it shows directly how the volume of the house responds to changes in temperature. It is the answer to our question.

As always, the multiplier in a microscope equation is a rate. The multiplier in the third microscope equation, 48 cubic inches/degree F, tells us the rate at which *volume changes with respect to temperature*. Thus, if the temperature increases by 10 degrees between night and day, the house will become about 480 cubic inches larger. Recall that Bodanis (see §4) said that the house might become only a few cubic inches larger—say, $\Delta V = 3$ cubic inches. If we solve the microscope equation

$$3 \approx 48 \cdot \Delta T$$

for ΔT, we see that the temperature would have risen only 1/16-th of a degree F!

The rate that appears as the multiplier in the third microscope equation is the product of the other two:

$$48 \,\frac{\text{cubic inches}}{\text{degree F}} = 120000 \,\frac{\text{cubic inches}}{\text{inch}} \times .0004 \,\frac{\text{inches}}{\text{degree F}}.$$

Each of these rates is a derivative:

$$\underbrace{48 \,\frac{\text{cubic inches}}{\text{degree F}}}_{dV/dT} = \underbrace{120000 \,\frac{\text{cubic inches}}{\text{inch}}}_{dV/ds} \times \underbrace{.0004 \,\frac{\text{inches}}{\text{degree F}}}_{ds/dT}.$$

We wrote the derivatives in Leibniz's notation because it's particularly helpful in keeping straight what is going on. For instance, dV/dT indicates very clearly the rate at which volume is changing with respect to *temperature*, and dV/ds the rate at which it is changing with respect to *length*. These rates are quite different—they even have different units—but the notation V' does not distinguish between them. In Leibniz's notation, the relation between the three rates takes this striking form:

$$\frac{dV}{dT} = \frac{dV}{ds} \cdot \frac{ds}{dT}.$$

This relation is called the **chain rule** for the variables T, s, and V. (We'll see in a moment what this has to do with chains.)

The chain rule is a consequence of the way the three microscope equations are related to each other. We can see how it emerges directly from the microscope equations if we replace the numbers that appear as multipliers in those equations by the three derivatives. To begin, we write

$$\Delta V \approx \frac{dV}{ds} \cdot \Delta s \quad \text{and} \quad \Delta s \approx \frac{ds}{dT} \cdot \Delta T.$$

Then, combining these equations, we get

$$\Delta V \approx \frac{dV}{ds} \cdot \frac{ds}{dT} \cdot \Delta T.$$

In fact, this is the microscope equation for V in terms of T, which can be written more directly as

$$\Delta V \approx \frac{dV}{dT} \cdot \Delta T.$$

In these two expressions we have the same microscope equation, so the multipliers must be equal. Thus, we recover the chain rule:

$$\frac{dV}{ds} \cdot \frac{ds}{dT} = \frac{dV}{dT}.$$

Recall that Leibniz worked directly with *differentials*, like dV and ds, so a derivative was a genuine fraction. For him, the chain rule is true simply because we can cancel the two appearances of "ds" in the derivatives. For us, though, a derivative is not really a fraction, so we need an argument like the one in the text to establish the rule.

Chains and the Chain Rule

Let's analyze the relationships between the three variables in the expanding house problem in more detail. There are three functions involved: volume is a function of length: $V = V(s)$; length is a function of temperature: $s = s(T)$; and finally, volume is a function of temperature, too: $V = V(s(T))$. To visualize these relationships better, we introduce the notion of an **input-output diagram**. The input-output diagram for the function $s = s(T)$ is just $T \to s$. It indicates that T is the input of a function whose output is s. Likewise $s \to V$ says that volume V is a function of length s. Since the output of $T \to s$ is the input of $s \to V$, we can make a *chain* of these two diagrams:

$$T \longrightarrow s \longrightarrow V.$$

An input-output chain

The result describes a function that has input T and output V. It is thus an input-output diagram for the third function $V = V(s(T))$.

We could also write the input-output diagram for the third function simply as $T \rightarrow V$; in other words,

$$T \longrightarrow V \qquad \text{equals} \qquad T \longrightarrow s \longrightarrow V.$$

We say that $T \rightarrow s \rightarrow V$ is a **chain** that is made up of the two **links** $T \rightarrow s$ and $s \rightarrow V$. Since each input-output diagram represents a function, we can attach a derivative that describes the rate of change of the output with respect to the input:

$$T \xrightarrow{\frac{ds}{dT}} s \qquad\qquad s \xrightarrow{\frac{dV}{ds}} V \qquad\qquad T \xrightarrow{\frac{dV}{dT}} V$$

Here is a single picture that shows all the relationships:

$$T \underset{\frac{ds}{dT}}{\longrightarrow} s \underset{\frac{dV}{ds}}{\longrightarrow} V$$

$$\overset{\frac{dV}{dT}}{\big\frown}$$

The **chain rule** tells us how the derivative dV/dT of the whole chain is related to the derivatives dV/ds and ds/dT of the individual links:

$$\boxed{\textbf{The chain rule:} \quad \frac{dV}{dT} = \frac{dV}{ds} \cdot \frac{ds}{dT}}$$

A simple example. We can sometimes use the chain rule without giving it much thought. For instance, suppose a bookstore makes an average profit of \$3 per book, and its sales are increasing at the rate of 40 books per month. At what rate is its monthly profit increasing, in dollars per month? Does it seem clear to you that the rate is \$120 per month?

Let's analyze the question in more detail. There are three variables here:

$$\begin{array}{lll} \textbf{time} & t & \text{measured in months;} \\ \textbf{sales} & s & \text{measured in books;} \\ \textbf{profit} & p & \text{measured in dollars.} \end{array}$$

The two known rates are

$$\frac{dp}{ds} = 3\ \frac{\text{dollars}}{\text{book}} \qquad \text{and} \qquad \frac{ds}{dt} = 40\ \frac{\text{books}}{\text{month}}.$$

The rate we seek is dp/dt, and we find it by the chain rule:

$$\begin{aligned}\frac{dp}{dt} &= \frac{dp}{ds} \cdot \frac{ds}{dt} \\[2mm] &= 3\ \frac{\text{dollars}}{\text{book}} \times 40\ \frac{\text{books}}{\text{month}} \\[2mm] &= 120\ \frac{\text{dollars}}{\text{month}}\end{aligned}$$

Chains, in general. The chain rule applies whenever the output of one function is the input of another. For example, suppose $u = f(x)$ and $y = g(u)$. Then $y = g(f(x))$, and we have:

$$x \xrightarrow[\frac{du}{dx}]{} u \xrightarrow[\frac{dy}{du}]{} y \qquad\qquad \frac{dy}{dx} = \frac{dy}{du} \cdot \frac{du}{dx}$$

Let's take

$$u = x^2 \qquad \text{and} \qquad y = \sin(u);$$

then $y = \sin(x^2)$, and it is not at all obvious what the derivative dy/dx ought to be. None of the basic rules in §4 covers this function. However, those rules *do* cover $u = x^2$ and $y = \sin(u)$:

$$\frac{du}{dx} = 2x \qquad \text{and} \qquad \frac{dy}{du} = \cos(u).$$

We can now get dy/dx by the chain rule:

$$\frac{dy}{dx} = \frac{dy}{du} \cdot \frac{du}{dx} = \cos(u) \cdot 2x.$$

Since we are interested in y as a function of x—rather than u—we should rewrite dy/dx so that it is expressed entirely in terms of x:

$$\text{If} \quad y = \sin(x^2), \quad \text{then} \quad \frac{dy}{dx} = 2x \cos(u) = 2x \cos(x^2).$$

Let's start over, using the function names f and g we introduced at the outset:

$$u = f(x) \quad \text{and} \quad y = g(u), \quad \text{so} \quad y = g(f(x)).$$

The third function, $y = g(f(x))$, needs a name of its own; let's call it h. Thus

Composition of functions

$$y = h(x) = g(f(x)).$$

We say that h is **composed** of g and f, and h is called the **composite**, or the **composition**, of g and f.

The problem is to find the derivative h' of the composite function, knowing g' and f'. Let's translate all the derivatives into Leibniz's notation.

$$h'(x) = \frac{dy}{dx} \qquad g'(u) = \frac{dy}{du} \qquad f'(x) = \frac{du}{dx}.$$

We can now invoke the chain rule:

$$h'(x) = \frac{dy}{dx} = \frac{dy}{du} \cdot \frac{du}{dx} = g'(u) \cdot f'(x).$$

Although h' is now expressed in terms of g' and f', we are not yet done. The variable u that appears in $g'(u)$ is out of place—because h is a function of x, not u. (We got to the same point in the example; the original form of the derivative of $\sin(x^2)$ was $2x\cos(u)$.) The remedy is to replace u by $f(x)$; we can do this because $u = f(x)$ is given.

The chain rule: $h'(x) = g'(f(x)) \cdot f'(x)$ when $h(x) = g(f(x))$

Using the Chain Rule

The chain rule will allow us to differentiate nearly any formula. The key is to recognize when a given formula can be written as a chain—and then, how to write it.

Example 1. Here is a problem first mentioned on page 147: What is the derivative of $y = \sin(5x)$? If we set

$$y = \sin(u) \qquad \text{where} \qquad u = 5x,$$

then we find immediately

$$\frac{dy}{du} = \cos(u) \qquad \text{and} \qquad \frac{du}{dx} = 5.$$

Thus, by the chain rule we see

$$\frac{dy}{dx} = \frac{dy}{du} \cdot \frac{du}{dx} = \cos(u) \cdot 5 = 5\cos(5x).$$

Example 2. $w = 2^{\cos z}$. Set

$$w = 2^u \qquad \text{and} \qquad u = \cos z.$$

Then, once again, the basic rules from §4 are sufficient to differentiate the individual links:

$$\frac{dw}{du} = k_2\, 2^u \qquad \text{and} \qquad \frac{du}{dz} = -\sin z.$$

The chain rule does the rest:

$$\frac{dw}{dz} = \frac{dw}{du} \cdot \frac{du}{dz} = k_2\, 2^u \cdot (-\sin z) = -k_2 \sin z\, 2^{\cos z}.$$

Example 3. $p = \sqrt{7t^3 + \sin^2 t}$. This presents several challenges. First let's make a chain:

$$p = \sqrt{u} \qquad \text{where} \qquad u = 7t^3 + \sin^2 t.$$

The basic rules give us $dp/du = 1/2\sqrt{u}$, but it is more difficult to deal with u. Let's at least introduce separate labels for the two terms in u:

$$q = 7t^3 \qquad \text{and} \qquad r = \sin^2 t.$$

Then

$$\frac{du}{dt} = \frac{dq}{dt} + \frac{dr}{dt} \qquad \text{and} \qquad \frac{dq}{dt} = 21t^2.$$

The remaining term $r = \sin^2 t = (\sin t)^2$ can itself be differentiated by the chain rule. Set

$$r = v^2 \qquad \text{where} \qquad v = \sin t.$$

Then

$$\frac{dr}{dv} = 2v \quad \text{and} \quad \frac{dv}{dt} = \cos(t),$$

so

$$\frac{dr}{dt} = \frac{dr}{dv} \cdot \frac{dv}{dt} = 2v \cos t = 2 \sin t \cos t.$$

The final step is to assemble all the pieces:

$$\frac{dp}{dt} = \frac{dp}{du} \cdot \frac{du}{dt} = \frac{1}{2\sqrt{u}} \cdot \left(21t^2 + 2 \sin t \cos t \right) = \frac{21t^2 + 2 \sin t \cos t}{2\sqrt{7t^3 + \sin^2 t}}$$

By breaking down a complicated expression into simple pieces, and applying the appropriate differentiation rule to each piece, it is possible to differentiate a vast array of formulas. You may meet two sorts of difficulties: you may not see how to break down the expression into simpler parts; and you may overlook a step. Practice helps overcome the first, and vigilance the second.

Here is an example of the second problem: find the derivative of $y = -3 \cos(2x)$. The derivative is *not* $3 \sin(2x)$; it is $6 \sin(2x)$. Besides remembering to deal with the constant multiplier -3, and with the fact that there is a minus sign in the derivative of $\cos u$, you must not overlook the link $u = 2x$ in the chain that connects y to x.

Exercises

1.   Use the chain rule to find dy/dx, when y is given as a function of x in the following way.

a) $y = 5u - 3$, where $u = 4 - 7x$.

b) $y = \sin u$, where $u = 4 - 7x$.

c) $y = \tan u$, where $u = x^3$.

d) $y = 10^u$, where $u = x^2$.

e) $y = u^4$, where $u = x^3 + 5$.

2.   Find the derivatives of the following functions.

a) $(9x + 6x^3)^5$

b) $\sqrt{4w^2 + 1}$

c) $\sqrt{(4w^2 + 1)^3}$

d) $\dfrac{1}{1-x} = (1-x)^{-1}$

e) $3\tan\left(\dfrac{1}{z}\right)$

f) $\sin^2(w^3+1)$

g) $\cos(2^t)$

h) $5^{1/x}$

3. If $f(x) = (1+x^2)^5$, what are the numerical values of $f'(0)$ and $f'(1)$?

4. If $h(t) = \cos(\sin t)$, what are the numerical values of $h'(0)$ and $h'(\pi)$?

5. If $f'(x) = g(x)$, which of the following defines a function which also has g as its derivative?

$$f(x+17) \qquad f(17x) \qquad 17f(x) \qquad 17+f(x) \qquad f(17)$$

6. Let $f(t) = t^2+2t$ and $g(t) = 5t^3-3$. Determine all of the following: $f'(t),\ g'(t),\ g(f(t)),\ f(g(t)),\ g'(f(t)),\ f'(g(t)),\ (f(g(t)))',\ (g(f(t)))'$.

7. a) Sketch the graphs of $f(x) = 2^{-x^2}$ and its derivative on the interval $-2 \le x \le 2$.

b) At what point, or points, is $f'(x) = 0$? What is true about the graph of f at these points?

c) Where does the graph of f have positive slope, and where does it have negative slope?

8. a) With a graphing utility, find the point x where the function $y = 1/(3x^2 - 5x + 7)$ takes its maximum value. Obtain the numerical value of x accurately to two decimal places.

b) Find the derivative of $y = 1/(3x^2 - 5x + 7)$, and determine where it takes the value 0.

[Answer: $y' = -(6x-5)(3x^2-5x+7)^{-2}$, and $y' = 0$ when $x = 5/6$.]

c) Using part (b), find the *exact* value of x where $y = 1/(3x^2 - 5x + 7)$ takes its maximum value.

d) At what point is the graph of $y = 1/(3x^2 - 5x + 7)$ rising most steeply? Describe how you determined the location of this point.

9. a) Write the microscope equation for the function $y = \sin\sqrt{x}$ at $x = 1$.

b) Using the microscope equation, estimate the following values: $\sin\sqrt{1.05}$, $\sin\sqrt{.9}$.

10. a) Write the microscope equation for $w = \sqrt{1+x}$ at $x = 0$.

b) Use the microscope equation to estimate the values of $\sqrt{1.1056}$ and $\sqrt{.9788}$. Compare your estimates with the values provided by a calculator.

11. When the sides of a cube are 5 inches, its surface area is changing at the rate of 60 square inches per inch increase in the side. If, at that moment, the sides are increasing at a rate of 3 inches per hour, at what rate is the area increasing: is it 60, 3, 63, 20, 180, 5, or 15 square inches per hour?

12. Find a function $f(x)$ for which $f'(x) = 3x^2(5 + x^3)^{10}$. Find a function $p(x)$ for which $p'(x) = x^2(5 + x^3)^{10}$.

13. Find a function $g(t)$ for which $g'(t) = t/\sqrt{1+t^2}$.

§7. Partial Derivatives

Let's return to the sunrise function once again. The time of sunrise depends not only on the date, but on our latitude. In fact, if we are far enough north or south, there are days when the sun never rises at all. We give in the table below the time of sunrise at eight different latitudes on March 15, 1990.

Latitude	36°N	38°N	40°N	42°N	44°N	46°N	48°N	50°N
Mar 15	6:10	6:11	6:12	6:13	6:13	6:13	6:14	6:14

Thus on March 15, the time of sunrise increases as latitude increases.

Clearly what this shows is that the time of sunrise is actually a function of two independent inputs: the date and the latitude. If T denotes the time of sunrise, then we will write $T = T(d, \lambda)$ to make explicit the dependence of T on both the date d and the latitude λ. To capture this double dependence, we need information like the following table:

The time of sunrise depends on latitude as well as on the date

Latitude	36°N	38°N	40°N	**42°N**	44°N	46°N	48°N	50°N
Mar 3	6:24	6:27	6:31	**6:33**	6:34	6:36	6:38	6:40
7	6:20	6:22	6:25	**6:26**	6:27	6:29	6:30	6:32
11	6:15	6:17	6:19	**6:19**	6:20	6:21	6:22	6:23
15	**6:10**	**6:11**	**6:12**	**6:13**	**6:13**	**6:13**	**6:14**	**6:14**
19	6:06	6:06	6:06	**6:06**	6:06	6:06	6:06	6:06
23	6:01	6:00	5:59	**5:59**	5:58	5:58	5:58	5:57
27	5:56	5:54	5:53	**5:52**	5:51	5:50	5:49	5:48

Thus we can say $T(74, 42°N) = 6:13$ (March 15 is the 74-th day of the year). Note, though, that at this date and place the time of sunrise is changing in two very different senses:

First: At 42°N, during the eight days between March 11 and March 19, the time of sunrise gets 13 minutes earlier. We thus would say that on March 15 at 42°N, sunrise is changing at −1.63 minutes/day.

Second: On the other hand, on March 15 we see that the time of sunrise varies by 1 minute as we go from 40°N to 44°N. We would thus say that at 42°N the rate of change of sunrise as the latitude varies is approximately 1 minute/4° = +.25 minutes/degree of latitude.

Two quite different rates are at work here, one with respect to time, the other with respect to latitude.

We need a notation which allows us to talk about the different rates

A function of several variables has several rates of change

at which a function can change, when that function depends on more than one variable. A rate of change is, of course, a derivative. But since a change in one input produces only part of the change that a function of several variables can experience, we call the rate of change with respect to any one of the inputs a **partial derivative**. If the value of z depends on the variables x and y according to the rule $z = F(x, y)$, then we denote the rate at which z is changing with respect to x when $x = a$ and $y = b$ by

$$F_x(a, b) \qquad \text{or by} \qquad \frac{\partial z}{\partial x}(a, b).$$

Partial derivatives

We call this rate the **partial derivative of F with respect to x**. Similarly, we define the partial derivative of F with respect to y to be the rate at which z is changing when y is varied. It is denoted

$$F_y(a, b) \qquad \text{or} \qquad \frac{\partial z}{\partial y}(a, b).$$

There is nothing conceptually new involved here; to calculate either of these partial derivatives you simply hold one variable constant and go through the same limiting process as before for the input variable of interest. Note that, to call attention to the fact that there is more than one input variable present, we write

$$\frac{\partial z}{\partial x} \qquad \text{rather than} \qquad \frac{dz}{dx},$$

as we did when x was the only input variable.

To calculate the partial derivative of F with respect to x at the point (a, b), we can use

$$F_x(a, b) = \frac{\partial z}{\partial x}(a, b) = \lim_{\Delta x \to 0} \frac{F(a + \Delta x, b) - F(a, b)}{\Delta x}.$$

Similarly,

$$F_y(a, b) = \frac{\partial z}{\partial y}(a, b) = \lim_{\Delta y \to 0} \frac{F(a, b + \Delta y) - F(a, b)}{\Delta y}.$$

By using this notation for partial derivatives, we can cast some of our earlier statements about the sunrise function $T = T(d, \lambda)$ in the following form:

$$T_d(74, 42°\text{N}) \approx -1.63 \text{ minutes per day};$$
$$T_\lambda(74, 42°\text{N}) \approx +.25 \text{ minutes per degree}.$$

Partial Derivatives as Multipliers

For any given date d and latitude λ we can write down two microscope equations for the sunrise function $T(d, \lambda)$. One describes how the time of sunrise responds to changes in the date, the other to changes in the latitude. Let's consider variations in the time of sunrise in the vicinity of March 15 and 42°N.

The partial derivative $T_d(74, 42°\text{N})$ of T with respect to d is the multiplier in the first of these microscope equations:

$$\Delta T \approx T_d(74, 42°\text{N}) \cdot \Delta d.$$

The microscope equation for dates

For example, from March 15 to March 17 ($\Delta d = 2$ days), we would expect the time of sunrise to change by

$$\Delta T \approx -1.63 \, \frac{\text{min}}{\text{day}} \times 2 \text{ days} = -3.3 \text{ minutes}.$$

Thus, we would expect the time of sunrise on March 17 at 42°N to be approximately

$$T(76, 42°\text{N}) \approx 6{:}09.7.$$

The partial derivative $T_\lambda(74, 42°\text{N})$ of T with respect to λ is the multiplier in the second microscope equation:

$$\Delta T \approx T_\lambda(74, 42°\text{N}) \cdot \Delta \lambda.$$

The microscope equation for latitudes

If, say, we moved 1° north, to 43°N, we would expect the time of sunrise on March 7 to change by

$$\Delta T \approx .25 \, \frac{\text{min}}{\text{deg}} \times 1 \text{ degree} = .25 \text{ minute}.$$

The time of sunrise on March 15 at 43°N would therefore be

$$T(74, 43°\text{N}) \approx 6{:}13.25.$$

We have seen what happens to the time of sunrise from March 15 to March 17 if we stay at 42°N, and we have seen what happens to the time on March 15 if we move from 42°N to 43°N. Can we put these two pieces of information together? That is, can we determine the time of sunrise on March 17 at 43°N? This involves changing *both* the date and the latitude.

The total change   To determine the total change we shall just combine the two changes ΔT we have already calculated. Making the date two days later moves the time of sunrise 3.3 minutes *earlier*, and travelling one degree north makes the time of sunrise .25 minutes *later*, so the net effect would be to change the time of sunrise by

$$\Delta T \approx -3.3 \text{ min} + .25 \text{ min} \approx -3 \text{ minutes.}$$

This puts the time of sunrise at $T(76, 43°N) \approx 6{:}10$.

We can formulate this idea more generally in the following way: partial derivatives are not only multipliers for gauging the separate effects that changes in each input have on the output, but they also serve as multipliers for gauging the *cumulative* effect that changes in all inputs have on the output. In general, if $z = F(x, y)$ is a function of two variables, then near the point (a, b), the combined change in z caused by small changes in x and y can be stated by the *full* microscope equation:

The full microscope equation:
$$\Delta z \approx F_x(a, b) \cdot \Delta x + F_y(a, b) \cdot \Delta y$$

As was the case for the functions of one variable, there is an important class of functions for which we may write "=" instead of "≈" in this relation, the linear functions. The most general form of a **linear function of two variables** is $z = F(x, y) = mx + ny + c$, for constants m, n, and c.

In the exercises you will have an opportunity to verify that for a linear function $z = F(x, y) = mx + ny + c$ and for all (a, b), we know that $F_x(a, b) = m$ and $F_y(a, b) = n$, and the full microscope equation $\Delta z = F_x(a, b) \cdot \Delta x + F_y(a, b) \cdot \Delta y$ is true for all values of Δx and Δy.

Formulas for Partial Derivatives

No new rules are needed to find the formulas for the partial derivatives of a function of two variables that is given by a formula. To find the partial derivative with respect to one of the variables, just treat the other variable as a constant and follow the rules for functions of a single variable. (The basic rules are described in §5 and the chain rule in §6.) We give two examples to illustrate the method.

To find a partial derivative, treat the other variable as a constant

Example 1. For $z = F(x, y) = 3x^2 y + 5y^2 \sqrt{x}$, we have

$$
\begin{aligned}
F_x(x, y) &= 3y(2x) + 5y^2 \frac{1}{2\sqrt{x}} = 6xy + \frac{5y^2}{2\sqrt{x}} \\
F_y(x, y) &= 3x^2 + 10y\sqrt{x}
\end{aligned}
$$

Example 2. For $w = G(u, v) = 3u^5 \sin v - \cos v + u$, we have

$$
\begin{aligned}
\frac{\partial w}{\partial u} &= 15u^4 \sin v + 1 \\
\frac{\partial w}{\partial v} &= 3u^5 \cos v + \sin v
\end{aligned}
$$

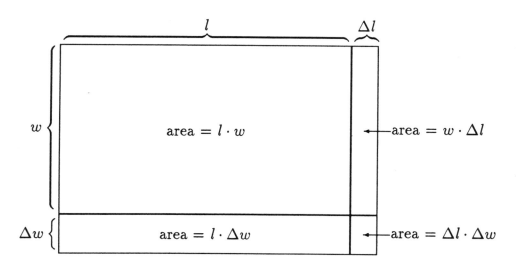

The formulas for derivatives and the combined multiplier effect of partial derivatives allow us to determine the effect of changes in length and in width on the area of a rectangle. The area A of the rectangle is a simple function of its dimensions l and w, $A = F(l, w) = lw$. The partial derivatives are

$$F_l(l, w) = w \quad \text{and} \quad F_w(l, w) = l.$$

Changes Δl and Δw in the dimensions produce a change

The full microscope equation for rectangular area

$$\Delta A \approx w \cdot \Delta l + l \cdot \Delta w$$

in the area. The picture on the opposite page shows that the *exact* value of ΔA includes an additional term—namely $\Delta l \cdot \Delta w$—that is not in the approximation $w \cdot \Delta l + l \cdot \Delta w$. The difference, $\Delta l \cdot \Delta w$, is *very* small when the changes Δl and Δw are small. In chapter 5 we will have a further look at the nature of this approximation.

Exercises

1.   Use differentiation formulas to find the partial derivatives of the following functions.

a) $x^2 y$.

b) $\sqrt{x + y}$

c) $x^2 y + 5x^3 - \sqrt{x + y}$

d) 10^{xy}

e) $\dfrac{y}{x}$

f) $\sin \dfrac{y}{x}$

g) $17\dfrac{x^2}{y^3} - x^2 \sin y + \pi$

h) $\dfrac{uv}{5} + \dfrac{5}{uv}$

i) $2\sqrt{x} \, \sqrt[3]{y} - 7 \cos x$

j) $x \tan y$

2.   On March 7 in the Northern Hemisphere, the farther south you are the earlier the sun rises. The sun rises at 6:25 on this date at 40°N. If we had been far enough south, we could have experienced a 6:25 sunrise on March 5. Near what latitude did this happen?

3. The volume V of a given quantity of gas is a function of the temperature T (in degrees Kelvin) and the pressure P. In a so-called ideal gas the functional relationship between volume and pressure is given by a particularly simple rule called the **ideal gas law**:

$$V(T, P) = R\frac{T}{P},$$

where R is a constant.

a) Find formulas for the partial derivatives $V_T(T, P)$ and $V_P(T, P)$

b) For a particular quantity of an ideal gas called a *mole*, the value of R can be expressed as 8.3×10^3 newton-meters per degree Kelvin. (The *newton* is the unit of force in the *MKS* unit system.) Check that the units in the ideal gas law are consistent if V is measured in cubic meters, T in degrees Kelvin, and P in newtons per square meter.

c) Suppose a mole of gas at 350 degrees Kelvin is under a pressure of 20 newtons per square meter. If the temperature of the gas increases by 10 degrees Kelvin and the volume increases by 1 cubic meter, will the pressure increase or decrease? By about how much?

4. Write the formula for a linear function $F(x, y)$ with the following properties:

$$
\begin{aligned}
F_x(x, y) &= .15 &\text{for all } x \text{ and } y \\
F_y(x, y) &= 2.31 &\text{for all } x \text{ and } y \\
F(4, 1) &= 8
\end{aligned}
$$

5. The purpose of this exercise is to verify the claims made in the text for the linear function $z = F(x, y) = mx + ny + c$, where m, n and c are constants.

a) Use the differentiation rules to find the partial derivatives of F.

b) Use the *definition* of the partial derivative $F_x(a, b)$ to show that $F_x(a, b) = m$ for any input (a, b). That is, show that the value of

$$\frac{F(a + \Delta x, b) - F(a, b)}{\Delta x}$$

exactly equals m, no matter what a and b are.

c) Compute the exact value of the change

$$\Delta z = F(a + \Delta x, b + \Delta y) - F(a, b)$$

corresponding to changing a by Δx and b by Δy.

6.   Suppose $w = G(u, v) = \dfrac{uv}{3 + v}$.

a) Approximate the value of the partial derivative $G_u(1, 2)$ by computing $\Delta w / \Delta u$ for $\Delta u = \pm.1, \pm.01, \ldots, \pm.00001$.

b) Approximate the value of $G_v(1, 2)$ by computing $\Delta w / \Delta v$ for $\Delta v = \pm.1, \pm.01, \ldots, \pm.00001$.

c) Write the full microscope equation for $G(u, v)$ at $(u, v) = (1, 2)$.

d) Use the full microscope equation to approximate $G(.8, 2.1)$. How close is your approximation to the true value of $G(.8, 2.1)$?

7.   a) A rectangular piece of land has been measured to be 51 feet by 2034 feet. What is its area?

b) The narrow dimension has been measured with an accuracy of 4 inches, but the long dimension is accurate only to 10 feet. What is the error, or uncertainty, in the calculated area? What is the percentage error?

[Answer: The uncertainty is about 1188 square feet in the total area of 103,734 square feet. The percentage error is just over 1%.]

8.   Suppose $z = f(x, y)$ and

$$f(3, 12) = 240 \qquad f_x(3, 12) = 7 \qquad f_y(3, 12) = 4.$$

a) Estimate $f(4, 12)$, $f(3, 13)$, $f(4, 13)$, $f(4, 10)$.

b) When $x = 3$ and $y = 12$, how much does a 1% increase in x cause z to change? How much does a 1% increase in y cause z to change? Which has the larger effect: a 1% increase in x or a 1% increase in y?

9.   Let $P(K, L)$ represent the monthly profit, in thousands of dollars, of a company that produces a product using capital whose monthly cost is K thousand dollars and labor whose monthly cost is L thousand dollars. The current levels of expense for capital and labor are $K = 23.5$ and $L = 39.0$. Suppose now that company managers have determined

$$\frac{\partial P}{\partial K}(23.5, 39.0) = -.12 \qquad \frac{\partial P}{\partial L}(23.5, 39.0) = -.20.$$

a) Estimate what happens to the monthly profit if monthly capital expenses increase to $24,000.

b) Each typical person added to the work force increases the monthly labor expense by $1,500. Estimate what happens to the monthly profit if one more person is added to the work force. What, therefore, is the rate of change of profit, in thousands of dollars per person? Is the rate positive or negative?

c) Suppose managers respond to increased demand for the product by adding three workers to the labor force. What does that do to monthly profit? If the managers want to keep the profit level unchanged, they could try to alter capital expenses. What change in K would leave profit unchanged after the three workers are added? (This is called a **trade-off**.)

10.   A forester who wants to know the height of a tree walks 100 feet from its base, sights to the top of the tree, and finds the resulting angle to be 57 degrees.

a) What height does this give for the tree?

b) If the 100-foot measurement is certain only to 1 foot and the angle measurement is certain only to 5 degrees, what can you say about the uncertainty of the height measured in part (a)? (Note: you need to express angles in *radians* to use calculus: π radians = 180 degrees.)

c) Which would be more effective: improving the accuracy of the angle measurement, or improving the accuracy of the distance measurement? Explain.

§8.   Chapter Summary

The Main Ideas

- The functions we study with the calculus have graphs that are **locally linear**; that is, they look approximately straight when magnified under a microscope.

- The **slope of the graph** at any point is the **limit** of the slopes seen under a microscope at that point.

- The **rate of change** of a function at a point is the slope of its graph at that point, and thus is also a **limit**. Its dimensional units are (units of output)/(unit of input).

- The **derivative** of $f(x)$ at $x = a$ is name given to both the rate of change of f at a and the slope of the graph of f at $(a, f(a))$.

- The derivative of $y = f(x)$ at $x = a$ is written $f'(a)$. The **Leibniz notation** for the derivative is dy/dx.

- To calculate the derivative $f'(a)$, make **successive approximations** using $\Delta y / \Delta x$:

$$f'(a) = \lim_{\Delta x \to 0} \frac{\Delta y}{\Delta x} = \lim_{h \to 0} \frac{f(a + h) - f(a - h)}{2h} = \lim_{h \to 0} \frac{f(a + h) - f(a)}{h}.$$

- The **microscope equation** $\Delta y \approx f'(a) \cdot \Delta x$ describes the relation between x and $y = f(x)$ as seen under a microscope at $(a, f(a))$; Δx and Δy are the **microscope coordinates**.

- The microscope equation describes how the output changes in response to small changes in the input. The response is proportional, and the derivative $f'(a)$ plays the role of **multiplier**, or scaling factor.

- The microscope equation expresses the **local linearity** of a function in analytic form. The microscope equation is exact for **linear** functions.

- The microscope equation describes **error propagation** when one quantity, known only approximately, is used to calculate another.

- The **derivative function** is the rule that assigns to any x the number $f'(x)$.

- The derivative of a function gives information about the shape of the graph of the function, and conversely.

- If a function is given by a formula, its derivative also has a formula. There are formulas for the derivatives of the **basic functions**, and there are **rules** for the derivatives of combinations of basic functions.

- The **chain rule** gives the formula for the derivative of a **chain**, or **composite** of functions.

- Functions that have more than one input variable have **partial derivatives**. A partial derivative is the rate at which the output changes with respect to one variable when we hold all the others constant.

- If a multi-input function is given by a formula, its partial derivatives also have formulas that can be found using the same rules that apply to single-input functions.

- A function $z = F(x, y)$ of two variables also has a **microscope equation**:
$$\Delta z \approx F_x(a, b) \cdot \Delta x + F_y(a, b) \cdot \Delta y.$$

The partial derivatives are the **multipliers** in the microscope equation.

Self-Testing

- You should be able to approximate $f'(a)$ by zooming in on the graph of f near a and calculating the slope of the graph on an interval on which the graph appears straight.

- You should be able to approximate $f'(a)$ using a table of values of f near a.

- From the microscope equation $\Delta y \approx f'(a) \cdot \Delta x$, you should be able to estimate any one of Δx, Δy and $f'(a)$ if given the other two.

- If $y = f(x)$ and there is an error in the measured value of x, you should be able to determine the absolute and relative error in y.

- You should be able to sketch the graph of f' if you are given the graph of f.

- You should be able to use the basic differentiation rules to find the derivative of a function given by a formula that involves sums of constant multiples of x^p, $\sin x$, $\cos x$, $\tan x$, or b^x.

- You should be able to break down a complicated formula into a chain of simple pieces.

- You should be able to use the chain rule to find the derivative of a chain of functions. This could involve several independent steps.

- For $z = F(x, y)$, you should be able to approximate any one of Δz, Δx, Δy, $F_x(a, b)$ and $F_y(a, b)$, if given the other four.

- You should be able to find formulas for partial derivatives using the basic rules and the chain rule for finding formulas for derivatives.

Chapter 4

Differential Equations

The rate equations with which we began our study of calculus are called **differential equations** when we identify the rates of change that appear within them as derivatives of functions. In this chapter we model new situations with differential equations. Euler's method remains a basic tool, but we also use the analytic and geometric power of the derivative to investigate the solutions of systems of differential equations. In addition, we *define* important classes of functions by means of differential equations.

§1. Modelling with Differential Equations

To analyze the way an infectious disease spreads through a population, we asked how three quantities S, I, and R would vary over time. This was difficult to answer; we found no simple, direct relation between S (or I or R) and t. What we *did* find, though, was a relation between the variables S, I, and R and their rates S', I', and R'. We expressed the relation as a set of rate equations. Then, given the rate equations and initial values for S, I, and R, we used Euler's method to estimate the values at any time in the future. By constructing a sequence of successive approximations, we were able to make these estimates as accurate as we wished.

There are two ideas here. The first is that we could write down equations for the rates of change that reflected important features of the process we sought to model. The second is that these equations *determined* the variables as functions of time, so we could make predictions about the real process we were modelling. Can we apply these ideas to other processes?

Differential equations and initial value problems

To answer this question, it will be helpful to introduce some new terms. What we have been calling rate equations are more commonly called **differential equations**. (The name is something of an historical accident. Since the equations involve functions and their derivatives, we might better call them *derivative* equations.) Euler's method treats the differential equations for a set of variables as a prescription for finding future values of those variables. However, in order to get started, we must always specify the initial values of the variables—their values at some given time. We call this specification an **initial condition**. The differential equations together with an initial condition is called an **initial value problem**. Each initial value problem determines a set of functions which we find by using Euler's method.

If we use Leibniz's notation for derivatives, a differential equation like $S' = -aSI$ takes the form $dS/dt = -aSI$. If we then treat dS/dt as a quotient of the individual *differentials* dS and dt (see page 122), we can even write the equation as $dS = -aSI\,dt$. Since this expresses the differential dS in terms of the differential dt, it was natural to call it a differential equation.

To illustrate how differential equations can be used to describe a wide range of processes in the physical, biological, and social sciences, we'll devote this section to a number of ways to model the long-term behavior of animal populations. To be specific, we will talk about rabbits and foxes, but the ideas can be adapted to the population dynamics of virtually all living things.

In each model, we will begin by identifying variables that describe what is happening. Then, we will try to establish how those variables change over time. Of course, no model can hope to capture every feature of the process we seek to describe, so we begin simply. We choose just one or two elements that seem particularly important. After examining the predictions of our simple model and checking how well they correspond to reality, we make modifications. We might include more features of the population dynamics, or we might describe the same features in different ways. Gradually, through a succession of refinements of our original simple model, we hope for descriptions that come closer and closer to the real situation we are studying.

Models can provide successive approximations to reality

Single-species Models: Rabbits

The problem. If we turn 2000 rabbits loose on a large, unpopulated island that has plenty of food for the rabbits, how might the number of rabbits vary over time? If we let $R = R(t)$ be the number of rabbits at time t (measured in months, let us say), we would like to be able to make some predictions about the function $R(t)$. It would be ideal to have a formula for $R(t)$—but this is not usually possible. Nevertheless, there may still be a great deal we can say about the behavior of R. To begin our explorations we will construct a model of the rabbit population that is obviously too simple. After we analyze the predictions it makes, we'll look at various ways to modify the model so that it approximates reality more closely.

The first model. Let's assume that, at any time t, the rate at which the rabbit population changes is simply proportional to the number of rabbits present at that time. For instance, if there were twice as many rabbits, then the *rate* at which new rabbits appear will also double. In mathematical terms, our assumption takes the form of the differential equation

(1)
$$\frac{dR}{dt} = k R \, \frac{\text{rabbits}}{\text{month}}.$$

Constant per capita growth

The multiplier k is called the **per capita growth rate** (or the **reproductive rate**), and its units are rabbits per month per rabbit. Per capita growth is discussed in problem 22 in chapter 1, §2.

For the sake of discussion, let's suppose that $k = .1$ rabbits per month per rabbit. (By cancelling the two occurences of *rabbit*, we can also just say $k = .1$ per month.) This assumption means that, on the average, one rabbit will produce .1 new rabbits every month. In the *S-I-R* model of chapter 1, the reciprocals of the coefficients in the differential equations had natural interpretations. The same is true here for the per capita growth rate. Specifically, we can say that $1/k = 10$ months is the average length of time required for a rabbit to produce one new rabbit.

Since there are 2000 rabbits at the start, we can now state a clearly defined initial value problem for the function $R(t)$:

$$\frac{dR}{dt} = .1\,R \qquad R(0) = 2000.$$

Use Euler's method to find $R(t)$

By modifying the program SIRPLOT, we can readily produce the graph of the function that is determined by this problem. Before we do that, though, let's first consider some of the implications that we can draw out of the problem without the graph.

Since $R'(t) = .1\,R(t)$ rabbits per month and $R(0) = 2000$ rabbits, we see that the initial rate of growth is $R'(0) = 200$ rabbits per month. If this rate were to persist for 20 years (= 240 months), R would have increased by

$$\Delta R = 240 \text{ months} \times 200 \;\frac{\text{rabbits}}{\text{month}} = 48000 \text{ rabbits,}$$

yielding altogether

$$R(240) = R(0) + \Delta R = 2000 + 48000 = 50000 \text{ rabbits}$$

at the end of the 20 years. However, since the population R is always getting larger, the differential equation tells us that the growth rate R' will *also* always be getting larger. Consequently, 50,000 is actually an underestimate of the number of rabbits predicted by this model.

Let's restate our conclusions in a graphical form. If R' were always 200 rabbits per month, the graph of R plotted against t would just be a straight line whose slope is 200 rabbits/month. But R' is always getting bigger, so the slope of the graph should increase from left to right. This will make the graph curve upward. In fact, SIRPLOT will produce the following graph of $R(t)$:

The graph of R curves up

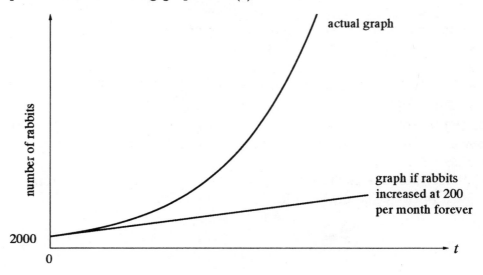

Later, we will see that the function $R(t)$ determined by this initial value problem is actually an exponential function of t, and we will even be able to write down a formula for $R(t)$, namely

$$R(t) = 2000\,(1.10517)^{t}.$$

This model is too simple to be able to describe what happens to a rabbit population very well. One of the obvious difficulties is that it predicts the rabbit population just keeps growing—forever. For example, if we used the formula for $R(t)$ given above, our model would predict that after 20 years ($t = 240$) there will be more than 50 *trillion* rabbits! While rabbit populations can, under good conditions, grow at a nearly constant per capita rate for a surprisingly long time (this happened in Australia during the 19th century), our model is ultimately unrealistic.

> It is a good idea to think qualitatively about the functions determined by a differential equation and make some rough estimates before doing extensive calculations. Your sketches may help you see ways in which the model doesn't correspond to reality. Or, you may be able to catch errors in your computations if they differ noticeably from what your estimates led you to expect.

The second model. One way out of the problem of unlimited growth is to modify equation (1) to take into account the fact that any given ecological system can support only some finite number of creatures over the long term. This number is called the **carrying capacity** of the system. We expect that when a population has reached the carrying capacity of the system, the population should neither grow nor shrink. At carrying capacity, a population should hold steady—its rate of change should be zero.

The carrying capacity of the environment

What we would like to do, then, is to find an expression for R' which is similar to equation (1) when the number of rabbits R near 2000, but which becomes 0 as R approaches 25,000. One possibility which captures these features is the **logistic equation**, first proposed by the Belgian mathematician Otto Verhulst in 1845:

$$(2) \qquad R' = k\,R\left(1 - \frac{R}{b}\right)\frac{\text{rabbits}}{\text{month}}.$$

Logistic growth

In this equation, the coefficient k is called the **natural growth rate**. It plays the same role as the per capita growth rate in equation (1), and it has the same units—rabbits per month per rabbit. The number

b is the **carrying capacity**; it is measured in rabbits. (We first saw the logistic equation on pages 81–86.) Notice also that we have written the derivative of R in the simpler form R', a practice we will continue for the rest of the section.

Suppose the carrying capacity of the island is 25,000 rabbits. If we keep the natural growth rate at .1 rabbits per month per rabbit, then the logistic equation for the rabbit population is

$$R' = .1\, R \left(1 - \frac{R}{25000}\right) \frac{\text{rabbits}}{\text{month}}.$$

Check to see that this equation really does have the behavior claimed for it—namely, that a population of 25,000 rabbits neither grows or declines. Notice also that R' is positive as long as R is less than 25000, so the population increases. However, as R approaches 25000, R' will get closer and closer to 0, so the graph will become nearly horizontal. (What would happen if the island ever had more than 25,000 rabbits?)

> The graph of $R(t)$ levels off near $R = 25000$

These observations about the qualitative behavior of $R(t)$ are consistent with the following graph, produced by a modified version of the program SIRPLOT. For comparison, we have also graphed the exponential function produced by the first model. Notice that the two graphs "share ink" when R near 2000.

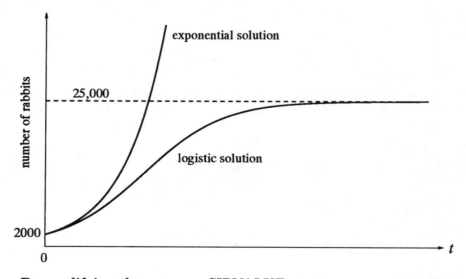

By modifying the program SIRVALUE, we can even get numerical answers to specific questions about the two models. For example, after 30 months under constant per capita growth, the rabbit population will be more than 40,000—well beyond the carrying capacity of the island. Under logistic growth, though, the population will be only about 16,000.

In the following figure we display several functions that are determined by the logistic equation

$$R' = .1\,R\left(1 - \frac{R}{25000}\right)$$

when different initial conditions are given. Each graph therefore predicts the future for a different initial population $R(0)$. One of the graphs is just the t-axis itself. What does this graph predict about the rabbit population?

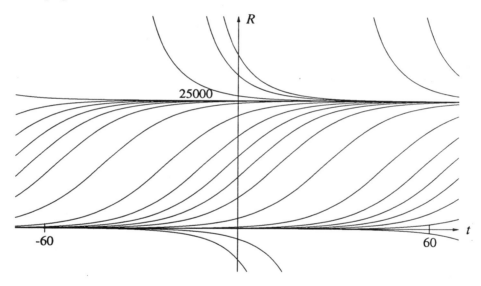

Solutions to the logistic equation $R' = .1R\,(1 - R/25000)$

Two-species Models: Rabbits and Foxes

No species lives alone in an environment, and the same is true of the rabbits on our island. The rabbit population will probably have to deal with predators of various sorts. Some are microscopic—disease organisms, for example—while others loom as obvious threats. We will enrich our population model by adding a second species—foxes—that will prey on the rabbits. We will continue to suppose that the rabbits live on abundant native vegetation, and we will now assume that the

Introduce
predators

rabbits are the sole food supply of the foxes. Can we say what will happen? Will the number of foxes and rabbits level off and reach a "steady state" where their numbers don't vary? Or will one species perhaps become extinct?

Let F denote the number of foxes, and R the number of rabbits. As before, measure the time t in months. Then F and R are functions of t: $F(t)$ and $R(t)$. We seek differential equations that describe how the growth rates F' and R' are related to the population sizes F and R. We make the following assumptions.

- In the absence of foxes, the rabbit population grows logistically.

- The population of rabbits declines at a rate proportional to the product $R \cdot F$. This is reasonable if we assume rabbits never die of old age—they just get a little too slow. Their death rate, which depends on the number of fatal encounters between rabbits and foxes, will then be approximately proportional to both R and F— and thus to their product. (This is the same kind of interaction effect we used in our epidemic model to predict the rate at which susceptibles become infected.)

- In the absence of rabbits, the foxes die off at a rate proportional to the number of foxes present.

- The fox population increases at a rate proportional to the number of encounters between rabbits and foxes. To a first approximation, this says that the birth rate in the fox population depends on maternal fox nutrition, and this depends on the number of rabbit-fox encounters, which is proportional to $R \cdot F$.

Our assumptions are about birth and death rates, so we can convert them quite naturally into differential equations. Pause here and check that the assumptions translate into these differential equations:

Lotka–Volterra equations with bounded growth

$$R' = aR\left(1 - \frac{R}{b}\right) - cRF = aR - \frac{a}{b}R^2 - cRF$$
$$F' = dRF - eF$$

These are the **Lotka–Volterra equations with bounded growth**. The coefficients a, b, c, d, and e are **parameters**—constants that have to be determined through field observations in particular circumstances.

An example. To see what kind of predictions the Lotka–Volterra equations make, we'll work through an example with specific values for the parameters. Let

$$
\begin{aligned}
a &= .1 \quad \text{rabbits per month per rabbit} \\
b &= 10000 \quad \text{rabbits} \\
c &= .005 \quad \text{rabbits per month per rabbit-fox} \\
d &= .00004 \quad \text{foxes per month per rabbit-fox} \\
e &= .04 \quad \text{foxes per month per fox}
\end{aligned}
$$

(Check that these five parameters have the right units.) These choices give us the specific differential equations

$$
\begin{aligned}
R' &= .1\,R - .00001\,R^2 - .005\,RF \\
F' &= .00004\,RF - .04\,F
\end{aligned}
$$

To use this model to follow R and F into the future, we need to know the initial sizes of the two populations. Let's suppose that there are 2000 rabbits and 10 foxes at time $t = 0$. Then the two populations will vary in the following way over the next 250 months.

The graphs of R and F

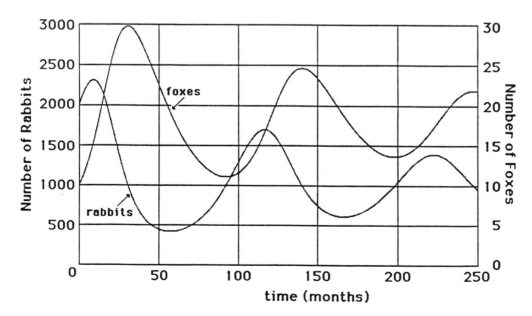

Rabbit and fox populations as a function of time

A variant of the program SIRPLOT was used to produce these
graphs. Notice that it plots $100F$ rather than F itself. This is because
the number of foxes is about 100 times smaller than the number of
rabbits. Consequently, $100F$ and R are about the same size, so their
graphs fit nicely together on the same screen.

The graphs have several interesting features. There are different
scales for the R and the F values, because the program plots $100F$
instead F. The peak fox population is about 30, while the peak rabbit
population is about 2300. The rabbit and fox populations rise and fall
in a regular manner. They rise and fall less with each repeat, though,
and if the graphs were continued far enough into the future we would
see R and F level off to nearly constant values.

How rabbits
respond to
changes in the
initial fox
population

The illustration below shows what happens to an initial rabbit population of 2000 in the presence of three different initial fox populations
$F(0)$. Note that the peak rabbit populations are different, and they
occur at different times. The size of the intervals between peaks also
depends on $F(0)$.

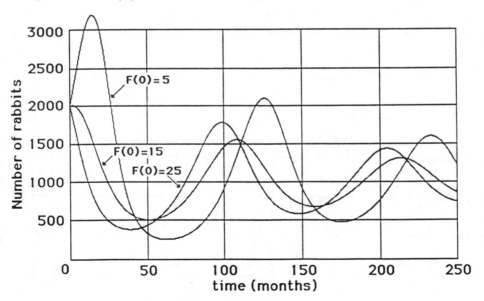

Rabbit populations for different initial fox populations

We have looked at three models, each a refinement of the preceding
one. The first was the simplest. It accounted only for the rabbits,
and it assumed the rabbit population grew at a constant per capita
rate. The second was also restrict to rabbits, but it assumed logistic

growth to take into account the carrying capacity of the environment. The third introduced the complexity of a second species preying on the rabbits. In the exercises you will have an opportunity to explore these and other models. Remember that when you use Euler's method to find the functions determined by an initial value problem, you must construct a sequence of successive approximations, until you obtain the level of accuracy desired.

Exercises

Single-species models

1. **Constant per capita Growth.** This question considers the initial value problem given in the text:

$$R' = 0.1\,R \text{ rabbits per month}; \qquad R(0) = 2000 \text{ rabbits}.$$

a) Use Euler's method to determine how many rabbits there are after 6 months. Present a table of successive approximations from which you can read the exact value to whole-number accuracy.

b) Determine, to whole-number accuracy, how many rabbits there are after 24 months.

c) How many months does it take for the rabbit population to reach 25,000?   [Answer: 25.258 months.]

2. **Logistic Growth.** The following questions concern a rabbit population described by the logistic model

$$R' = 0.1\,R \left(1 - \frac{R}{25000} \right) \text{ rabbits per month}.$$

a) What happens to a population of 2000 rabbits after 6 months, after 24 months, and after 5 years? To answer each question, present a table of successive approximations that allows you to give the exact value to the nearest whole number.

b) Sketch the functions determined by the logistic equation if you start with either 2000 or 40000 rabbits. (Suggestion: you can modify the program SIRPLOT to answer this question.) Compare the two functions. How do they differ? In what ways are they similar?

3.   **World population**. The world's population in 1990 was about 5 billion, and data show that birth rates range from 35 to 40 per thousand per year and death rates from 15 to 20.   Take this to imply a net annual growth rate of 20 per thousand. One model for world population assumes constant per capita growth, with a per capita growth rate of $20/1000 = 0.02$.

a) Write a differential equation for P that expresses this assumption. Use P to denote the world population, measured in billions.

b) According to the differential equation in (a), at what rate (in billions of persons per year) was the world population growing in 1990?

c) By applying Euler's method to this model, using the initial value of 5 billion in 1990, estimate the world population in the years 1980, 2000, 2040, and 2230. Present a table of successive approximations that stabilizes with one decimal place of accuracy (in billions). What step size did you have to use to obtain this accuracy?

4.   **Supergrowth**. Another model for the world population assumes "supergrowth"—the rate P' is proportional to a *higher power* of P, rather than to P itself. The model is

$$P' = .015\, P^{1.2}.$$

As in the previous exercise, assume that P is measured in billions, and the population in 1990 was about 5 billion.

a) According to this model, at what rate (in billions of persons per year) was the population growing in 1990?

b) Using Euler's method, estimate the world population in the years 1980, 2000 and 2040. Use successive approximations until you have one decimal place of accuracy (in billions). What step size did you have to use to obtain this accuracy?

c) Use an Euler approximation with a step size of 0.1 to estimate the world population in the year 2230. What happens if you repeat your calculation with a step size of 0.01? [Comment: Something strange is going on here. We will look again at this model in the next section.]

Two-species models

Here are some other differential equations that model a predator-prey interaction between two species.

5.   **The May Model.** This model has been proposed by the contemporary ecologist, R.M. May, to incorporate more realistic assumptions about the encounters between predators (foxes) and their prey (rabbits). So that you can work with quantities that are about the same size, let y be the number of foxes and let x be the number of rabbits *divided by* 100. May makes the following assumptions.

- In the absence of foxes, the rabbits grow logistically.

- Rabbits are captured at a rate proportional to $x \cdot y$ when x is small but only to y when x is large. (If there are many rabbits, it is much easier for the foxes to catch them. Because each fox can catch as many rabbits as it can eat, the capture rate will depend only on the number of foxes. If there are few rabbits, then the capture rate will depend on the number of rabbits, too.)

- The fox population is governed by the logistic equation, and the carrying capacity is proportional to the number of rabbits.

a) Explain how the following system of equations incorporates these assumptions.

$$x' = a x \left(1 - \frac{x}{b}\right) - \frac{c x y}{x + d}$$
$$y' = e y \left(1 - \frac{y}{f x}\right)$$

The parameters a, b, c, d, e and f are all positive.

b) Assume you begin with 2000 rabbits and 10 foxes. (Be careful: $x(0) \neq 2000$.) What does May's model predict will happen to the rabbits and foxes over time if the values of the parameters are $a = .6$, $b = 10$, $c = .5$, $d = 1$, $e = .1$ and $f = 2$? Use a suitable modification of the program SIRPLOT.

c) Using the same parameters, describe what happens if you begin with 2000 rabbits and 20 foxes; with 1000 rabbits and 10 foxes; with 1000 rabbits and 20 foxes. Does the eventual long-term behavior depend on the initial condition? How does the long-term behavior here compare with the long-term behavior of the two populations in the Lotka–Volterra model of the text?

6. **The Lotka–Volterra Equations**. This model for predator and prey interactions is slightly simpler than the "bounded growth" version we consider in the text. It is important historically, though, because it was first proposed as a way of understanding why the harvests of certain species of fish in the Adriatic Sea exhibited cyclical behavior over the years. For the sake of variety, let's take the prey to be hares and the predators to be lynxes.

Let $H(t)$ denote the number of hares at time t and $L(t)$ the number of lynxes. This model, the basic Lotka–Volterra model, differs from the bounded growth model in only one respect: it assumes the hares would experience constant per capita growth if there were no lynxes.

a) Explain why the following system of equations incorporates the assumptions of the basic model. (The parameters a, b, c, and d are all positive.)

$$H' = aH - bHL$$
$$L' = cHL - dL$$

(These are called the **Lotka–Volterra equations**. They were developed independently by the Italian mathematical physicist Vito Volterra in 1925–26, and by the mathematical ecologist and demographer Alfred James Lotka a few years earlier. Though very simplified, they form one of the principal starting points in ecological modeling.)

b) Explain why a and b have the units hares per month per hare and hares per month per hare-lynx, respectively. What are the units of c and d? Explain why.

Suppose time t is measured in months, and suppose the parameters have values

$$
\begin{aligned}
a &= .1 && \text{hares per month per hare} \\
b &= .005 && \text{hares per month per hare-lynx} \\
c &= .00004 && \text{lynxes per month per hare-lynx} \\
d &= .04 && \text{lynxes per month per lynx}
\end{aligned}
$$

This leads to the system of differential equations

$$H' = .1H - .005HL$$
$$L' = .00004HL - .04L.$$

c) Suppose that you start with 2000 hares and 10 lynxes—that is, $H(0) = 2000$ and $L(0) = 10$. Describe what happens to the two populations.

A good way to do this is to draw graphs of the functions $H(t)$ and $L(t)$. It will be convenient to have the Hare scale run from 0 to 3000, and the Lynx scale from 0 to 50. If you modify the program SIRPLOT, have it plot H and $60L$.

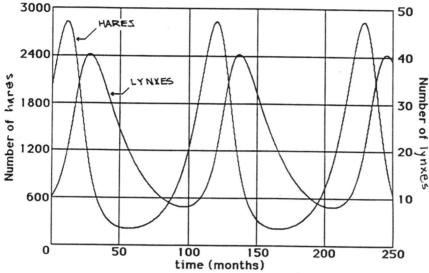

Hare and lynx populations as a function of time

You should get graphs like those above. Notice that the hare and lynx populations rise and fall in a fashion similar to the rabbits and foxes, but here they oscillate—returning *periodically* to their original values.

d) What happens if you keep the same initial hare population of 2000, but use different initial lynx populations? Try $L(0) = 20$ and $L(0) = 50$. (In each case, use a step size of .1 month.)

e) Start with 2000 hares and 10 lynxes. From part (c), you know the solutions are periodic. The goal of this part is to analyze this periodic behavior. You can do this with your program in part (c), but you may prefer to replace the FOR-NEXT loop in your program by a variety of DO-WHILE loops (see page 77). First find the maximum number of hares. What is the length of one **period** for the hare population? That is, how long does it take the hare population to complete one cycle (e.g., to go from one maximum to the next)? Find the length of one period for the lynxes. Do the hare and lynx populations have the same periods?

f) Plot the hare populations over time when you start with 2000 hares and, successively, 10, 20, and 50 lynxes. Is the hare population periodic in each case? What is the period? Does it vary with the size of the initial lynx population?

Fermentation

Wine is made by yeast; yeast digest the sugars in grape juice and produce alcohol as a waste product. This process is called fermentation. The alcohol is toxic to the yeast, though, and the yeast is eventually killed by the alcohol. This stops fermentation, and the liquid has become wine, with about 8–12% alcohol.

Although alcohol isn't a "species," it acts like a predator on yeast. Unlike the other predator-prey problems we have considered, though, the yeast does not have an unlimited food supply. The following exercises develop a sequence of models to take into account the interactions between sugar, yeast, and alcohol.

7. a) In the first model assume that the sugar supply is not depleted, that no alcohol appears, and that the yeast simply grows *logistically*. Begin by adding 0.5 lb of yeast to a large vat of grape juice whose carrying capacity is 10 lbs of yeast. Assume that the natural growth rate of the yeast is 0.2 lbs of yeast per hour, per pound of yeast. Let $Y(t)$ be the number of pounds of live yeast present after t hours; what differential equation describes the growth of Y?

b) Graph the solution $Y(t)$, for example by using a suitable modification of the program SIRPLOT. Indicate on your graph approximately when the yeast reaches one-half the carrying capacity of the vat, and when it gets to within 1% of the carrying capacity.

c) Suppose you use a second strain of yeast whose natural growth rate is only half that of the first strain of yeast. If you put 0.5 lb of *this* yeast into the vat of grape juice, when will it reach one-half the carrying capacity of the vat, and when will it get to within 1% of the carrying capacity? Compare these values to the values produced by the first strain of yeast: are they larger, or smaller? Sketch, on the same graph as in part (b), the way this yeast grows over time.

8. a) Now consider how the yeast produces alcohol. Suppose that waste products are generated at a rate proportional to the amount of yeast present; specifically, suppose each pound of yeast produces 0.05 lbs of alcohol per hour. (The other major waste product is carbon

dioxide gas, which bubbles out of the liquid.) Let $A(t)$ denote the amount of alcohol generated after t hours. Write a differential equation that describes the growth of A.

b) Consider the toxic effect of the alcohol on the yeast. Assume that yeast cells die at a rate proportional to the amount of alcohol present, and also to the amount of yeast present. Specifically, assume that, in each pound of yeast, a pound of alcohol will kill 0.1 lb of yeast per hour. Then, if there are Y lbs of yeast and A lbs of alcohol, how many pounds of yeast will die in one hour? Modify the logistic equation for Y to take this effect into account. The modification involves subtracting off a new term that describes the rate at which alcohol kills yeast. What is the new differential equation?

c) You should now have two differential equations describing the rates of growth of yeast and alcohol. The equations are **coupled**, in the sense that the yeast equation involves alcohol, and the alcohol equation involves yeast. Assuming that the vat contains, initially, 0.5 lb of yeast and no alcohol, describe by means of a graph what happens to the yeast. How close does the yeast get to carrying capacity, and when does this happen? Does the fermentation end? If so, when; and how much alcohol has been produced by that time?

9.   What happens if the rates of toxicity and alcohol production are different? Specifically, increase the rate of alcohol production by a factor of five—from 0.05 to 0.25 lbs of alcohol per hour, per pound of yeast—and at the same time reduce the toxicity rate by the same factor—from 0.10 to 0.02 lb of yeast per hour, per pound of alcohol and pound of yeast. How do these changes affect the time it takes for fermentation to end? How do they affect the amount of alcohol produced? What happens if only the rate of alcohol production is changed? What happens if only the toxicity rate is reduced?

10.   a) The third model will take into account that the sugar in the grape juice is consumed. Suppose the yeast consumes .15 lb of sugar per hour, per lb of yeast. Let $S(t)$ be the amount of sugar in the vat after t hours. Write a differential equation that describes what happens to S over time.

b) Since the carrying capacity of the vat depend on the amount of sugar in it, the carrying capacity must now vary. Assume that the carrying capacity of S lbs of sugar is .4 S lbs of yeast. How much sugar is needed to maintain a carrying capacity of 10 lbs of yeast? How much is needed to maintain a carrying capacity of 1 lb of yeast? Rewrite the logistic equation for yeast so that the carrying capacity is $.4\,S$ lbs, instead of 10 lbs, of yeast.

c) There are now three differential equations. Using them, describe what happens to .5 lbs of yeast that is put into a vat of grape juice that contains 25 lbs of sugar at the start. Does all the sugar disappear? Does all the yeast disappear? How long does it take before there is only .01 lb of yeast? How much sugar is left then? How much alcohol has been produced by that time?

Newton's law of cooling

Suppose that we start off with a freshly brewed cup of coffee at 90°C and set it down in a room where the temperature is 20°C. What will the temperature of the coffee be in 20 minutes? How long will it take the coffee to cool to 30°C?

If we let the temperature of the coffee be Q (in °C), then Q is a function of the time t, measured in minutes. We have $Q(0) = 90°C$, and we would like to find the value t_1 for which $Q(t_1) = 30°C$.

It is not immediately apparent how to give Q as a function of t. However, we can describe the *rate* at which a liquid cools off, using **Newton's law of cooling**: the rate at which an object cools (or warms up, if it's cooler than its surroundings) is proportional to the *difference* between its temperature and that of its surroundings.

11.   In our example, the temperature of the room is 20°C, so Newton's law of cooling states that $Q'(t)$ is proportional to $Q - 20$. In symbols, we have

$$Q' = -k\,(Q - 20)$$

where k is some positive constant.

a) Why is there a minus sign in the equation?

The particular value of k would need to be determined experimentally. It will depend on things like the size and shape of the cup, how much sugar and cream you use, and whether you stir the liquid. Suppose that k has the value of .1° per minute per °C of temperature difference. Then the differential equation becomes:

$$Q' = -.1(Q - 20) \quad °C \text{ per minute.}$$

b) Use Euler's method to determine the temperature Q after 20 minutes. Write a table of successive approximations with smaller and smaller step sizes. The values in your table should stabilize to the second decimal place.

c) How long does it take for the temperature Q to drop to 30°C? Use a DO-WHILE loop to construct a table of successive approximations that stabilize to the second decimal place.

12. On a hot day, a cold drink warms up at a rate approximately proportional to the difference in temperature between the drink and its surroundings. Suppose the air temperature is 90°F and the drink is initially at 36°F. If Q is the temperature of the drink at any time, we shall suppose that it warms up at the rate

$$Q' = -0.2(Q - 90) \quad \text{°F per minute.}$$

According to this model, what will the temperature of the drink be after 5 minutes, and after 10 minutes. In both cases, produce values that are accurate to two decimal places.

13. In our discussion of cooling coffee, we assumed that the coffee did not heat up the room. This is reasonable because the room is large, compared to the cup of coffee. Suppose, in an effort to keep it warmer, we put the coffee into a small insulated container—such as a microwave oven (which is turned off). We must assume that the coffee *does* heat up the air inside the container. Let A be the air temperature in the container and Q the temperature of the coffee. Then both A and Q change over time, and Newton's law of cooling tells us the *rates* at which they change. In fact, the law says that both Q' and A' are proportional to $Q - A$. Thus,

$$\begin{aligned} Q' &= -k_1(Q - A) \\ A' &= k_2(Q - A), \end{aligned}$$

where k_1 and k_2 are positive constants.

a) Explain the signs that appear in these differential equations.

b) Suppose $k_1 = .3$ and $k_2 = .1$. If $Q(0) = 90°C$ and $A(0) = 20°C$, when will the temperature of the coffee be 40°C? What is the temperature of the air at this time? Your answers should be accurate to one decimal place.

c) What does the temperature of the coffee become eventually? How long does it take to reach that temperature?

S-I-R revisited

Consider the spread of an infectious disease that is modelled by the S-I-R differential equations

$$
\begin{aligned}
S' &= -.00001\, SI \\
I' &= .00001\, SI - .08\, I \\
R' &= .08\, I
\end{aligned}
$$

Take the initial condition of the three populations to be

$$
\begin{aligned}
S(0) &= 35,400 \text{ persons} \\
I(0) &= 13,500 \text{ persons} \\
R(0) &= 22,100 \text{ persons}
\end{aligned}
$$

14. How many susceptibles are left after 40 days? When is the largest number of people infected? How many susceptibles are there at that time? Explain how you could determine the last number *without* using Euler's method.

[Answer: There are 105 susceptibles after 40 days. The largest number of people are infected when $t = 5.3$ days. At that time, there are 8000 susceptibles.]

15. What happens as the epidemic "runs its course"? That is, as more and more time goes by, what happens to the numbers of infecteds and susceptibles?

16. One of the principal uses of a mathematical model is to get a qualitative idea how a system will behave with different initial conditions. For instance, suppose we introduce 100 infected individuals into a population. How will the spread of the infection depend on the size of the population? Assume the same S-I-R differential equations that were used in the previous exercise, and draw the graphs of $S(t)$ for initial susceptible population sizes $S(0)$ ranging from 0 to 45,000 in increments of 5000 (that is, take $S(0) = 0, 5000, 10000, \ldots, 45000$). In each case assume that $R(0) = 0$ and $I(0) = 100$. Use these graphs

to argue that the larger the initial susceptible population, the more rapidly the epidemic runs its course.

17.   Draw the graphs of $I(t)$ for the same initial conditions as in the previous problem. Using these graphs you can demonstrate that the larger the susceptible population, the larger will be the fraction of the population that is infected during the worst stages of the epidemic. Do this by constructing a table displaying I_{max}, t_{max}, and P_{max}, where I_{max} is the maximum value of $I(t)$, t_{max} is the time at which this maximum occurs (that is, $I_{max} = I(t_{max})$), and P_{max} is the ratio of I_{max} to the initial susceptible population: $P_{max} = I_{max}/S(0)$. The table below gives the first three sets of values.

$S(0)$	I_{max}	P_{max}	t_{max}
5 000	100	0.02	0
10 000	315	0.03	> 100
15 000	2071	0.14	66
⋮	⋮	⋮	⋮

Your table should show that there is a time when over half the population is infected if $S(0) = 45000$, while there is never a time when more than one-fourth of the population is infected if $S(0) = 20000$.

Constructing models

Questions in which we know a number of quantities at a given time and would like to know their values at a future time (or know at what future time they will attain given values) occur in many different contexts. The following are some questions for discussion. Can any of these be modelled as initial value problems? What information would you need to resolve the question?

18.   We deposit a fixed sum of money in a bank, and we'd like to know how much will be there in ten years.

19.   We know the diameter of the mold spot growing on a cheese sandwich is 1/4 inch, and we'd like to know when its diameter will be one inch.

20.   We know the fecal bacterial and coliform concentrations in a local swimming hole, and we'd like to know when they fall below certain prescribed levels (which the Board of Health deems safe).

21.   We know what the temperature and rainfall is today, and we'd like to know what both will be one week from today.

22.   We know what the winning lottery number was yesterday, and we'd like to know what the winning number will be the day after to-morrow.

23.   We know where the earth, sun, and moon are in relation to each other now, and how fast and in what direction they are moving. We would like to be able to predict where they are going to be at any time in the future. We know the gravity of each affects the motions of the others by determining the way their velocities are changing.

§2.   Solutions of Differential Equations

Differential Equations are Equations

Differential equations give instructions for Euler's method

Until now, we have viewed a system of differential equations as a set of *instructions* for "stepping into the future" (or the past). Put another way, an initial value problem is a prescription for using Euler's method to determine a set of functions. In this section we take a new point of view: we will think of differential equations as *equations* for which those functions are *solutions*.

To see what this means, let's look first at equations in algebra. Consider the equation $x^2 = x + 6$. As it stands, this is neither true nor false. We make it true or false, though, when we substitute a particular number for x. For example, $x = 3$ makes the equation true, because $3^2 = 3 + 6$. On the other hand, $x = 1$ makes the equation false, because

Equations have solutions

$1^2 \neq 1 + 6$. Any number that makes an equation true is called a *solution* to that equation. In fact, $x^2 = x + 6$ has exactly two solutions: $x = 3$ and $x = -2$.

We can view differential equations the same way. Consider, for example, the differential equation

$$\frac{dy}{dt} = \frac{1}{2y}.$$

Because it involves the expression dy/dt, we understand that y is a function of t. As it stands, the differential equation is neither true nor false. We make it true or false, though, when we substitute a particular *function* for y. For example, $y = \sqrt{t} = t^{1/2}$ makes the differential equation true. To see this, first look at the left-hand side of the equation:

$$\frac{dy}{dt} = \tfrac{1}{2}t^{-1/2} = \frac{1}{2\sqrt{t}}.$$

Substitute $y = \sqrt{t}$ into the differential equation

Now look at the right-hand side:

$$\frac{1}{2y} = \frac{1}{2\sqrt{t}}.$$

The two sides of the equation are equal, so the substitution $y = \sqrt{t}$ makes the equation true.

The function $y = t^2$, however, makes the differential equation *false*. The left-hand side is

$$\frac{dy}{dt} = 2t,$$

but the right-hand side is

$$\frac{1}{2y} = \frac{1}{2t^2}.$$

Since $2t$ and $1/2t^2$ are different functions, the two sides are unequal and the equation is therefore false.

We say that $y = \sqrt{t}$ is a **solution** to this differential equation. The function $y = t^2$ is *not* a solution. To decide whether a particular function is a solution when the function is given by a formula, notice that we need to be able to differentiate the formula.

A solution makes the equation true

> If we view differential equations simply as instructions for carrying out Euler's method, we need only the microscope equation $\Delta y \approx y' \cdot \Delta t$ in order to find functions. However, if we want to find functions that are solutions to differential equations from our new point of view, we first need to introduce the idea of the derivative and the rules for differentiating functions.

Just as an algebraic equation can have more than one solution, so can a differential equation. In fact, we can show that $y = \sqrt{t + C}$ is a solution to the differential equation

$$\frac{dy}{dt} = \frac{1}{2y},$$

for any value of the constant C. To evaluate the left-hand side dy/dt, we need the chain rule (chapter 3, §6). Let's write

$$y = \sqrt{u} \quad \text{where} \quad u = t + C.$$

Then the left-hand side is the function

$$\frac{dy}{dt} = \frac{dy}{du} \cdot \frac{du}{dt} = \frac{1}{2\sqrt{u}} \cdot 1 = \frac{1}{2\sqrt{t+C}}.$$

Since the right-hand side of the differential equation is

$$\frac{1}{2y} = \frac{1}{2\sqrt{t+C}},$$

A differential equation can have infinitely many solutions

the two sides are equal—no matter what value C happens to have. This proves that every function of the form $y = \sqrt{t+C}$ is a solution to the differential equation. Since there are infinitely many different values that C can take, the differential equation has infinitely many different solutions!

If a differential equation arises in modelling a physical or biological process, the variables involved must also satisfy an initial condition. Suppose we add an initial condition to our differential equation:

$$\frac{dy}{dt} = \frac{1}{2y} \quad \text{and} \quad y(0) = 5.$$

Does *this* problem have a solution—that is, can we find a function $y(t)$ that is a solution to the differential equation and also satisfies the condition $y(0) = 5$?

Notice $y = \sqrt{t}$ is not a solution to this new problem. Although it satisfies the differential equation, it fails to satisfy the initial condition:

$$y(0) = \sqrt{0} = 0 \neq 5.$$

Perhaps one of the other solutions to the differential equation will work. When we evaluate the solution $y = \sqrt{t+C}$ at $t = 0$ we get

$$y(0) = \sqrt{0+C} = \sqrt{C}.$$

An initial value problem has only one solution

We want this to equal 5, and it will if $C = 25$. Thus, $y = \sqrt{t+25}$ is a solution to the initial value problem. Furthermore, the only value of C which will make $y(0) = 5$ is $C = 25$, so the initial value problem has only one solution of the form $\sqrt{t+C}$. The graph of this solution is at the top of the opposite page.

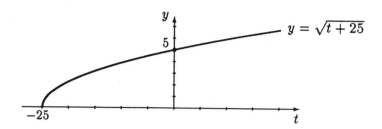

As always, you can use Euler's method to find the function determined by an initial value problem, and you can graph that function using the program SIRPLOT, for example. How will *that* graph compare with this one? In the exercises you can explore this question.

Checking solutions versus finding solutions. Notice that we have only *checked* whether a given function solves an initial value problem; we have not *constructed* a formula to solve the problem. This was true even for the algebraic equation $x^2 = x + 6$. Of course, there are methods to find solutions. One possibility is to rewrite $x^2 = x + 6$ in the form $x^2 - x - 6 = 0$. By factoring $x^2 - x - 6$ as

$$x^2 - x - 6 = (x - 3)(x + 2)$$

we can see that either $x - 3 = 0$ (so $x = 3$), or $x + 2 = 0$ (so $x = -2$). Another method is to use the **quadratic formula**

$$x = \frac{-b \pm \sqrt{b^2 - 4ac}}{2a}$$

for the roots of the quadratic function $ax^2 + bx + c$. In our case, the quadratic formula yields

$$x = \frac{-(-1) \pm \sqrt{1 - 4 \cdot 1 \cdot (-6)}}{2 \cdot 1} = \frac{1 \pm \sqrt{1 + 24}}{2} = \frac{1 \pm 5}{2},$$

so again we find that x must be either 3 or -2.

The methods we use to solve an algebraic equation depend very much on the equation we face. For example, there is no way to find a solution to $\sin x = 2^x$ by factoring, or by using a "magic formula" like the quadratic formula. There are methods that *do* work, though. In chapters 1 and 2 we dealt with similar problems by using a computer graphing utility that could zoom in on the point of intersection

There are special methods to solve particular equations, but other methods work generally

of two graphs. In chapter 6 we will introduce another tool, the Newton–Raphson method, for finding roots. These are both powerful methods, because they will work with nearly all algebraic equations.

The methods we use to solve a differential equation also depend on the equation we face. A course in differential equations provides methods for finding formulas that solve many different kinds of differential equations. The methods are like the quadratic formula in algebra, though—they give a complete solution, but they work only with differential equations that have a very specific form. This course will not attempt to survey the methods that find such formulas. Instead, it will continue to rely on Euler's method, which works with nearly all differential equations.

However, there are clear benefits to having a formula for the solution to a differential equation. In this section, we will look at some of those benefits.

World Population Growth

Two models

In the exercises in the last section, we looked at two different models that seek to describe how the world population will grow. One model assumed constant per capita growth—rate of change proportional to population size. The other assumed "supergrowth"—rate of change proportional to a *higher* power of the population size. Let's write P for the population size in the constant per capita growth model and Q for the population size in the supergrowth model. In both cases, the population is expressed in billions of persons and time is measured in years, with $t = 0$ in 1990. In this notation, the two models are

Models for world population growth

$$\text{constant per capita:} \quad \frac{dP}{dt} = .02\,P \qquad P(0) = 5$$

$$\text{supergrowth:} \quad \frac{dQ}{dt} = .015\,Q^{1.2} \qquad Q(0) = 5$$

By using Euler's method, we discover that the two models predict fairly similar results over sixty years, although the supergrowth model lives up to its name by predicting larger populations than the constant per capita growth model as time passes:

t	P	Q
-10	4.09	4.08
0	5.00	5.00
10	6.11	6.18
50	13.59	15.94

These estimates are accurate to one decimal place, and that level of accuracy was obtained with the step size $\Delta t = .1$.

However, the predictions made by the models differ widely over longer time spans. If we use Euler's method to estimate the populations after 240 years, we get

Δt	$P(240)$	$Q(240)$
.1	6.046×10^2	1.979×10^{10}
.01	6.073×10^2	2.573×10^{11}
.001	6.075×10^2	3.825×10^{11}

As the step size decreases from 0.1 to 0.01 to 0.001, the estimates of the constant per capita growth model $P(240)$ behave as we have come to expect: already three digits have stabilized. But in the estimates of the supergrowth model, not even one digit of $Q(240)$ has stabilized.

In this section we will see that there are actually *formulas* for the functions $P(t)$ and $Q(t)$. These formulas will illuminate the reason behind the differences in speed of stabilization in the estimates.

A formula for the supergrowth model

Without asking how the following formula might have been derived, let's check that it is indeed a solution to the supergrowth initial value problem.

$$Q(t) = \left(\frac{1}{\sqrt[5]{5}} - .003\,t \right)^{-5}$$

First of all, the formula satisfies the initial condition $Q(0) = 5$:

Checking the initial condition

$$Q(0) = \left(\frac{1}{\sqrt[5]{5}} \right)^{-5} = (\sqrt[5]{5})^5 = 5.$$

Checking the differential equation

To check that it also satisfies the differential equation, we must evaluate the two sides of the differential equation

$$\frac{dQ}{dt} = .015\,Q^{1.2}.$$

Left-hand side

Let's begin by evaluating the left-hand side. To differentiate $Q(t)$, we will write Q as a *chain* of functions:

$$Q = u^{-5} \qquad \text{where} \qquad u = \frac{1}{\sqrt[5]{5}} - .003\,t.$$

Since $Q = u^{-5}$, $dQ/du = (-5)u^{-6}$. Also, since u is just a linear function of t in which the multiplier is $-.003$, we have $du/dt = -.003$. Consequently,

$$\begin{aligned}
\frac{dQ}{dt} &= \frac{dQ}{du} \cdot \frac{du}{dt} \\
&= (-5)\,u^{-6} \cdot (-.003) \\
&= .015\,u^{-6}
\end{aligned}$$

Ordinarily, we would "finish the job" by substituting for u its formula in terms of t. However, let's just leave the left-hand side in this form for the moment.

Right-hand side

To evaluate the right-hand side (which is the expression $.015\,Q^{1.2}$), we would expect to substitute for Q its formula in terms of t. But notice that, on the left-hand side, we stopped when we got to u. We'll do the same thing here. Since $Q = u^{-5}$,

$$Q^{1.2} = Q^{6/5} = (u^{-5})^{6/5} = u^{-5 \cdot 6/5} = u^{-6}.$$

Therefore, the right-hand side is equal to $.015\,u^{-6}$. But so is the left-hand side, so $Q(t)$ is indeed a solution to the differential equation

$$\frac{dQ}{dt} = .015\,Q^{1.2}.$$

Notice two things about this result. First, when we work with formulas we have greater need for algebra to manipulate them. For example, we needed one of the laws of exponents, $(a^b)^c = a^{bc}$, to evaluate the right-hand side. Second, we found it simpler to express Q in terms of the intermediate variable u, instead of the original input variable t. In another computation, it might be preferable to replace u by its formula in terms of t. You need to choose your algebraic strategy to fit the circumstances.

Behavior of the supergrowth solution

It was convenient to use a negative exponent in the formula for $Q(t)$ when we wanted to differentiate Q. However, to understand what the formula tells us about supergrowth, it will be more useful to write Q as

$$Q(t) = \left(\frac{1}{1/\sqrt[5]{5} - .003\, t} \right)^5 .$$

This way makes it clear that Q is a fraction, and we can see its denominator. In particular, this fraction is not defined when the denominator is zero—that is, when

$$\frac{1}{\sqrt[5]{5}} - .003\, t = 0, \quad \text{or} \quad t = \frac{1/\sqrt[5]{5}}{.003} = 241.6\ldots \quad \text{years after 1990.}$$

Consider what happens, though, as t approaches this special value 241.6.... The denominator isn't *yet* zero, but it is *approaching* zero, so the fraction Q is becoming infinite. This means that the supergrowth model predicts the world population will become infinite in about 240 years!

Q "blows up" as $t \to 241.6\ldots$

Let's see what the predicted population size is when $t = 240$ (which is the year 2230 A.D.), shortly before Q becomes infinite. We have

$$Q(240) = \left(\frac{1}{\sqrt[5]{5}} - (.003)(240) \right)^{-5} \approx 4.0088 \times 10^{11}.$$

Remember that Q expresses the population in *billions* of people, so the supergrowth model predicts about 4×10^{20} people (i.e. 400 quintillion!) in the year 2230. Refer back to our estimates of $Q(240)$ using Euler's method (page 203). Although not even one digit of the estimates had stabilized, at least the final one (with a step size of .001) had reached the right power of ten. In fact, estimates made with still smaller step sizes do eventually approach the value given by the formula for Q:

step size	$Q(240)$
.1	0.1979×10^{11}
.01	2.5727×10^{11}
.001	2.8249×10^{11}
.0001	3.9999×10^{11}
.00001	4.0069×10^{11}

Let's look at the relationship between the Euler approximations of Q and the formula for Q graphically. Here are graphs produced by a modification of the program EULER.

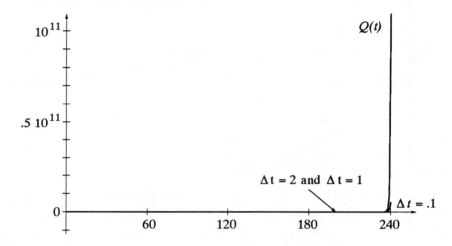

The range of values of Q for $0 \le t \le 240$ is so immense that these graphs are useless. In a case like this, it is helpful to rescale the vertical axis so that the space between one power of 10 and the next is the same. In other words, instead of seeing 1, 2, 3, ..., we see 10^1, 10^2, 10^3, ... This is called a **logarithmic** scale. Here's what happens to the graphs if we put a logarithmic scale on the vertical axis:

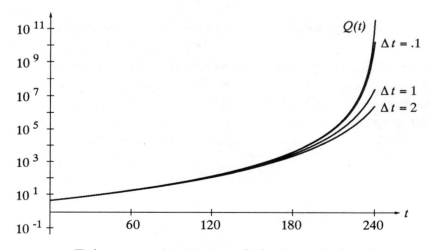

Euler approximations and the formula for Q

The second graph makes it clearer that the Euler approximations do indeed approach the graph of the function given by our formula, but they approach more and more slowly, the closer t approaches 241.6....

> Graphs are made with logarithmic scales particularly when the numbers being plotted cover a wide range of values. When just one axis is logarithmic, the result is called a *semi–log* plot; when both axes are logarithmic, the result is called a *log–log* plot.

Since $Q(t)$ becomes infinite when $t = 241.6...$, we must conclude that the solution to the original initial value problem is meaningful only for $t < 241.6...$. Of course, the formula for Q works quite well when $t > 241.6...$. It just has no meaning as the size of a population. For instance, when $t = 260$ we get

The initial value problem has a solution only for $t < 241.6...$

$$Q(260) = \left(\frac{1}{\sqrt[5]{5}} - (.003)(260) \right)^{-5} \approx (-.05522)^{-5} \approx -1.948 \times 10^6.$$

In other words, the function determined by the initial value problem is defined only on intervals around $t = 0$ that do not contain $t = 241.6...$.

The formula for $Q'(t)$ is informative too:

$$Q'(t) = .015 \left(\frac{1}{1/\sqrt[5]{5} - .003t} \right)^6$$

Since Q' has the same denominator as Q, it becomes infinite the same way Q does: $Q'(t) \to \infty$ as $t \to 241.6...$. Because Euler's method uses the microscope equation $\Delta Q \approx Q' \cdot \Delta t$ to predict the next value of Q, we can now understand why the estimates of $Q(240)$ were so slow to stabilize: as $Q' \to \infty$, $\Delta Q \to \infty$, too.

Q' blows up as $t \to 241.6...$

A formula for constant per capita growth

The constant per capita growth model for the world population that we are considering is

$$\frac{dP}{dt} = .02\,P \qquad P(0) = 5.$$

This differential equation has a very simple form; if $P(t)$ is a solution, then the derivative of P is just a multiple of P. We have already seen in chapter 3 that *exponential* functions behave this way (exercises 5–7 in §3). For example, if $P(t) = 2^t$, then

$$\frac{dP}{dt} = .69 \cdot 2^t = .69\, P.$$

Of course, the multiplier that appears here is .69, not .02, so $P(t) = 2^t$ is not a solution to our problem.

Exponential functions satisfy $dy/dt = ky$

However, the multiplier that appears when we differentiate an exponential function changes when we change the base. That is, if $P(t) = b^t$, then $P'(t) = k_b \cdot b^t$, where k_b depends on b. Here is a sample of values of k_b for different bases b:

b	k_b
.5	$-.693147\ldots$
2	$.693147\ldots$
3	$1.098612\ldots$
10	$2.302585\ldots$

Notice that k_b gets larger as b does. Since .02 lies between $-.693147$ and $+.693147$, the table suggests that the value of b we want lies somewhere between .5 and 2.

We can say even more about the multiplier. Since $P'(t) = k_b \cdot P(t)$ and $P(t) = b^t$, we find

$$P'(0) = k_b \cdot P(0) = k_b \cdot b^0 = k_b \cdot 1 = k_b.$$

In other words, k_b *is the slope of the graph of $P(t) = b^t$ at the origin.*

The correct exponential function has slope .02 at the origin

Thus, we will be able to solve the differential equation $dP/dt = .02\, P$ if we can find an exponential function $P(t) = b^t$ whose graph has slope .02 at the origin. This is a problem that we can solve with a computer microscope. Pick a value of b and graph b^t. Zoom in on the graph at the origin and measure the slope. If the slope is more than .02, choose a smaller value for b; if the slope is less than .02, choose a larger value for b. Repeat this process, narrowing down the possibilities for b until the slope is as close to .02 as you wish. Eventually, we get

$$P(t) = (1.0202)^t.$$

You should check that $P'(0) = .02000\ldots$; see the exercises.

Thus $P(t) = (1.0202)^t$ solves the differential equation $P' = .02\,P$. However, it does not satisfy the initial condition, because

$$P(0) = (1.0202)^0 = 1 \neq 5.$$

This is easy to fix; $P(t) = 5 \cdot (1.0202)^t$ satisfies both conditions. More generally, $P(t) = C\,(1.0202)^t$ satisfies the initial condition $P(0) = C$ as well as the differential equation $P'(t) = .02\,P(t)$. To check the initial condition, we compute

$$P(0) = C\,(1.0202)^0 = C \cdot 1 = C.$$

The differential equation is also satisfied:

$$P'(t) = (C\,(1.0202)^t)' = C \cdot ((1.0202)^t)' = C \cdot (.02\,(1.0202)^t) = .02\,P(t).$$

So we have verified that the solution to our problem is

$$P(t) = 5\,(1.0202)^t.$$

The formula
for P

Because exponential functions are involved, constant per capita growth is commonly called **exponential growth**. In the figure below we compare exponential growth $P(t)$ to "supergrowth" $Q(t)$. The two graphs agree quite well when $t < 50$. Notice that population is plotted on a logarithmic scale (a *semi–log* plot). This makes the graph of P a straight line!

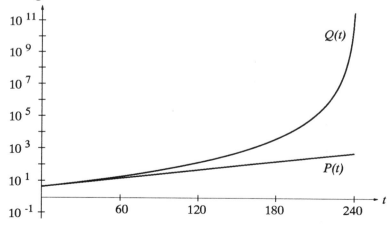

The graphs of $P(t)$ and $Q(t)$

Differential Equations Involving Parameters

The *S-I-R* model contained two parameters—the transmission and recovery coefficients a and b. When we used Euler's method to analyze S, I, and R, we were working numerically. To do the computations, **How do parameters affect solutions?** we had to give the parameters definite numerical values. That made it more difficult to deal with our questions about the effects of *changing* the parameters. As a result, we took other approaches to explore those questions. For example, we used algebra to see that there was a threshold for the spread of the disease: if there were fewer than b/a people in the susceptible population, the infection would fade away.

This is the situation generally. Euler's method can be used to produce solutions to a very broad range of initial value problems. However, if the model includes parameters, then we usually want to know how the solutions are affected when the parameters change. Euler's method is a rather clumsy tool for investigating this question. Other methods—ones that don't require the values of the parameters to be fixed—work better. One possibility is to start with a formula.

Supergrowth parameters The supergrowth problem illustrates both how questions about parameters can arise and how useful a formula for the solution can be to answer the questions. One of the most striking features of the supergrowth model is that it predicts the population becomes infinite in 241.6... years. That prediction was based on an initial population of 5 billion and a growth constant of .015. Suppose those values turn out to be incorrect, and we need to start with different values. Will that change the prediction? If so, how?

We should treat the initial population and the growth constant as parameters—that is, as quantities that *can* vary, although they will have fixed values in any specific situation that we consider. Suppose we let A denote the size of the initial population, and k the growth constant. If we incorporate these parameters into the supergrowth model, the initial value problem takes this form:

$$\frac{dQ}{dt} = k\,Q^{1.2} \qquad Q(0) = A$$

Here is the formula for a function that solves this problem:

The supergrowth solution with parameters

$$Q(t) = \left(\frac{1}{\sqrt[5]{A}} - .2kt \right)^{-5}$$

Notice that, when $A = 5$ and $k = .015$, this formula reduces to the one we considered earlier.

Let's check that the formula does indeed solve the initial value problem. First, the initial condition:

$$Q(0) = \left(\frac{1}{\sqrt[5]{A}} - .2k \cdot 0 \right)^{-5} = \left(\frac{1}{\sqrt[5]{A}} \right)^{-5} = (\sqrt[5]{A})^5 = A.$$

Next, the differential equation. To differentiate $Q(t)$ we introduce the chain

$$Q = u^{-5} \qquad \text{where} \qquad u = \frac{1}{\sqrt[5]{A}} - .2kt.$$

We see that $dQ/du = -5\,u^{-6}$. Since u is a linear function of t in which the multiplier is $-.2k$, we also have $du/dt = -.2k$. Thus, by the chain rule,

$$\frac{dQ}{dt} = \frac{dQ}{du} \cdot \frac{du}{dt} = -5\,u^{-6} \cdot (-.2k) = k\,u^{-6}.$$

That is the left-hand side of the differential equation. To evaluate the right-hand side, we use the fact that $Q = u^{-5}$. Thus

$$k\,Q^{1.2} = k\,Q^{6/5} = k(u^{-5})^{6/5} = k\,u^{-5 \cdot 6/5} = k\,u^{-6}.$$

Since both sides equal $k\,u^{-6}$, they equal each other, proving that $Q(t)$ is a solution to the differential equation.

Next, we ask when the population becomes infinite. Exactly as before, this will happen when the denominator of the formula for $Q(t)$ becomes zero:

$$\frac{1}{\sqrt[5]{A}} - .2\,kt = 0, \qquad \text{or} \qquad t = \frac{1}{.2\,k\sqrt[5]{A}}.$$

Here, in fact, is a *formula* that tells us how each of the parameters A and k affects the time it takes for the population to become infinite.

Let's use τ (the Greek letter "tau") to denote the "time to infinity." For example, if we double the initial population, so $A = 10$ billion people, while keeping the original growth constant $k = .015$, then the time to infinity is

$$\tau = \frac{1}{.003 \times \sqrt[5]{10}} \approx 210.3 \text{ years.}$$

By contrast, if we double the growth rate, to $k = .030$, while keeping the original $A = 5$, then the time to infinity is only

$$\tau = \frac{1}{.006 \times \sqrt[5]{5}} \approx 120.8 \text{ years}$$

Conclusion: doubling the growth rate has a much greater impact than doubling the initial population.

Uncertainty in the size of τ

For any specific growth rate and initial population, we can always calculate the time to infinity. But we can actually do more; the formula for τ allows us to do an *error analysis* along the patterns described in chapter 3, §4. For example, suppose we are uncertain of our value of the growth rate k; there may be an error of size Δk. How uncertain does that make us about the calculated value of τ? Likewise, if the current world population A is known only with an error of ΔA, how uncertain does *that* make τ? Also, how are the *relative* errors related? Let's do this analysis, assuming that $k = .015$ and $A = 5$.

How an error in k propagates

Our tool is the error propagation equation—which is the microscope equation. If we deal with k first, then

$$\Delta \tau \approx \frac{\partial \tau}{\partial k} \cdot \Delta k.$$

We have used partial derivatives because τ is a function of *two* variables, A as well as k. If we write

$$\tau = \frac{1}{.2\sqrt[5]{A}} k^{-1},$$

then the differentiation rules yield

$$\frac{\partial \tau}{\partial k} = -1 \cdot \frac{1}{.2\sqrt[5]{A}} k^{-2} = \frac{-1}{.2\sqrt[5]{5}} \times (.015)^{-2} \approx -16106.$$

Thus $\Delta \tau \approx -16106 \cdot \Delta k$. For example, if the uncertainty in the value of $k = .015$ is $\Delta k = \pm.001$, then the uncertainty in τ is about ∓ 16 years.

How an error in A propagates

To determine how an error in A propagates to τ, we first write

$$\tau = \frac{1}{.2\,k} A^{-1/5}.$$

Then

$$\frac{\partial \tau}{\partial k} = -\frac{1}{5} \cdot \frac{1}{.2\,k} A^{-6/5} = \frac{-1}{5 \times .2 \times .015} \times 5^{-6/5} \approx -9.7.$$

The error propagation equation is thus $\Delta \tau \approx -9.7 \cdot \Delta A$. If the uncertainty in the world population is about 100 million persons, so $\Delta A = \pm .1$, then the uncertainty in τ is less than 1 year.

To complete the analysis, let's compare relative errors. This involves a lot of algebra. Since

Relative errors

$$\Delta \tau \approx -\frac{\Delta k}{.2\, k^2 \sqrt[5]{A}} \qquad \text{and} \qquad \tau = \frac{1}{.2\, k \sqrt[5]{A}},$$

we can compute that a given relative error in k propagates as

$$\frac{\Delta \tau}{\tau} \approx -\frac{\Delta k}{.2\, k^2 \sqrt[5]{A}} \cdot \frac{.2\, k \sqrt[5]{A}}{1} = -\frac{\Delta k}{k}.$$

Thus, a 1% error in k leads to a 1% error in τ, although the sign is reversed.

To analyze how a given relative error in A propagates, we start with

$$\Delta \tau \approx -\frac{1}{5} \cdot \frac{\Delta A}{.2\, k A^{6/5}}.$$

Then

$$\frac{\Delta \tau}{\tau} \approx -\frac{1}{5} \cdot \frac{\Delta A}{.2\, k A^{6/5}} \cdot \frac{.2\, k \sqrt[5]{A}}{1} = -\frac{1}{5} \cdot \frac{\Delta A}{A}.$$

This says that it takes a 5% error in A to produce a 1% error in τ. Consequently, the time to infinity τ *is 5 times more sensitive to errors in k than to errors in A.*

How sensitive τ is to errors in k and A

The exercises in this section will give you an opportunity to check that a particular formula is a solution to an initial value problem that arises in a variety of contexts. Later in this chapter, we will make a modest beginning on the much harder task of *finding* solutions given by formulas for special initial value problems. There are more sophisticated methods for finding formulas, when the formulas exist, and they provide powerful tools for some important problems, especially in physics. However, most initial value problems we encounter cannot be solved by formulas. This is particularly true when two or more variables are needed to describe the process being modelled. The tool of widest applicability is Euler's method. This isn't so different from the situation in algebra, where exact solutions given by formulas (e.g. the quadratic formula) are also relatively rare, and numerical methods

Special methods give formulas, general methods are numerical

play an important role. (Chapter 5, §5, presents the Newton–Raphson method for solving algebraic equations by successive approximation.) In most cases that will interest us, there are simply no formulas to be found—the limitation lies in the mathematics, not the mathematicians.

Exercises

In exercises 1–4, verify that the given formula is a *solution* to the initial value problem.

1. **Powers of y.**

a) $y' = y^2$, $y(0) = 5$: $y(t) = 1/(\frac{1}{5} - t)$

b) $y' = y^3$, $y(0) = 5$: $y(t) = 1/\sqrt{\frac{1}{25} - 2t}$

c) $y' = y^4$, $y(0) = 5$: $y(t) = 1/\sqrt[3]{\frac{1}{125} - 3t}$

d) Write a general formula for the solution of the initial value problem $y' = y^n$, $y(0) = 5$, for any integer $n > 1$.

2. **Powers of t.**

a) $y' = t^2$, $y(0) = 5$: $y(t) = \frac{1}{3}t^3 + 5$

b) $y' = t^3$, $y(0) = 5$: $y(t) = \frac{1}{4}t^4 + 5$

c) $y' = t^4$, $y(0) = 5$: $y(t) = \frac{1}{5}t^5 + 5$

d) Write a general formula for the solution of the initial value problem $y' = t^n$, $y(0) = 5$ for any integer $n > 1$.

3. **Sines and cosines.**

a) $x' = -y$, $y' = x$, $x(0) = 1$, $y(0) = 0$: $x(t) = \cos t$, $y(t) = \sin t$

b) $x' = -y$, $y' = x$, $x(0) = 0$, $y(0) = 1$: $x(t) = \cos(t + \pi/2)$, $y(t) = \sin(t + \pi/2)$

4. **Exponential functions.**

a) $y' = 2.3\,y$, $y(0) = 5$: $y(t) = 5 \cdot 10^t$

b) $y' = 2.3\,y$, $y(0) = C$: $y(t) = C \cdot 10^t$

c) $y' = -2.3\,y$, $y(0) = 5$: $y(t) = 5 \cdot 10^{-t}$

d) $y' = 4.6\,ty$, $y(0) = 5$: $y(t) = 5 \cdot 10^{t^2}$

5. **Initial Conditions.**

a) Choose C so that $y(t) = \sqrt{t+C}$ is a solution to the initial value problem

$$y' = \frac{1}{2y} \quad y(3) = 17.$$

[Answer: $C = 286$.]

b) Choose C so that $y(t) = -1/(t+C)$ is a solution to the initial value problem

$$y' = y^2 \quad y(0) = -5.$$

c) Choose C so that $y(t) = -1/(t+C)$ is a solution to the initial value problem

$$y' = y^2 \quad y(2) = 3.$$

World population growth with parameters

6. a) Using a graphing utility or a calculator, show that the derivative of $P(t) = (1.0202)^t$ at the origin is approximately .02: $P'(0) \approx .02$. Since quick convergence is desirable, use

$$\frac{\Delta P}{\Delta t} = \frac{P(0+h) - P(0-h)}{2h} = \frac{(1.0202)^h - (1.02020)^{-h}}{2h}$$

b) By using more decimal places to get higher precision, show that $P(t) = (1.0202013)^t$ satisfies $P'(0) = .02$ even more exactly.

7. a) Show that the function $y = 2^{t/.69}$ satisfies the differential equation $dy/dt = y$. (Use the chain rule: $y = 2^u$, $u = t/.69$.)

b) Show that the function $y = 2^{kt/.69}$ satisfies the differential equation $dy/dt = k\,y$.

c) Show that the function $P(t) = A \cdot 2^{kt/.69}$ is a solution to the initial value problem

$$\frac{dP}{dt} = k\,P \qquad P(0) = A.$$

Note that this describes a population that grows at the constant per capita rate k from an initial size of A.

8. a) Show that the function $y = 10^{t/2.3}$ satisfies the differential equation $dy/dt = y$. (Use the chain rule: $y = 10^u$, $u = t/2.3$.)

b) Show that the function $y = 10^{kt/2.3}$ satisfies the differential equation $dy/dt = k\,y$.

c) Show that the function $P(t) = A \cdot 10^{kt/2.3}$ is a solution to the initial value problem

$$\frac{dP}{dt} = k\,P \qquad P(0) = A.$$

This formula provides an alternative way to describe a population that grows at the constant per capita rate k from an initial size of A.

9.   a) The formula $P(t) = 5 \cdot 2^{k\,t/.69}$ describes how an initial population of 5 billion will grow at a constant per capita rate of k persons per year per person. Use this formula to determine how many years t it will take for the population to double, to 10 billion persons.

[Answer: The doubling time is $t = .69/k$.]

b) Suppose the initial population is A billion, instead of 5 billion. What is the doubling time then?

c) Suppose the initial population is 5 billion, and the per capita growth rate is .02, but that value is certain only with an error of Δk. How much uncertainty is there in the doubling time that you found in part (a)?

Newton's law of cooling

There are formulas that describe how a body cools, or heats up, to match the temperature of its surroundings. See the exercises on Newton's law of cooling in §1. Consider first the model

$$\frac{dT}{dt} = -.1(T - 20) \qquad T(0) = 90,$$

introduced on page 194 to describe how a cup of coffee cools.

10.   Show that the function $y = 2^{-.1\,t/.69}$ is a solution to the differential equation $dy/dt = -.1\,y$. (Use the chain rule: $y = 2^u$, $u = -.1\,t/.69$.)

11.   a) Show that the function

$$T = 70 \cdot 2^{-.1\,t/.69} + 20$$

is a solution to the initial value problem $dT/dt = -.1(T - 20)$, $T(0) = 90$. This is the temperature T of a cup of coffee, initially at 90°C, after t minutes have passed in a room whose temperature is 20°C.

b) Use the formula in part (a) to find the temperature of the coffee after 20 minutes. Compare this result with the value you found in exercise 11 (b), page 195.

c) Use the formula in part (a) to determine how many minutes it takes for the coffee to cool to 30°C. In doing the calculations you will find it helpful to know that $1/7 = 2^{-2.8}$. Compare this result with the value you found in exercise 11 (c), page 195.

12.   a) A cold drink is initially at $Q = 36°$F when the air temperature is 90°F. If the temperature changes according to the differential equation

$$\frac{dQ}{dt} = -.2(Q - 90)°\text{F per minute,}$$

show that the function $Q(t) = 90 - 54 \cdot 2^{-.2\,t/.69}$ describes the temperature after t minutes.

b) Use the formula to find the twmperature of the drink after 5 minutes and after 10 minutes. Compare your results with the values you found in exercise 12, page 195.

13.   Find a formula for a function that solves the initial value problem

$$\frac{dQ}{dt} = -k(Q - A) \qquad Q(0) = B.$$

A leaking tank

The rate at which water leaks from a small hole at the bottom of a tank is proportional to the square root of the height of the water surface above the bottom of the tank. Consider a cylindrical tank that is 10 feet tall and stands on one of its circular ends, which is 3 feet in diameter. Suppose the tank is currently half full, and is leaking at a rate of 2 cubic feet per hour.

14.   a) Let $V(t)$ be the volume of water in the tank t hours from now. Explain why the leakage rate can be written as the differential equation

$$V'(t) = -k\sqrt{V(t)},$$

for some positive constant k. (The issue to deal with is this: why is it permissible to use the square root of the *volume* here, when the rate is known to depend on the square root of the *height*?)

b) Determine the value of k. [Answer: $k \approx .3364$; you need to explain *why* this is the value.]

15. How much water leaks out of the tank in 12 hours; in 24 hours? Use Euler's method, and compute successive approximations until your results stabilize.

a) How many hours does it take for the tank to empty?

16. a) Use the differentiation rules to show that any function of the form

$$V(t) = \begin{cases} \dfrac{k^2}{4}(C - t)^2 & \text{if } 0 \leq t \leq C \\ 0 & \text{if } C < t \end{cases}$$

satisfies the differential equation.

b) For the situation we are considering, what is the value of C? According to this solution, how long does it take for the tank to empty? Compare this result with your answer using Euler's method.

c) Sketch the graph of $V(t)$ for $0 \leq t \leq 2C$, taking particular care to display the value $V(0)$ in terms of k and C.

Motion

Newton created the calculus to study the motion of the planets. He said that all motion obeys certain basic laws. One law says that the velocity of an object changes only if a force acts on the body. Furthermore, the *rate* at which the velocity changes is proportional to the force. By knowing the forces that act on a body we can construct—and then solve—a differential equation for the velocity.

Falling bodies–with gravity. A body falling through the air starts up slowly but picks up speed as it falls. Its velocity is thus changing, so there must be a force acting. We call the force that pulls objects to the earth **gravity**. Near the earth, the strength of gravity is essentially constant.

Suppose an object is x meters above the surface of the earth after t seconds have passed. Then, by definition, its velocity is

$$v = \frac{dx}{dt} \text{ meters/second.}$$

According to Newton's laws of motion, the force of gravity causes the velocity to change, and we can write

$$\frac{dv}{dt} = -g.$$

Here g is a constant whose numerical value is about 9.8 meters/second per second. Since x and v are positive when measured *upwards*, but gravity acts *downwards*, a minus sign is needed in the equation for dv/dt. (The derivative of velocity is commonly called **acceleration**, and g is called the **acceleration due to gravity**.)

17.   Verify that $v(t) = -g\,t + v_0$ is a solution to the differential equation $dv/dt = -g$ with initial velocity v_0.

18.   Since $dx/dt = v$, and since $v(t) = -g\,t + v_0$, the *position* x of the body satisfies the differential equation

$$\frac{dx}{dt} = -g\,t + v_0 \text{ meters/second.}$$

Find a formula for $x(t)$ that solves this differential equation. This function describes how a body moves under the force of gravity.

19.   Suppose the initial position of the body is x_0, so that the position x is a solution to the initial value problem

$$\frac{dx}{dt} = -g\,t + v_0 \qquad x(0) = x_0 \text{ meters.}$$

Find a formula for $x(t)$.

20.   a) Suppose a body is held motionless 200 meters above the ground, and then released. What values do x_0 and v_0 have? What is the formula for the motion of this body as it falls to the ground?

b) How far has the body fallen in 1 second? In 2 seconds?

c) How long does it take for the body to reach the ground?

Falling bodies–with gravity and air resistance. As a body falls, air pushes against it. Air resistance is thus another force acting on a falling body. Since air resistance is slight when an object moves slowly but increases as the object speeds up, the simplest assumption we can make is that the force of air resistance is proportional to the velocity: force $= -bv$. The multiplier b is positive, and the minus sign tells us that the direction of the force is always opposite the velocity.

The forces of gravity and air resistance combine to change the velocity:

$$\frac{dv}{dt} = -g - bv \text{ meters/second per second.}$$

21. Show that

$$v(t) = \frac{g}{b} \left(2^{-bt/.69} - 1 \right) \text{ meters/second}$$

is a solution to this differential equation that also satisfies the initial condition $v(0) = 0$ meters/second.

22. a) Show that the position $x(t)$ of a body that falls against air resistance from an initial height of x_0 meters is given by the formula

$$x(t) = x_0 - \frac{g}{b}t - \frac{g}{b^2} \left(2^{-bt/.69} - 1 \right) \text{ meters.}$$

b) Suppose the coefficient of air resistance is $b = .2$ per second. If a body is held motionless 200 meters above the ground, and then released, how far will it fall in 1 second? In 2 seconds? Compare these values with those you obtained assuming these was no air resistance.

c) How long does it take for the body to reach the ground? Compare this value with the one you obtained assuming these was no air resistance. How much does air resistance add to the time?

23. a) According to the equation $dv/dt = -g - bv$, there is a velocity v_T at which the force of air resistance exactly balances the force of gravity, and the velocity doesn't change. What is v_T, expressed as a function of g and b. Note: v_T is called the **terminal velocity** of the body. Once the body reaches its terminal velocity, it continues to fall at that velocity.

b) What is the terminal velocity of the body in the previous exercise?

The oscillating spring. Springs can smooth out life's little irregularities (as in the suspension of a car) or amplify and measure them (as

in earthquake detection devices). Suppose a
spring that hangs from a hook has a weight at its
end. Let the weight come to rest. Then, when
the weight moves, let x denote the position of the
weight above the rest position. (If x is negative,
this means the weight is below the rest position.)
If you pull down on the weight, the spring pulls
it back up. If you push up on the weight, the
spring (and gravity) push it back down. This
push is the **spring force**.

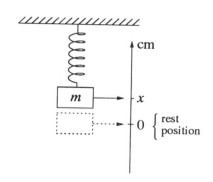

The simplest assumption is that the spring force is proportional to
the amount x that the spring has been stretched: force $= -c^2 x$. The
constant c^2 is customarily written as a square to emphasize that it is
positive. The minus sign tells us the force pushes down if $x > 0$ (so the
weight is above the rest position), but it pushes up if $x < 0$.

If $v = dx/dt$ is the velocity of the weight, then Newton's law of
motion says

$$\frac{dv}{dt} = -c^2 x.$$

Suppose we move the weight to the point $x = a$ on the scale, hold
it motionless momentarily, and then release it at time $t = 0$. This
determines the initial value problem

$$x' = v \qquad\qquad x(0) = a$$
$$v' = -c^2 x \qquad\quad v(0) = 0.$$

24. a) Show that

$$x(t) = a \cos(ct) \qquad v(t) = -ac \sin(ct)$$

is a solution to the initial value problem.

b) What range of values does x take on; that is, how far does the
weight move from its rest position?

25. a) Use a graphing utility to compare the graphs of $y = \cos(x)$, $y = \cos(2x)$, $y = \cos(3x)$, and $y = \cos(.5x)$. Based on your observations,
explain how the value of c affects the nature of the motion $x(t) = a \cos(ct)$ for a fixed value of a.

b) How long does it take the weight to complete one cycle (from $x = a$ back to $x = a$) when $c = 1$? The motion of the weight is said to be **periodic**, and the time it takes to complete one cycle is called its **period**.

c) What is the period of the motion when $c = 2$? When $c = 3$? Does the period depend on the initial position a?

d) Write a formula that expresses the period of the motion in terms of the parameters a and c.

26.   a) The parameter c depends on two things: the mass m of the weight, and the stiffness k of the spring:

$$c = \sqrt{\frac{k}{m}}.$$

Write a formula that expresses the period of the motion of the weight in terms of m and k.

b) Suppose you double the weight on the spring. Does that *increase* or *decrease* the period of the motion? Does your answer agree with your intuitions?

c) Suppose you put the first weight on a second spring that is twice as stiff as the first (i.e., double the value of k). Does that *increase* or *decrease* the period of the motion? Does your answer agree with your intuitions?

d) When you calculate the period of the motion using your formula form part (a), suppose you know the actual value of the mass only to within 5%. How accurately do you know the period—as a percentage of the calculated value?

§3. The Exponential Function

The Equation $y' = ky$

As we have seen, initial value problems *define* functions—as their solutions. They therefore provide us with a vast, if somewhat bewildering, array of new functions. Fortunately, a few differential equations—in fact, the very simplest—arise over and over again in an astonishing variety of contexts. The functions they define are among the most important in mathematics.

One of the simplest differential equations is $dy/dt = ky$, where k is a constant. It is also one of the most useful. We used it in chapter 1 to model the populations of Poland and Afghanistan, as well as bacterial growth and radioactive decay. In this chapter, it was our initial model of a rabbit population and one of our models of the world population. Later, we will use it to describe how money accrues interest in a bank and how radiation penetrates solid objects.

A simple and natural model of growth and decay

In chapter 3 we established that the solutions to $dy/dt = ky$ are exponential functions. Specifically, for each base b, the exponential function $y = b^t$ was a solution to $dy/dt = k_b \cdot b^t = k_b \cdot y$, where k_b was the slope of the graph of $y = b^t$ at the origin. In this approach, if the constant k changes, we must change the base b so that $k_b = k$.

Exponential solutions— variable base ...

Exercise 7 in the last section (page 215) opened up a new possibility: for the *fixed* base 2, the function

...and fixed base

$$y = 2^{kt/.69}$$

was a solution to the differential equation $dy/dt = kt$, no matter what value k took. There was nothing special about the base 2, of course. In the next exercise, we saw that the functions

$$y = 10^{kt/2.3}$$

would serve equally well as solutions.

In fact, we can show that, for any base b, the functions

$$y = b^{kt/k_b}$$

are also solutions to $dy/dt = ky$. Construct the chain

$$y = b^u \qquad \text{where} \qquad u = kt/k_b.$$

Then $dy/du = k_b \cdot b^u = k_b \cdot y$, while $du/dt = k/k_b$. Thus, by the chain rule we have

$$\frac{dy}{dt} = \frac{dy}{du} \cdot \frac{du}{dt} = k_b\, y \cdot \frac{k}{k_b} = ky.$$

Advantages of a fixed base

If we express solutions to $dy/dt = ky$ by exponential functions with a *fixed* base, it is easy to alter the solution if the growth constant k changes. We just change the value of k in the exponent of b^{kt/k_b}. Let's see how this works when $b = 2$ and $b = 10$:

differential equation	solution base 2	solution base 10
$\dfrac{dy}{dt} = .16\,y$	$2^{.232\,t}$	$10^{.070\,t}$
$\dfrac{dy}{dt} = .18\,y$	$2^{.261\,t}$	$10^{.078\,t}$

Notice that the growth constant k gets "swallowed up" in the exponent of the solution when k has a specific numerical value. The number that appears in the exponent is k divided by $k_2 = .69$ (when the base is 2) and by $k_{10} = 2.30$ (when the base is 10).

The base e

The most vivid solution to $dy/dt = ky$ would use the base b for which $k_b = 1$ *exactly*. There is such a base, and it is always denoted e. (We will determine the value of e in a moment.) Since $k_e = 1$, k would stand out in the exponent:

differential equation	solution base e
$\dfrac{dy}{dt} = .16\,y$	$e^{.16\,t}$
$\dfrac{dy}{dt} = .18\,y$	$e^{.18\,t}$

The simplicity and clarity of this expression have led to the universal adoption of the base e for describing exponential growth and decay— that is, for describing solutions to $dy/dt = ky$.

The use of the symbol e to denote the base dates back to a paper that Euler wrote at age 21, entitled *Meditatio in experimenta explosione tormentorum nuper instituta* (Meditation upon recent experiments on the firing of cannons), where the symbol e was used sixteen times. It is now in universal use. The number e is, like π, one of the most important and ubiquitous numbers in mathematics.

By design, $y = e^t$ is a solution to the differential equation $dy/dt = y$. In particular, the slope of the graph of $y = e^t$ at the origin is exactly 1. As we have just seen, the function $y = e^{kt}$ is a solution to the differential equations $dy/dt = ky$ whose growth constant is k. Finally:

> $y = C \cdot e^{kt}$ is the solution to the initial value problem
> $$\frac{dy}{dt} = ky \qquad y(0) = C.$$

The general initial value problem for exponential functions

We can check this quickly. The initial condition is satisfied because $e^0 = 1$, so $y(0) = C \cdot e^{k \cdot 0} = C \cdot 1 = C$. The differential equation is satisfied because

$$\left(C \cdot e^{kt}\right)' = C \cdot \left(e^{kt}\right)' = C \cdot k\, e^{kt} = k\, y.$$

We used the differentiation rule for a constant multiple of a function, and we used the fact that the derivative of e^{kt} was already established to be $k\, e^{kt}$.

The Number e

The number e is determined by the property that $k_e = 1$. Since this number is the slope of the graph of $y = e^t$ at the origin, one way to find e is with a computer microscope. Pick a value for e and graph $y = e^t$. Zoom in the graph at the origin, and measure the slope. If the slope is more than 1, choose a smaller value for e; if the slope is less than 1, choose a larger value. Repeat this process, narrowing down the value of e until you know its value to as many decimal places as you wish.

Finding e with a computer microscope

We already know $e = 2$ is too small, because the slope of $y = 2^t$ at the origin is .69. Likewise, $e = 3$ is too large, because the slope of $y = 3^t$ at the origin is 1.09. Thus $2 < e < 3$, and is closer to 3 than to 2. At the next stage we learn that $e = 2.7$ is too small but $e = 2.8$ is too large. Thus, at least we know $e = 2.7\ldots$. Several stages later we would learn $e = 2.71828\ldots$.

Under successive magnifications: $e = 2.71828\ldots$

We can take a very different approach to finding the numerical value of e by using the fact that e is defined by an initial value problem. Here is the idea: e is the value of the function e^t when $t = 1$, and $y(t) = e^t$ is the solution to the initial value problem

$$y' = y \qquad y(0) = 1.$$

Find $e = y(1)$ by solving this initial value problem using Euler's method.

Suppose we take n steps to go from $t = 0$ to $t = 1$. Then the step size is $\Delta t = 1/n$. The following table shows the calculations:

Finding $y(1)$ by Euler's method when $y' = y$ and $y(0) = 1$

t	y	$y' = y$	$\Delta y = y' \cdot \Delta t$
0	1	1	$1 \cdot 1/n$
$1/n$	$1 + 1/n$	$1 + 1/n$	$(1 + 1/n) \cdot 1/n$
$2/n$	$(1 + 1/n)^2$	$(1 + 1/n)^2$	$(1 + 1/n)^2 \cdot 1/n$
$3/n$	$(1 + 1/n)^3$	$(1 + 1/n)^3$	$(1 + 1/n)^3 \cdot 1/n$
$\vdots$	$\vdots$	$\vdots$	$\vdots$
n/n	$(1 + 1/n)^n$		

The entries in the y column need to be explained. The first two should be clear: $y(0)$ is the initial value 1, and $y(1/n) = y(0) + \Delta y = 1 + 1/n$. To get from any entry to the next we must do the following:

$$
\begin{aligned}
\text{new } y &= \text{ current } y + \Delta y \\
&= \text{ current } y + \text{current } y \cdot \Delta t \\
&= \text{ current } y \cdot (1 + \Delta t) \\
&= \text{ current } y \cdot (1 + 1/n)
\end{aligned}
$$

The new y is the current y multiplied by $(1 + 1/n)$. Since the second y is itself $(1 + 1/n)$, the third will be $(1 + 1/n)^2$, the fourth will be $(1 + 1/n)^3$, and so on.

Euler's method with n steps therefore gives us the following estimate for $e = y(1) = y(n/n)$:

$$ e \approx (1 + 1/n)^n $$

We can calculate this number on a computer. To get the highest precision possible we choose n to be a power of 2 (That avoids round-off errors because computer arithmetic is done in base 2.) In the table on the next page, eleven digits of e have stabilized.

Expressing e as a limit   The *true* value of e is the limit of these approximations as we take n arbitrarily large:

$$ e = \lim_{n \to \infty} (1 + 1/n)^n = 2.7182818284\ldots $$

n	$(1+1/n)^n$
2^0	2.0
2^4	2.638
2^8	2.712 992
2^{12}	2.717 950 081
2^{16}	2.718 261 089 905
2^{20}	2.718 280 532 282
2^{24}	2.718 281 747 448
2^{28}	2.718 281 823 396
2^{32}	2.718 281 828 142
2^{36}	2.718 281 828 439
2^{40}	2.718 281 828 458

Differential Equations Define Functions

It seems we should be finished. We know everything we need to about the solutions of the initial value problems of the form $y' = ky$ and $y(0) = C$:

A new start

- We have a formula for the solution: $y = Ce^{kt}$.

- We can use Euler's method to approximate values of Ce^{kt}, including the value of the number e.

But we can ask another question. Since the initial value problem itself defines the function giving its solution, could we have deduced *directly* from the differential equation that the solution was an exponential function? We "stumbled on" the formula $y = Ce^{kt}$ by noticing that every exponential function $y = b^t$ satisfies a differential equation of the form $y' = ky$. What if we hadn't known that at the start? There is another route to the formula for the solution to this initial value problem, and we will examine it here.

You may be asking yourself: Why are we solving the same problem twice? Here are two reasons.

- Often, a second approach casts a different light on the problem and leads to a deeper understanding.

- This particular second solution requires a chain of deductions. Many people find this kind of reasoning satisfying and even beautiful. You may be one of them!

So, we make a fresh start. This time, we will assume *nothing* about the function $y = e^t$. Instead, we begin simply with the observation that each initial value problem defines a function—its solution. Therefore, the specific problem

$$y' = y \qquad y(0) = 1$$

Defining exp(t) defines a function; we call it $y = \exp(t)$. At the outset, all we know about the function $\exp(t)$ is that

$$\exp'(t) = \exp(t) \qquad \exp(0) = 1.$$

From these facts alone we want to *deduce* that $\exp(t)$ is the ordinary exponential function e^t. (Of course, the name "exp" is chosen because we *will* eventually show that $\exp(t)$ is an exponential function.) The following theorem is the first of several facts we wish to establish about the function $\exp(t)$.

Theorem 1. *For any real numbers r and s,*

$$\exp(r + s) = \exp(r) \cdot \exp(s).$$

If we already knew that $\exp(t) = e^t$, then Theorem 1 would just be the ordinary law of exponents:

$$e^{r + s} = e^r \cdot e^s.$$

However, we *don't* know $\exp(t) = e^t$—at least, not yet. So we must give an argument to establish the theorem. Before we do that, though, we'll see how we can use Theorem 1. For example, notice that

$$\exp(2) = \exp(1 + 1) = \exp(1) \cdot \exp(1) = (\exp(1))^2.$$

Notice that we invoked Theorem 1 to equate $\exp(1 + 1)$ with $\exp(1) \cdot \exp(1)$. In a similar way,

$$\exp(3) = \exp(2 + 1) = \exp(2) \cdot \exp(1) = (\exp(1))^2 \cdot \exp(1) = (\exp(1))^3.$$

Set $e = \exp(1)$ Let's define $\exp(1) = e$. For the moment, this e has nothing to do with the base e we introduced in the last section; it is just our name for $\exp(1)$. But at least we know

$$\exp(1) = e^1 \qquad \exp(2) = e^2 \qquad \exp(3) = e^3.$$

In fact, we can repeat our argument for any positive integer m, and get

$$\exp(m) = (\exp(1))^m = e^m.$$

We can even express $\exp(t)$ in terms of $e = \exp(1)$ when t is a *negative* integer. We begin with another consequence of Theorem 1: **Negative integers**

$$1 = \exp(0) = \exp(-1+1) = \exp(-1) \cdot \exp(1).$$

This says $e = \exp(1)$ is the reciprocal of $\exp(-1)$:

$$\exp(-1) = (\exp(1))^{-1} = e^{-1}.$$

Since $-2 = -1-1$, $-3 = -2-1$, and so forth, we can eventually show that, for any negative integer $-m$,

$$\exp(-m) = (\exp(1))^{-m} = e^{-m}.$$

We can even do the same thing with fractions. Here's how to deal with $\exp(1/3)$, for example: **Rational numbers**

$$\begin{aligned}
\exp(1) &= \exp\left(\tfrac{1}{3} + \tfrac{1}{3} + \tfrac{1}{3}\right) \\
&= \exp(1/3) \cdot \exp(1/3) \cdot \exp(1/3) \\
&= (\exp(1/3))^3,
\end{aligned}$$

so $\exp(1/3)$ is the cube root of $\exp(1)$:

$$\exp(1/3) = (\exp(1))^{1/3} = e^{1/3}.$$

A similar argument will show that $\exp(1/n) = e^{1/n}$. Finally, we can deal with any rational number m/n:

$$\exp(m/n) = (\exp(1/n))^m = \left(e^{1/n}\right)^m = e^{m/n}.$$

In other words, Theorem 1 implies that the function $\exp(t)$ is the genuine exponential function e^t—at least when t is a rational number m/n. The base of this exponential function is, *by definition*, the value of $\exp(t)$ when $t = 1$: $e = \exp(1)$. We summarize these conclusions in the following theorem.

Theorem 2. *For any rational number r, $\exp(r) = (\exp(1))^r = e^r$.*

A sketch that $\exp(t) = e^t$ when t is irrational

Can we even equate $\exp(t)$ and e^t when t is irrational? We can, but a proof would involve knowing that these functions are *continuous*. We won't pursue this, but here is the idea. First of all, any irrational number t can be approximated by a sequence of rational numbers that approach it in the limit: $r_1, r_2, r_3, \ldots \to t$. For example, we can approximate π by the sequence

$$
\begin{aligned}
r_1 &= 3 \\
r_2 &= 3.1 &= 31/10 \\
r_3 &= 3.14 &= 314/100 \\
r_4 &= 3.141 &= 3141/1000 \\
&\vdots
\end{aligned}
$$

Thus, if we take $r_1, r_2, r_3, \ldots \to t$, we can write

$$
\begin{aligned}
\exp(r_1) &= e^{r_1} \\
\exp(r_2) &= e^{r_2} \\
\exp(r_3) &= e^{r_3} \\
\downarrow & \qquad \downarrow \\
\exp(t) & \qquad e^t
\end{aligned}
$$

This implies $\exp(t) = e^t$.

The key step was to be able to conclude that

$$
f(r_1), \ f(r_2), \ f(r_3), \ \ldots \ \to \ f(t)
$$

when $r_1, r_2, r_3, \ldots \to t$, for each of our functions $f(t) = \exp(t)$ and $f(t) = e^t$. This property of f is called *continuity*. We say that **f is continuous at t** if $f(t_k) \to f(t)$ whenever $t_k \to t$.

Continuity is a technical condition, and it is crucial to resolving the questions we have raised. This is one place where, in developing an idea, we encounter a mathematical issue that we will defer to another time.

There are two items of unfinished business. One of them is to prove Theorem 1. The other is to show that the number e defined in this section is the same as the number e defined in the last section. We'll attend to the second item first.

In this section $e = \exp(1)$, where $y = \exp(t)$ is the solution to the initial value problem $y' = y$, $y(0) = 1$. In the last section e was the base of the exponential function whose derivative at the origin had the value 1. But we now know that $y = \exp(t)$ is an exponential function whose base is $\exp(1)$, and its derivative at the origin is $y'(0) = y(0) = 1$. So the base $\exp(1)$ must be the same as the base e from the last section.

Theorem 1 involves two fixed real numbers, r and s. We use one of them to define two new functions of t:

The proof of
Theorem 1

$$P(t) = \exp(r + t) \qquad Q(t) = \exp(r) \cdot \exp(t).$$

We shall show that *both* of these functions are solutions to the same initial value problem:

$$\frac{dy}{dt} = y \qquad y(0) = \exp(r).$$

(Remember, $\exp(r)$ is a constant, because r is fixed.) But an initial value problem has only one solution, so these functions must be equal: $P(t) = Q(t)$, for all t. In particular, they must be equal when t has the fixed value s, so

$$\exp(r + s) = P(s) = Q(s) = \exp(r) \cdot \exp(s).$$

This is exactly the statement of Theorem 1.

The only thing left to do is to show that $P(t)$ and $Q(t)$ both solve the initial value problem. Look first at the initial condition $y(0) = \exp(r)$. We find

$$P(0) = \exp(r + 0) = \exp(r)$$
$$Q(0) = \exp(r) \cdot \exp(0) = \exp(r) \cdot 1 = \exp(r)$$

Now consider the differential equation $y' = y$.

$$Q'(t) = (\exp(r) \cdot \exp(t))' = \exp(r) \cdot (\exp(t))' = \exp(r) \cdot \exp(t) = Q(t).$$

To differentiate $P(t)$ we construct a chain:

$$P = \exp(u) \qquad \text{where} \qquad u = r + t.$$

Then $dP/du = \exp(u)$ and $du/dt = 1$, so

$$P'(t) = \frac{dP}{du} \cdot \frac{du}{dt} = \exp(u) \cdot 1 = P(t) \cdot 1 = P(t).$$

The crux of this argument is that a given initial value problem has one, *and only one*, solution. We have observed many times that an initial value problem is a prescription for using Euler's method to determine a specific set of functions, the unique solution of the problem. In more advanced courses you will see that this is true in general, provided some mild conditions are satisfied.

Existence and Uniqueness Principle
Under most conditions, an initial value problem
has one and only one solution.

The existence and uniqueness principle is one of the most important mathematical results in the theory of differential equations.

Now that we have established $\exp(x) = e^x$, we will call $\exp(x)$ **the exponential function** and we will use the forms e^x and $\exp(x)$ interchangeably. The following theorem summarizes several more properties of the exponential function.

Theorem 3. *For any real numbers r and s,*

$$\exp(s) \; > \; 0$$

$$\exp(-s) \; = \; \frac{1}{\exp(s)}$$

$$\exp(r - s) \; = \; \frac{\exp(r)}{\exp(s)}$$

$$\exp(rs) \; = \; (\exp(r))^s \; = \; (\exp(s))^r \,.$$

To make the statements in this theorem seem more natural, you should stop and translate them from $\exp(x)$ to e^x. Proofs will be covered in the exercises.

Exponential Growth

The function $\exp(x) = e^x$, like polynomials and the sine and cosine functions, is defined for all real numbers. Nevertheless, it behaves in a way that is quite different from any of those functions.

One difference occurs when x is large, either positive or negative. The sine function and the cosine function stay bounded between $+1$ and -1 over their entire domain. By contrast, every polynomial "blows up" as $x \to \pm\infty$. In this regard, the exponential function is a hybrid. As $x \to -\infty$, $\exp(x) \to 0$. As $x \to +\infty$, however, $\exp(x) \to +\infty$.

Let's look more closely at what happens to power functions x^n and the exponential function e^x as $x \to \infty$. Both kinds of functions "blow up" but they do so at quite different rates, as we shall see. Before we compare power and exponential functions directly, let's compare one power of x with another—say x^2 with x^5. As $x \to \infty$, both x^2 and x^5 get very large. However, x^2 is only a small fraction of the size of x^5, and that fraction gets smaller, the larger x is. The table at the top of the opposite page demonstrates this. Even though x^2 grows enormous, we interpret the fact that $x^2/x^5 \to 0$ to mean that x^2 *grows more slowly than* x^5.

How fast do x^n and e^x become infinite?

x	x^2	x^5	x^2/x^5
10	10^2	10^5	10^{-3}
100	10^4	10^{10}	10^{-6}
1000	10^6	10^{15}	10^{-9}
↓	↓	↓	↓
∞	∞	∞	0

It should be clear to you that we can compare *any* two powers of x this way. We will find that x^p grows more slowly than x^q if, and only if, $p < q$. To prove this, we must see what happens to the ratio x^p/x^q, as $x \to +\infty$. We can write $x^p/x^q = 1/x^{q-p}$, and the exponent $q - p$ that appears here is positive, because $q > p$. Consequently, as $x \to \infty$, $x^{q-p} \to \infty$ as well, and therefore $1/x^{q-p} \to 0$. This completes the proof.

x^p grows more slowly than x^q if $p < q$

x	x^{50}	e^x	x^{50}/e^x
100	$\sim 10^{100}$	$\sim 10^{43}$	$\sim 10^{56}$
200	10^{115}	10^{86}	10^{28}
300	10^{123}	10^{130}	10^{-7}
400	10^{130}	10^{173}	10^{-44}
500	10^{134}	10^{217}	10^{-83}
↓	↓	↓	↓
∞	∞	∞	0

How does e^x compare to x^p? To make it tough on e^x, let's compare it to x^{50}. We know already that x^{50} grows faster than any lower power of x. The table above compares x^{50} to e^x. However, the numbers involved are so large that the table shows only their *order of magnitude*—that is, the number of digits they contain. At the start, x^{50} is *much* larger than e^x. However, by the time $x = 500$, the ratio x^{50}/e^x is so small its first 82 decimal places are 0!

So x^{50} grows more slowly than e^x, and so does any lower power of x. Perhaps a higher power of x would do better. It does, but ultimately the ratio $x^p/e^x \to 0$, no matter how large the power p is. We don't yet have all the tools needed to prove this, but we will after we introduce the logarithm function in the next section.

e^x grows more rapidly than any power of x

The speed of exponential growth has had an impact in computer science. In many cases, the number of operations needed to calculate a particular quantity is a power of the number of digits of precision required in the answer. Sometimes, though, the number of operations is an *exponential* function of the number of digits. When that happens, the number of operations can quickly exceed the capacity of the computer. In this way, some problems that can be solved by an algorithm that is straightforward in theoretical terms are intractable in practical terms.

Exercises

The exponential functions b^t

1. Use a graphing utility or a calculator to show the following.

a) If $f(t) = (2.71)^t$, then $f'(0) < 1$.

b) If $g(t) = (2.72)^t$, then $g'(0) > 1$.

c) Use parts (a) and (b) to explain why $2.71 < e < 2.72$.

2. a) Use a calculator or a graphing utility to find the value of the parameter k_b for the bases $b = .5, .75$, and $.9$.

b) What is the shape of the graph of $y = b^t$ when $0 < b < 1$? What does that imply about the *sign* of k_b for $0 < b < 1$? Explain your reasoning.

Differentiating exponential functions

3. Differentiate the following functions.

a) $7e^{3x}$

b) Ce^{kx}, where C and k are constants.

c) $1.5e^t$

d) $1.5e^{2t}$

e) $2e^{3x} - 3e^{2x}$

f) $e^{\cos t}$

4. Find partial derivatives of the following functions.

a) e^{xy}

b) $3x^2 e^{2y}$

c) $e^u \sin v$

d) $e^{u \sin(v)}$

Powers of e

5. Simplify the following and rewrite as powers of e. For each, explain your work, citing any theorems you use.

a) $\exp(2x + 3)$

b) $(\exp(x))^2$

c) $\exp(17x)/\exp(5x)$

6. Use the second property in Theorem 3 to explain why

$$\lim_{t \to -\infty} \exp(t) = 0.$$

7. This purpose of this exercise is to prove the fourth property listed in Theorem 3: $\exp(rs) = (\exp(s))^r$, for all real numbers r and s. The idea of the proof is the same as for Theorem 1: show that two different-looking functions solve the same initial value problem, thus demonstrating that the functions must be the same. The initial value problem is

$$y' = ry \qquad y(0) = 1.$$

a) Show that $P(t) = \exp(rt)$ solves the initial value problem. (You need to use the chain rule.)

b) Show that $Q(t) = (\exp(t))^r$ solves the initial value problem. (Here use the chain $Q = u^r$, where $u = \exp(t)$. There is a bit of algebra involved.)

c) From parts (a) and (b), and the fact that an initial value problem has a unique solution, it follows that $P(t) = Q(t)$, for every t. Explain how this establishes the result.

Solving $y' = ky$ using e^t

8. **Poland and Afghanistan.** Refer to problem 21 in chapter 1, §2.

a) Write out the initial value problems that summarize the information about the populations P and A given in parts (a) and (b) of problem 21.

b) Write formulas for the solutions P and A of these initial value problems.

c) Use your formulas in part (b) (and a calculator) to find the population of each country in the year 2005. What were the populations in 1965?

9. **Bacterial growth.** Refer to problem 22 in chapter 1, §2.

a) Assuming that we begin with the colony of bacteria weighing 32 grams, write out the initial value problem that summarizes the information about the weight P of the colony.

b) Write a formula for the solution P of this initial value problem.

c) How much does the colony weigh after 30 minutes? after 2 hours?

10. **Radioactivity.** Refer to problem 23 in chapter 1, §2.

a) Assuming that when we begin the sample of radium weighs 1 gram, write out the initial value problem that summarizes the information about the weight R of the sample.

b) How much did the sample weigh 2 years ago? How much will it weigh 2 years hence?

11. **Intensity of radiation.** As gamma rays travel through an object, their intensity I decreases with the distance x that they have travelled. This is called **absorption**. The absorption rate dI/dx is proportional to the intensity. For some materials the multiplier in this proportion is large; they are used as radiation shields.

a) Write down a differential equation which models the intensity of gamma rays $I(x)$ as a function of distance x.

b) Some materials, such as lead, are better shields than others, such as air. How would this difference be expressed in your differential equation?

c) Assume the unshielded intensity of the gamma rays is I_0. Write a formula for the intensity I in terms of the distance x and verify that it gives a solution of the initial value problem.

12.   In this problem you will find a solution for the initial value problem $y' = ky$ and $y(t_0) = C$. (Notice that this isn't the original initial value problem, because t_0 was 0 originally.)

a) Explain why you may assume $y = Ae^{kt}$ for some constant A.

b) Find A in terms of k, C and t_0.

Solving other differential equations

13.   a) **Newton's law of cooling.** Verify that

$$Q(t) = 70e^{-.1t} + 20$$

is a solution to the initial value problem $Q'(t) = -k(Q-20)$, $Q(0) = 90$. What is the relationship between this formula and the one found in problem 11 in §2?

b) Verify that

$$Q(t) = (Q_0 - A)e^{-kt} + A$$

is a solution to the the initial value problem $Q'(t) = -k(Q - A)$. What is the relationship between this formula and the one found in problem 11 in §2?

14.   In *An Essay on the Principle of Population*, written in 1798, the British economist Thomas Robert Malthus (1766–1834) argued that food supplies grow at a constant rate, while human populations naturally grow at a constant *per capita* rate. He therefore predicted that human populations would inevitably run out of food (unless population growth was suppressed by unnatural means).

a) Write differential equations for the size P of a human population and the size F of the food supply that reflect Malthus' assumptions about growth rates.

b) Keep track of the population in millions, and measure the food supply in millions of units, where one unit of food feeds one person for one year. Malthus' data suggested to him that the food supply in Great Britain was growing at about .28 million units per year and the per capita growth rate of the population was 2.8% per year. Let $t = 0$ be the year 1798, when Malthus estimated the population of the British Isles was $P = 7$ million people. He assumed his countrymen were on average adequately nourished, so he estimated that the food supply was $F = 7$ million units of food. Using these values, write formulas for the solutions $P = P(t)$ and $F = F(t)$ of the differential equations in (a).

c) Use the formulas in (b) to calculate the amount of food and the population at 25 year intervals for 100 years. Use these values to help you sketch graphs of $P = P(t)$ and $F = F(t)$ on the same axes.

d) The per capita food supply in any year equals the ratio $F(t)/P(t)$. What happens to this ratio as t grows larger and larger? (Use your graphs in (c) to assist your explanation.) Do your results support Malthus's prediction? Explain.

15.   a) **Falling bodies**. Using the base e instead of the base 2, modify the solution $v(t)$ to the initial value problem

$$\frac{dv}{dt} = -g - bv \qquad v(0) = 0$$

that appears in exercise 21 on page 220. Show that the modified expression is still a solution.

b) If an object that falls against air resistance is $x(t)$ meters above the ground after t seconds, and it started x_0 meters above the ground, then it is the solution of the initial value problem

$$\frac{dx}{dt} = v(t) \qquad x(0) = x_0,$$

where $v(t)$ is the velocity function from the previous exercise. Find a formula for $x(t)$ using the exponential function with base e. (Compare this formula with the one in exercise 22 (a), page 220.)

c) Suppose the coefficient of air resistance is .2 per second. If a body is held motionless 200 meters above the ground, and then released, how far will it fall in 1 second? In 2 seconds? Use your formula from part (b). Compare these values with those you obtained in exercise 22 (b), page 220.

Interest rates

Bank advertisements sometimes look like this:

Civic Bank and Trust

- Annual rate of interest 6%.

- Compounded monthly.

- Effective rate of interest 6.17%.

The first item seems very straightforward. The bank pays 6% interest per year. Thus if you deposit $100.00 for one year then at the end of the year you would expect to have $106.00. Mathematically this is the simplest way to compute interest; each year add 6% to the account. The biggest problem with this is that people often make deposits for odd fractions of a year, so if interest were paid only once each year then a depositor who withdrew her money after 11 months would receive no interest. To avoid this problem banks usually compute and pay interest more frequently. The Civic Bank and Trust advertises interest **compounded** monthly. This means that the bank computes interest each month and credits it (that is, adds it) to the account.

Since this particular account pays interest at the rate of 6% per year and there are 12 months in a year the interest rate is 6%/12 = 0.5% per month. The following table shows the interest computations for one year for a bank account earning 6% annual interest compounded monthly.

Month	Start	Interest	End
1	$100.0000	.5000	$100.5000
2	$100.5000	.5025	$101.0025
3	$101.0025	.5050	$101.5075
4	$101.5075	.5075	$102.0151
5	$102.0151	.5101	$102.5251
6	$102.5251	.5126	$103.0378
7	$103.0378	.5152	$103.5529
8	$103.5529	.5178	$104.0707
9	$104.0707	.5204	$104.5911
10	$104.5911	.5230	$105.1140
11	$105.1140	.5256	$105.6396
12	$105.6396	.5282	$106.1678

Notice that at the end of the year the account contains $106.17. It has effectively earned 6.17% interest. This is the meaning of the advertised *effective* rate of interest. The reason that the effective rate of interest is higher than the original rate of interest is that the interest earned each month itself earns interest in each succeeding month. (We first encountered this phenomenon when we were trying to follow the values of S, I, and R into the future.) The difference between the original rate of interest and the effective rate can be very significant. Banks routinely advertise the effective rate to attract depositors. Of course, banks do the same computations for loans. They rarely advertise the effective rate of interest for loans because customers might be repelled by the true cost of borrowing.

The effective rate of interest can be computed much more quickly than we did in the previous table. Let R denote the annual interest rate as a decimal. For example, if the interest rate is 6% then $R = 0.06$. If interest is compounded n times per year then each time it is compounded the interest rate is R/n. Thus each time you compound the interest you compute

$$V + \left(\frac{R}{n}\right) V = \left(1 + \frac{R}{n}\right) V$$

where V is the value of the current deposit. This computation is done n times during the course of a year. So, if the original deposit has value V, after one year it will be worth

$$\left(1 + \frac{R}{n}\right)^n V.$$

For our example above this works out to

$$\left(1 + \frac{0.06}{12}\right)^{12} V = 1.061678\, V$$

and the effective interest rate is 6.1678%.

Many banks now compound interest daily. Some even compound interest *continuously*. The value of a deposit in an account with interest compounded continuously at the rate of 6% per year, for example, grows according to the differential equation

$$V' = 0.06V.$$

16.   Many credit cards charge interest at an annual rate of 18%. If this rate were compounded monthly what would the effective annual rate be?

17.   In fact many credit cards compound interest daily. What is the effective rate of interest for 18% interest compounded daily? Assume that there are 365 days in a year.

18.   The assumption that a year has 365 days is, in fact, *not* made by banks. They figure every one of the 12 months has 30 days, so their year is 360 days long. This practice stem from the time when interest computations were done by hand or by tables, so simplicity won out over precision. Therefore when banks compute interest they find the daily rate of interest by dividing the annual rate of interest by 360. For example, if the annual rate of interest is 18% then the daily rate of interest is 0.05%. Find the effective rate of interest for 18% compounded 360 times per year.

19.   In fact, once they've obtained the daily rate as 1/360-th of the annual rate, banks then compute the interest *every* day of the year. They compound the interest 365 times. Find the effective rate of interest if the annual rate of interest is 18% and the computations are done by banks. First, compute the daily rate by dividing the annual rate by 360 and then compute interest using this daily rate 365 times.

20.   Consider the following advertisement.

Civic Bank and Trust

- Annual rate of interest 6%.

- Compounded daily.

- Effective rate of interest 6.2716%.

Find the effective rate of interest for an annual rate of 6% compounded daily in the straightforward way—using 1/365-th of the annual rate 365 times. Then do the computations the way they are done in a bank. Compare your two answers.

21.   There are two advertisements in the newspaper for savings accounts in two different banks. The first offers 6% interest compounded quarterly (that is, four times per year). The second offers 5.5% interest compounded continuously. Which account is better? Explain.

§4.   The Logarithm Function

Suppose a population is growing at the net rate of 3 births per thousand persons per year. If there are 100,000 persons now, how many will there be 37 years from now? How long will it take the population to double?

Translating into mathematics, we want to find the function $P(t)$ that solves the initial value problem

$$P'(t) = .003P(t) \quad \text{and} \quad P(0) = 100000.$$

Using the results of §3 we know that the solution is the exponential function

$$P(t) = 100000 \, e^{.003\,t}.$$

The size of the population 37 years from now will therefore be

$$\begin{aligned} P(37) &= 100000 \, e^{.111} \\ &= 100000 \times 1.117395 \\ &\approx 111740 \text{ people} \end{aligned}$$

The doubling time of a population

To find out how long it will take the population to double, we want to find a value for t so that $P(t) = 200000$. In other words, we need to solve for t in the equation

$$100000 \, e^{.003\,t} = 200000.$$

Dividing both sides by 100,000, we have

$$e^{.003t} = 2.$$

We can't proceed because one side is expressed in exponential form while the other isn't. One remedy is to express 2 in exponential form. In fact, $2 = e^{.693147}$, as you should verify with a calculator. Then

$$e^{.003t} = 2 = e^{.693147} \quad \text{implies} \quad .003\,t = .693147,$$

so $t = .693147/.003 = 231.049$. Thus it will take about 231 years for the population to double.

To determine the doubling time of the population we had to know the number b for which

$$\exp(b) = e^b = 2.$$

This is an aspect of a very general question: given a positive number a, find a number b for which

$$e^b = a.$$

Solving an exponential equation

A glance at the graph of the exponential function below shows that, by working backwards from any point a on the vertical axis, we can indeed find a unique point b on the horizontal axis which gives us $\exp(b) = a$.

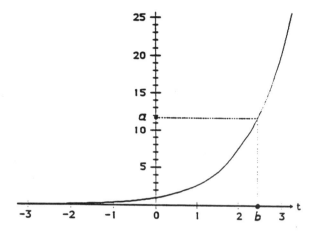

This process of obtaining the number b that satisfies $\exp(b) = a$ for any given positive number a is a clear and unambiguous rule. Thus, it defines a function. This function is called **the natural logarithm**, and it is denoted $\ln(a)$, or sometimes $\log(a)$. That is,

The natural logarithm function

$$\ln(a) = \log(a) = \{\text{the number } b \text{ for which } \exp(b) = a\}.$$

In other words, the two statements

$$\ln(a) = b \quad \text{and} \quad \exp(b) = a$$

express exactly the same relation between the quantities a and b.

The question that led to the introduction of the logarithm function was: what number gives the exponent to which e must be raised in order to produce the value 2? This number is $\ln(2)$, and we verified that $\ln(2) = .693147$. Quite generally we can say that the number $\ln(x)$ gives the exponent to which e must be raised in order to produce the value x:

$$e^{\ln(x)} = x.$$

If we set $y = \ln(x)$, then $x = e^y$ and we can restate the last equation as a pair of companion equations:

$$e^{\ln(x)} = x \quad \text{and} \quad \ln(e^y) = y.$$

The logarithm and exponential functions are inverses

The first equations says *the exponential function "undoes" the effect of the logarithm function* and the second one says *the logarithm function "undoes" the effect of the exponential function*. For this reason the exponential and logarithm functions are said to be **inverses** of each other.

Many of the other pairs of functions—sine and arscsine, squareroot and squaring—that share a key on a calculator have this property. There are even functions (at least one can be found on any calculator) that are their own inverses—apply such a function to any number, then apply this same function to the result, and you're back at the original number. What functions do this? We will say more about inverse functions later in this section.

Properties of the Logarithm Function

The inverse relationship allow us to translate each of the properties of the exponential function into a corresponding statement about the logarithm function. We list the major pairs of properties below.

exponential version	**logarithmic version**
$e^0 = 1$	$\ln(1) = 0$
$e^{a+b} = e^a \cdot e^b$	$\ln(m \cdot n) = \ln(m) + \ln(n)$
$e^{a-b} = e^a / e^b$	$\ln(m/n) = \ln(m) - \ln(n)$
$(e^a)^s = e^{as}$	$\ln(m^s) = s \cdot \ln(m)$
range of e^x is all positive reals	domain of $\ln(x)$ is all positive reals
domain of e^x is all real numbers	range of $\ln(x)$ is all real numbers
$e^x \to 0$ as $x \to -\infty$	$\ln(x) \to -\infty$ as $x \to 0$
e^x grows faster than x^n, any $n > 0$	$\ln(x)$ goes to infinity slower than $x^{1/n}$, any $n > 0$

For each pair, we can use the exponential property and the inverse relationship between exp and ln to establish the logarithmic property. As an example, we will establish the second property. You should be able to demonstrate the others.

Proof of the second property. Remember that to show $\ln(a) = b$, we need to show $e^b = a$. In our case a and b are more complicated. We have

$$
\begin{aligned}
a &= m \cdot n \\
b &= \ln(m) + \ln(n)
\end{aligned}
$$

Thus, we need to show

$$e^{\ln(m) + \ln(n)} = m \cdot n.$$

But, by the exponential version of property 2,

$$e^{\ln(m) + \ln(n)} = e^{\ln(m)} \cdot e^{\ln(n)} = m \cdot n,$$

and our proof is complete.

The Derivative of the Logarithm Function

Since the natural logarithm is a function in its own right, it is reasonable to ask: what is the derivative of this function? Since the derivative describes the slope of the graph, let us begin by examining the graph of ln. Can we to take advantage of the relationship between ln and exp—a function whose graph we know well—as we do this? Indeed we can, by making the following observations.

The graph of ln

- We know the point (a, b) is on the graph of $y = \ln(x)$ if and only if $b = \ln(a)$.

- We know $b = \ln(a)$ says the same thing as $a = e^b$.

- Finally, we know $a = e^b$ is true if and only if the point (b, a) is on the graph of $y = e^x$.

Putting our observations together, we have

$$(a, b) \text{ is on the graph of } y = \ln(x)$$
$$\text{if and only if}$$
$$(b, a) \text{ is on the graph of } y = e^x.$$

Reflection across the 45° line The picture below demonstrates that the point (a, b) and the point (b, a) are reflections of each other about the 45° line. (Remember that points on the 45° line have the same x and y coordinates.) This is because these two points are the endpoints of the diagonal of a square whose other diagonal is the 45° line.

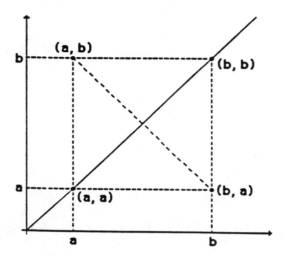

Since we have just seen that every point (a, b) on the graph of $y = \ln(x)$ corresponds to a point (b, a) on the graph of $y = e^x$, we see that *the graphs of $y = \ln(x)$ and $y = e^x$ are the reflections of each other about the 45° line.*

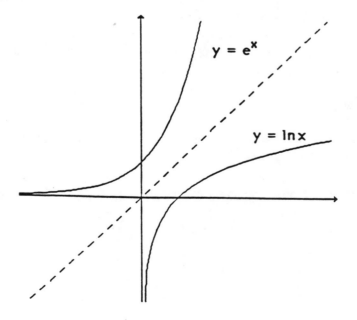

Finally, since the two graphs are reflections of one another, a microscopic view of $\ln(x)$ at any point (b, a) will be the mirror image of the microscopic view of of e^x at the point (a, b). Any change in the y-value on one of these lines will correspond to an equal change in the x-value in its mirror image, and vice versa. The figure below shows what microscopic views of a pair of corresponding points looks like, showing small changes in the x and y coordinates of each.

Microscopic views at mirror image points

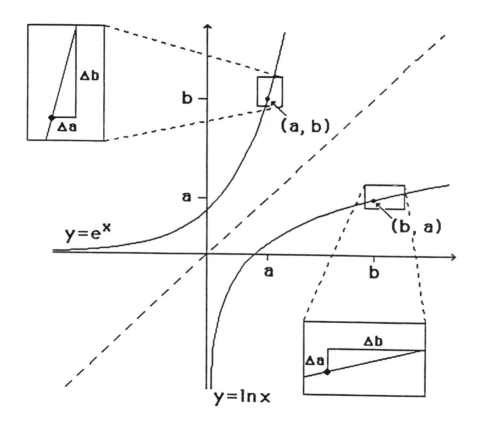

It follows that the slopes of the two lines must be reciprocals of each other. This says that the rate of change of $\ln(x)$ at $x = b$ is just the reciprocal of the rate of change of e^x at $x = a$, where $a = \ln(b)$. But the rate of change of e^x at $x = \ln(b)$ is just $e^{\ln(b)} = b$. Therefore the rate of change of $\ln(x)$ at $x = b$ is the reciprocal of this value, namely $1/b$. We have thus proved the following result:

The slopes are reciprocals

Theorem 1. $(\ln(x))' = 1/x$.

Exponential Growth

The logarithm gives us a useful tool for comparing the growth rates of exponential and power functions. In the last section we claimed that e^x grows faster than any power x^p of x, as $x \to +\infty$. We interpreted that to mean

$$\lim_{x \to +\infty} \frac{x^p}{e^x} = 0,$$

for any number p. Using the natural logarithm, we can now show why it is true.

To analyze the quotient $Q = x^p/e^x$, we first replace it by its logarithm

$$\ln Q = \ln\left(x^p/e^x\right) = \ln\left(x^p\right) - \ln\left(e^x\right) = p \ln x - x.$$

Several properties of the logarithm function were invoked here to reduce $\ln Q$ to $p \ln x - x$. By another property of the logarithm function, if we can show $\ln Q \to -\infty$ we will have established our original claim that $Q \to 0$.

Let $y = \ln Q = p \ln x - x$. We know y is increasing when $dy/dx > 0$ and decreasing where $dy/dx < 0$. Using the rules of differentiation, we find

$$\frac{dy}{dx} = \frac{p}{x} - 1.$$

The expression $p/x - 1$ is positive when x is less than p and negative when x is greater than p. For x near p, the graph of y must therefore look like this:

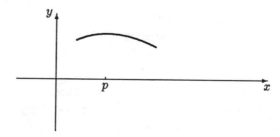

Since dy/dx remains negative as x gets large, y will continue to decrease. This does not, in itself, imply that $y \to -\infty$, however. It's conceivable that y might "level off" even as it continues to decrease—as it does in the graph at the top of the opposite page.

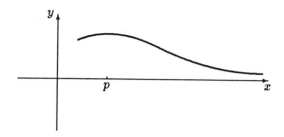

However, we can show that y does *not* "level off" in this way; it continues to plunge down to $-\infty$. We start by assuming that x has already become larger than $2p$: $x > 2p$. Then $1/x < 1/2p$ (the bigger number has the smaller reciprocal), and thus $p/x < p/2p = 1/2$. Thus, when $x > 2p$,

$$\frac{dy}{dx} = \frac{p}{x} - 1 < \frac{1}{2} - 1 = -\frac{1}{2}.$$

In other words, the slope of the graph of y is more negative than $-1/2$. The graph of y must therefore lie *below* the straight line with slope $-1/2$ that we see below:

y lies below a line that slopes down to $-\infty$

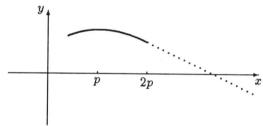

This guarantees that $y = \ln Q \to -\infty$ as $x \to \infty$. Hence $Q \to 0$, and since $Q = x^p/e^x$, we have shown that e^x grows faster than any power of x.

The Exponential Functions b^x

We have come to adopt the exponential function $\exp(x) = e^x$ as the natural one for calculus, and especially for dealing with differential equations of the form $dy/dx = ky$. Initially, though, all exponential functions b^x were on an equal footing. With the natural logarithm function, however, a *single* exponential function will meet our needs. Let's see why.

One exponential function is enough

If b is any positive real number, then $b = e^{\ln b}$. Consequently,

$$b^x = (e^{\ln b})^x = e^{\ln b \cdot x}.$$

$b^x = e^{\ln b \cdot x}$

In other words, $b^x = e^{cx}$, where $c = \ln b$. Thus, every exponential function can be expressed in terms of exp in a simple way.

This expression gives us a new way to find the derivative of b^x. We already know that

$$(e^{cx})' = c \cdot e^{cx},$$

for any constant c. This follows from the chain rule. When $c = \ln b$, we get

$$(b^x)' = (e^{\ln(b) \cdot x})' = \ln(b) \cdot e^{\ln(b) \cdot x} = \ln(b) \cdot b^x.$$

Thus, $y = b^x$ is a solution to the differential equation

$$\frac{dy}{dx} = \ln(b) \cdot y.$$

In chapter 3, we wrote this differential equation as

$$\frac{dy}{dx} = k_b \cdot y.$$

$k_b = \ln(b)$

We see now that $k_b = \ln(b)$.

We can use the connection between k_b and the natural logarithm, and between the natural logarithm and the exponential function, to gain new insights. For example, on page 208 we argued that there must be a value of b for which $k_b = .02$. This simply means

$$\ln b = .02 \qquad \text{or} \qquad b = e^{.02}.$$

In other words, we now have an explicit formula that tells us the value of b for which $k_b = .02$:

$$b = e^{.02} = 1.02020134\ldots$$

Inverse Functions

Most of what we have said about the exponential and logarithm functions carries over directly to *any* pair of inverse functions. We begin by saying precisely what it means for two functions f and g to be inverses of each other.

Definition. Two functions f and g are **inverses** if

$$\begin{aligned} f(g(a)) &= a \\ \text{and } g(f(b)) &= b \end{aligned}$$

for every a in the domain of g and every b in the domain of f.

Let's re-examine the examples we mentioned earlier to see how they fit this definition.

Example 1. Suppose $f(x) = \exp(x)$ and $g(x) = \ln(x)$. Then the equations

$$f(g(a)) \;=\; \exp(\ln(a)) = e^{\ln(a)} = a \text{ for } a > 0$$
$$\text{and } g(f(b)) \;=\; \ln(\exp(b)) = \ln(e^b) = b$$

hold for all real numbers b and for all *positive* real numbers a. The domain of the exponential function is all real numbers and the domain of the natural logarithm function is all positive real numbers.

Example 2. Suppose $f(x) = x^2$ and $g(x) = \sqrt{x}$. The squaring function is not invertible on its natural domain (since a number and its negative have the same square), but it is invertible if we restrict it to non-negative real numbers. Then

x^2 is invertible on $x \geq 0$

$$f(g(a)) \;=\; (\sqrt{a})^2 = a \;\;(\text{for } a \geq 0)$$
$$\text{and } g(f(b)) \;=\; \sqrt{b^2} = b \;\;(\text{for } b \geq 0).$$

The domain of the square root function is all $b \geq 0$.

Example 3. Suppose $f(x) = \sin(x)$ and $g(x) = \arcsin(x)$. By convention, the domain of $\sin(x)$ is taken to be $-\pi/2 \leq x \leq \pi/2$.

$\sin x$ is invertible on $-\pi/2 \leq x \leq \pi/2$

$$f(g(a)) \;=\; \sin(\arcsin(a)) = a \;\;(\text{for } -1 \leq a \leq 1) \text{ and}$$
$$g(f(b)) \;=\; \arcsin(\sin(b)) = b \;\;(\text{for } -\pi/2 \leq b \leq \pi/2).$$

Each pair of inverse functions share corresponding properties, just as the logarithm and exponential functions do—the particular properties depending on the particular functions. But two they all share are

- The range of f is the domain of g.

- The domain of f is the range of g.

The exercises check this for examples 2 and 3.

Finally, the graphs—and therefore the derivatives—of a function and of its inverse are mirror images, exactly like those of the exponential and logarithm functions. We begin with the same list of observations.

- We know the point (a, b) is on the graph of $y = g(x)$ if and only if $b = g(a)$.

- We know $b = g(a)$ says the same thing as $a = f(b)$.

- Finally, we know $a = f(b)$ is true if and only if the point (b, a) is on the graph of $y = f(x)$.

As before, putting our observations together, we have

$$(a, b) \text{ is on the graph of } y = g(x)$$
$$\text{if and only if}$$
$$(b, a) \text{ is on the graph of } y = f(x).$$

The graphs of inverse functions are mirror images ...

Exactly as before, we have that the point (a, b) and the point (b, a) are reflections of each other about the 45° line. Since we have just seen that every point (a, b) on the graph of $y = g(x)$ corresponds to a point (b, a) on the graph of $y = f(x)$, we again see that *the graphs of $y = g(x)$ and $y = f(x)$ are the reflections of each other about the 45° line.*

Finally, since the two graphs are reflections of one another, the local linear approximation of $g(x)$ at any point (a, b) will be the mirror image of the local linear approximation of $f(x)$ at the point (b, a). Any change in the y-value on one of these local lines will correspond to an equal change in the x-value in its mirror image, and vice versa. Just as before, it follows that the slopes of the two lines must be reciprocals of each other. This says that the rate of change of $g(x)$ at $x = b$ is the reciprocal of the rate of change of $f(x)$ at $x = a$, where $a = g(b)$. But the rate of change of $f(x)$ at $x = g(b)$ is just $f'(g(b))$. Therefore the rate of change of $g(x)$ at $x = b$ is the reciprocal of this value, namely $1/f'(g(b))$. We have thus proved the following result:

...and their derivatives are reciprocals

Theorem 2. *If the functions f and g are inverses, then g is locally linear at (b, a) if and only if f is locally linear at (a, b). When local linearity holds,*

$$g'(b) = \frac{1}{f'(a)}.$$

Exercises

1. Determine the numerical value of each of the following.

a) $\ln(2e)$ b) $\ln(e^3)$ c) e^{-1} d) $\ln(\sqrt{e})$

e) $e^{\ln 2}$ f) $e^{3\ln 2}$ g) $(e^{\ln 2})^3$ h) $e^{2\ln 3}$

i) $\ln 10$ j) $\ln 10^3$ k) $e^{\ln 10}$ l) $e^{\ln 1000}$

m) $\ln(1/e)$ n) $\ln(1/2)$ o) $e^{-\ln 2}$ p) $e^{-3\ln 2}$

2. The rate of growth of the population of a particular country is proportional to the population. The last two censuses determined that the population in 1980 was 40,000,000, and in 1985 it was 45,000,000. What will the population be in 1995?

3. Find the derivatives of the following functions.

a) $\ln(3x)$ d) $\ln(2^t)$

b) $17\ln(x)$ e) $\pi\ln(3e^{4s})$

c) $\ln(e^w)$

4. Suppose a bacterial population grows so that its mass is

$$P(t) = 200e^{.12t} \quad \text{grams}$$

after t hours. Its initial mass is $P(0) = 200$ grams. When will its mass double, to 400 grams? How much longer will it take to double again, to 800 grams? After the population reaches 800 grams, how long will it take for yet another doubling to happen? What is the *doubling time* of this population?

5. Suppose a beam of X-rays whose intensity is A rads (the "rad" is a unit of radiation) falls perpendicularly on a heavy concrete wall. After the rays have penetrated s feet of the wall, the radiation intensity has fallen to

$$R(s) = Ae^{-.35s} \quad \text{rads.}$$

What is the radiation intensity 3 inches inside the wall; 18 inches? (Your answers will be expressed in terms of A.) How far into the wall must the rays travel before their intensity is cut in half, to $A/2$? How much further before the intensity is $A/4$?

6. Use properties of exp to prove the following properties of the logarithm. (Remember that $\ln a = b$ means $a = \exp b$.)

a) $\ln(1) = 0$.

b) $\ln(m/n) = \ln(m) - \ln(n)$.

c) $\ln(m^n) = n \ln(m)$.

7. a) Use a graphing program to find a good numerical approximation to $(\ln x)'$ at $x = 2$. Make a short table, for decreasing interval sizes Δx, of the quantity $\Delta(\ln x)/\Delta x$.

b) Use a graphing program to find a good numerical approximation to $(e^x)'$ at $x = \ln(2)$. Make a short table for decreasing interval sizes Δx, of the quantity $\Delta(e^x)/\Delta x$.

8. Find a solution (using $\ln x$) to the differential equation

$$f'(x) = 3/x \quad \text{satisfying} \quad f(1) = 2 \ .$$

9. a) Find a formula using the natural logarithm function giving the solution of $y' = a/x$ with $y(1) = b$.

b) Solve $P' = 2/t$ with $P(1) = 5$.

10. Find the domain and range of each of the following pairs of inverse functions.

a) $f(x) = x^2$ (restricted to $x \geq 0$) and $g(x) = \sqrt{x}$.

b) $f(x) = \sin(x)$ (restricted to $-\pi/2 \leq x \leq \pi/2$) and $g(x) = \arcsin(x)$.

11. Show that $f(x) = 1/x$ equals its own inverse. What are the domain and range of f?

12. Use the relationship between the derivatives of a function and its inverse to find the indicated derivatives.

a) $g'(100)$ for $g(x) = \sqrt{x}$.

b) $g'(\sqrt{2}/2)$ for $g(x) = \arcsin(x)$.

c) $g'(1/2)$ for $g(x) = 1/x$.

13. Compare the rates of growth of e^x and b^x for both $e < b$ and $1 < b < e$.

§5.   The Equation $y' = f(t)$

Most differential equations we have encountered express the rate of growth of a quantity *in terms of the quantity itself.* The simplest models for biological growth had this form: $y' = ky$ and $y' = ky^p$. Even when several variables were present—as in the *S-I-R* model and the predator-prey models—it was most natural to express the rates at which those variables change in terms of the variables themselves. Even the motion of a spring (pages 220–222) was described that way: the rate of change of position equalled the velocity, and the rate of change of velocity was proportional to the position.

Sometimes, though, a differential equation will express the rate of change of a variable *directly in terms of the input variable.* For example, on page 219 we saw that the velocity dx/dt of a body falling under the sole influence of gravity is a linear function of the time:

The motion of a falling body . . .

$$\frac{dx}{dt} = -gt + v_0.$$

Here x is the height of the body above the ground, g is the acceleration due to gravity, and v_0 is the velocity at time $t = 0$. This equation has the general form

$$\frac{dx}{dt} = f(t),$$

where $f(t)$ is a given function of t. We will now consider special methods that can be used to study this differential equation.

Antiderivatives

To solve the equation of motion of a body falling under gravity, we must find a function $x(t)$ whose derivative is given as

. . . and its solution

$$x' = -gt + v_0.$$

We can call upon our knowledge of the rules of differentiation to find x. Consider $-gt$ first. What function has $-gt$ as its derivative? We can start with t^2, whose derivative is $2t$. Since we want the derivative to turn out to be $-gt$, we can reason this way:

$-gt$ is the derivative of $-gt^2/2$

$$-gt = -\frac{g}{2} \cdot 2t = -\frac{g}{2} \times \text{the derivative of } t^2.$$

This leads us to identify $-gt^2/2$ as a function whose derivative is $-gt$. Check for yourself that this is correct by differentiating $-gt^2/2$.

Now consider v_0, the other part of dx/dt. What function has the constant v_0 as its derivative? A derivative is a rate of growth, and we know that the linear functions are precisely the ones that have constant growth rates. Furthermore, the rate is the multiplier for a linear function, so we conclude that any linear function of the form $v_0 t + b$ has derivative v_0.

v_0 is the derivative of $v_0 t$ and of $v_0 t + b$

If we put the two pieces together, we find that

$$x(t) = -\frac{g}{2} t^2 + v_0 t + b$$

is a solution to the differential equation, for any value of b. (Recall from §2 that a differential equation can have many solutions.) We constructed this formula for $x(t)$ by "undoing" the process of differentiation, a process sometimes called **antidifferentiation**. The function produced is called an **antiderivative**. Thus:

The antiderivative of a function

$-\frac{g}{2}t^2 + v_0 t + b$ is an *antiderivative* of $-gt + v_0$

because

$-gt + v_0$ is the *derivative* of $-\frac{g}{2}t^2 + v_0 t + b$

All the functions $F(x) + C$ are antiderivatives of $F'(x)$

Note that a function has only one derivative, but it has many antiderivatives. If $F(t)$ is an antiderivative of $f(t)$, then so is $F(t) + C$, where C is any constant.

The list of functions and their derivatives that we compiled in chapter 3 (see page 146) can be "turned around" to become a list of functions and their antiderivatives.

function	antiderivative
0	c
x^p	$\frac{1}{p+1} x^{p+1}$
x^{-1}	$\ln x$
$\sin x$	$-\cos x$
$\cos x$	$\sin x$
$\exp x = e^x$	$\exp x = e^x$
b^x	$\frac{1}{\ln b} b^x$

Notice the formula for the antiderivative of x^p requires $p + 1 \neq 0$, that is, $p \neq -1$. This leaves out x^{-1}. However, the antiderivative of x^{-1} is $\ln x$, so no power of x is excluded from the table.

Every power of x has an antiderivative

We also had differentiation rules that told us how to deal with different *combinations* of functions. Each of these rules has an analogue in antidifferentiation. The simplest combinations are a sum and a constant multiple.

function	antiderivative
$f(x)$	$F(x)$
$g(x)$	$G(x)$
$c \cdot f(x)$	$c \cdot F(x)$
$f(x) + g(x)$	$F(x) + G(x)$

We defer a discussion of the analogue of the chain rule to Calculus II.

With just these rules we can find the antiderivative of any polynomial, for instance. (Recall that a polynomial is a sum of constant multiples of powers of the input variable.) Here is a collection of sample antiderivatives that illustrate the various rules. You should compare this table with the one on page 148.

function	antiderivative
$5x^4 - 2x^3$	$x^5 - \frac{1}{2}x^4$
$5x^4 - 2x^3 + 17x$	$x^5 - \frac{1}{2}x^4 + \frac{17}{2}x^2$
$6 \cdot 10^z + 17/z^5$	$6 \cdot 10^z / \ln 10 - 17/6z^6$
$3 \sin t - 2t^3$	$-3 \cos t - \frac{1}{2}t^4$
$\pi \cos x + \pi^2$	$\pi \sin x + \pi^2 x$

Euler's Method Revisited

If we know the formula for an antiderivative of $f(t)$, then we can write down a solution to the differential equation $dy/dt = f(t)$. For example, the general solution to

$$\frac{dy}{dt} = 12t^2 + \sin t$$

is $y = 4t^3 - \cos t + C$. But it is not often that we know a formula for an antiderivative of $f(t)$—even when $f(t)$ itself has a simple formula. There is no formula for the antiderivative of $\cos(t^2)$, or $\sin t/t$, or $\sqrt{1 + t^3}$, for instance. In other cases, $f(t)$ may not even be given by a formula. It may be a data function, given as a graph made by a pen tracing on a moving sheet of graph paper.

When we cannot find a formula for an antiderivative of $f(t)$, we can still solve the differential equation $dy/dt = f(t)$ by Euler's method. Because the differential equation is special, Euler's method takes a special form. Let's investigate this in the following context.

The volume of a reservoir varies over time

Let V be the volume of water in a reservoir serving a small town, measured in millions of gallons. Then V is a function of the time t, measured in days. Rainfall adds water to the reservoir, while evaporation and consumption by the townspeople take it away. Let f be the *net rate* at which water is flowing into the reservoir, in millions of gallons per day. Sometimes f will be positive—when rainfall exceeds evaporation and consumption—and sometimes f will be negative. The net inflow rate varies from day to day; that is, f is a function of time: $f = f(t)$. Our model of the reservoir is the differential equation

$$\frac{dV}{dt} = f(t) \text{ millions of gallons per day.}$$

The net inflow rate

Suppose $f(t)$ is measured every two days, and those measurements are recorded in the following table.

time t (days)	rate $f(t)$ ($10^6 \times$ gals. per day)
0	.34
2	.11
4	−.07
6	−.23
8	−.14
10	.03
12	.08

If we assume the value of $f(t)$ remains constant for the two days after each measurement is made, we can determine the total change in V over these 14 days.

The following table tells us two things: first, how much V changes over each two-day period; second, the *total* change in V that has accumulated by the end of each period. Since $\Delta t = 2$ days, we calculate ΔV by

The accumulated change in V

$$\Delta V = V' \cdot \Delta t = f(t) \cdot \Delta t = 2 \cdot f(t).$$

starting t	current ΔV	accumulated ΔV	ending t
0	.68	.68	2
2	.22	.90	4
4	−.14	.76	6
6	−.46	.30	8
8	−.28	.02	10
10	.06	.08	12
12	.16	.24	14

At the end of the 14 days, V has accumulated a total change of .24 million gallons. Notice this does not depend on the initial size of V. If V had been 92.64 million gallons at the start, it would be 92.64 + .24 = 92.88 million gallons at the end. If it had been only 2 million gallons at the start, it would be 2 + .24 = 2.24 million gallons at the end. Other models do not behave this way: in two weeks, a rabbit population of 900 will change much more than a population of 90. The total change in V is independent of V because the *rate* at which V changes is independent of V.

We can therefore use Euler's method to solve any differential equation of the form $dy/dt = f(t)$ *independently* of an initial value for y. We just calculate the total accumulated change in y, and add that total to any given initial y. Here is how it works when the initial value of t is a, and the time step is Δt.

Euler's method can calculate just the accumulated change in y

starting t	current Δy	accumulated Δy	ending t
a	$f(a) \cdot \Delta t$	previously	$a + \Delta t$
$a + \Delta t$	$f(a + \Delta t) \cdot \Delta t$	accumulated	$a + 2\Delta t$
$a + 2\Delta t$	$f(a + 2\Delta t) \cdot \Delta t$	Δy	$a + 3\Delta t$
$a + 3\Delta t$	$f(a + 3\Delta t) \cdot \Delta t$		$a + 4\Delta t$
$\vdots$	$\vdots$	$+$	$\vdots$
		current	
$a + (n-1)\Delta t$	$f(a + (n-1)\Delta t) \cdot \Delta t$	Δy	$a + n\,\Delta t$

The third column is too small to hold the values of "accumulated Δy." Instead, it contains the instructions for obtaining those values. It says: to get the current value of "accumulated Δy," add the "current Δy" to the previous value of "accumulated Δy."

Let's use Euler's method to find the accumulated Δy when $t = 4$, given that

$$\frac{dy}{dt} = \cos(t^2)$$

and t is initially 0. If we use 8 steps, then $\Delta t = .5$ and we obtain the following:

starting t	current Δy	accumulated Δy	ending t
0	.5000	.5000	.5
.5	.4845	.9845	1.0
1.0	.2702	1.2546	1.5
1.5	−.3141	.9405	2.0
2.0	−.3268	.6137	2.5
2.5	.4997	1.1134	3.0
3.0	−.4556	.6579	3.5
3.5	.4752	1.1330	4.0

The following program generated the last three columns of this table.

Program: TABLE

```
DEF fnf (t) = COS(t ^ 2)
tinitial = 0
tfinal = 4
numberofsteps = 2 ^ 3
deltat = (tfinal - tinitial) / numberofsteps
t = tinitial
accumulation = 0
FOR k = 1 TO numberofsteps
        deltay = fnf(t) * deltat
        accumulation = accumulation + deltay
        t = t + deltat
        PRINT deltay, accumulation, t
NEXT k
```

TABLE is a modification of the program SIRVALUE (page 65).

As usual, to find the exact value of the accumulated Δy, it is necessary to recalculate, using more steps and smaller step sizes Δt. If we use TABLE to do this, we find

number of steps	accumulated Δy
2^3	1.13304
2^6	.65639
2^9	.60212
2^{12}	.59542
2^{15}	.59458
2^{18}	.59448

Thus we can say that if $dy/dt = \cos(t^2)$, then y increases by .594... when t increases from 0 to 4.

In the same way we changed SIRVALUE to produce the program SIRPLOT (page 69), we can change the program TABLE into one that will *plot* the values of y. In the following program all those changes are made, and one more besides: we have increased the number of steps to 400 to get a closer approximation to the true values of y.

Program: PLOT

```
Set up GRAPHICS
DEF fnf (t) = COS(t ^ 2)
tinitial = 0
tfinal = 4
numberofsteps = 400
deltat = (tfinal - tinitial) / numberofsteps
t = tinitial
accumulation = 0
FOR k = 1 TO numberofsteps
      deltay = fnf(t) * deltat
      Plot the line from (t, accumulation)
          to (t + deltat, accumulation + deltay)
      accumulation = accumulation + deltay
      t = t + deltat
NEXT k
```

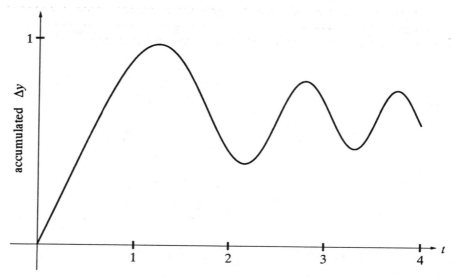

The accumulated Δy when $dy/dt = \cos(t^2)$

Let's compare our reservoir model with population growth. The rate at which a population grows depends, in an obvious way, on the size of the population. By contrast, the rate at which the reservoir fills does **Exogenous and** *not* depend on how much water there is in the reservoir. It depends on **endogenous** factors *outside* the reservoir: rainfall and consumption. These factors **factors** are said to be **exogenous** (from the Greek *exo-*, "outside" and *-gen*, "produced," or "born"). The opposite is called an **endogenous** factor (from the Greek *endo-*, "within"). Evaporation is an endogenous factor for the reservoir model; population size is certainly an endogenous factor for a population model.

Precisely because exogenous factors are "outside the system," we need to be *given* the information how they vary over time. In the reservoir model, this information appears in the function $f(t)$ that describes the rate at which V changes. In general, if y depends on exogenous factors that vary over time, we can expect the differential equation for y to involve a function of time:

$$\frac{dy}{dt} = f(t)$$

Thus, we can view this section as dealing with models that involve exogenous factors.

The differential equation of motion for a falling body, $dx/dt = -gt + v_0$, indicates that gravity is an exogenous factor. In Greek and medieval European science, the reason an object fell to the ground was assumed to lie within the object itself—it was the object's "heaviness." By making the cause of motion exogenous, rather than endogenous, Galileo and Newton started a scientific revolution.

Exercises

1. Find a formula $y = F(t)$ for a solution to the differential equation $dy/dt = f(t)$ when $f(t)$ is

a) $5t - 3$

b) $t^6 - 8t^5 + 22\pi^3$

c) $5e^t - 3\sin t$

d) $12\sqrt{t}$

e) $2^t + 7/t^9$

2. Find $G(5)$ if $y = G(x)$ is the solution to the initial value problem

$$\frac{dy}{dx} = \frac{1}{x^2} \qquad y(2) = 3.$$

3. Find $F(2)$ if $y = F(x)$ is the solution to the initial value problem

$$\frac{dy}{dx} = \frac{1}{x} \qquad y(1) = 5.$$

4. Sketch the graph of the solution to the initial value problem

$$\frac{dy}{dx} = \sin x \qquad y(0) = 0$$

over the interval $0 \le x \le 4\pi$.

5. a) Sketch the graph of the solution to the initial value problem

$$\frac{dy}{dx} = \sin(x^2) \qquad y(0) = 0$$

over the interval $0 \le x \le 5$.

b) What is the slope of the solution graph at $x = 0$? Does your graph show this?

c) How many peaks (local maxima) does the solution have on the interval $0 \le x \le 5$?

d) What is the maximum value that the solution achieves on the interval $0 \le x \le 5$? For which value of x does this happen?

6.   a) What is the accumulated change in y if $dy/dt = 3t^2 - 2t$ and t increases from 0 to 1? What if t increases from 1 to 2? What if t increases from 0 to 2?

b) Sketch the graph of the accumulated change in y as a function of t. Let $0 \le t \le 2$.

7.   a) Sketch the graph of the solution to the initial value problem

$$\frac{dy}{dx} = \frac{\sin x}{x} \qquad y(0) = 0$$

on the interval $0 \le x \le 40$. [Note: $\sin x / x$ is not defined when $x = 0$, so take the initial value of x to be .00001. That is, use $y(.00001) = 0$.]

b) How many peaks (local maxima) does the solution have on the interval $0 \le x \le 40$?

c) What is the maximum value of the solution on the interval $0 \le x \le 40$? For which x is this maximum achieved?

§6.   Chapter Summary

The Main Ideas

- A **system of differential equations** expresses the derivatives of a set of functions in terms of those functions and the input variable.

- An **initial value problem** is a system of differential equations together with values of the functions for some specified value of the input variable.

- Many processes in the physical, biological, and social sciences are **modelled** as initial value problems.

- A **solution to a system of differential equations** is a set of functions which make the equations *true* when they and their derivatives are substituted into the equations.

- A **solution to an initial value problem** is a set of functions that solve the differential equations and satisfy the initial conditions. Typically, a solution is unique.

- **Euler's method** provides a recipe to find the solution to an initial value problem.

- In special circumstances it is possible to find **formulas** for the solution to a system of differential equations. If the differential equations involve **parameters**, the solutions will, too.

- Systems of differential equations define functions as their solutions. Among the most important are the **exponential** and **logarithm functions**.

- The **natural logarithm** function is the inverse of the exponential function.

- The **graphs** and the **derivatives** of a function and its inverse are connected geometrically to each other by **reflection**.

- Exponential functions b^x **grow to infinity** faster than any power of x.

- The solution to $dy/dx = f(x)$ is an **antiderivative** of f—that is, a function whose derivative is f.

Self-Testing

- You should be able to use computer programs to produce tables and graphs of solutions to initial value problems.

- You should be able to check whether a system of differential equations reflects the hypotheses being made in constructing a model of a process.

- You should be able to verify whether a set of functions given by formulas is a solution to a system of differential equations.

- You should be familiar with the basic properties of the exponential and logarithm functions.

- You should be able to express solutions to initial value problems involving exponential growth or decay in terms of the exponential function.

- You should be able to solve $dy/dx = f(x)$ by antidifferentiation when $f(x)$ is a basic function or a simple combination of them.

- You should be able to analyze and graph the inverse of a given function.

Chapter 5

Techniques of Differentiation

In this chapter we focus on functions given by *formulas*. The derivatives of such functions are then also given by formulas. In chapter 4 we used derivatives to *define* functions; now we go *from* the function *to* its derivative. We develop the rules for *differentiating* a function: computing the formula for its derivative from the formula for the function. Then we use differentiation to investigate the properties of functions, expecially their *extreme values*. Finally we examine a powerful method for solving equations that depends on being able to find a formula for a derivative.

§1. The Differentiation Rules

There are three kinds of differentiation rules. First, any basic function has a specific rule giving its derivative. Second, the *chain rule* will find the derivative of a *chain* of functions. Third, there are general rules that allow us to calculate the derivatives of algebraic combinations—e.g., sums, products, and quotients—of any functions provided we know the derivatives of each of the component functions. To obtain all three kinds of rules we will typically start with the analytic definition of the derivative as the limit of a quotient of differences:

Definition. The **derivative** of the function f at x is the value of the limit

$$\lim_{\Delta x \to 0} \frac{f(x + \Delta x) - f(x)}{\Delta x} = f'(x).$$

In this chapter we will look at the cases where this limit can be evaluated exactly. Although using this definition of derivative usually leads to many algebraic manipulations, the other interpretations of derivatives as slopes, rates, and multipliers will still be helpful in visualizing what's going on. The process of calculating the derivative of a function is called **differentiation**. For this reason, functions which are locally linear and not locally vertical (so they do have slopes, and hence derivatives at every point) are called **differentiable** functions. Our goal in this chapter is to differentiate functions given by formulas.

Derivatives of Basic Functions

Functions given by formulas have derivatives given by formulas

When a function is given by a *formula*, there is in fact a formula for its derivative. We have already seen several examples in chapters 3 and 4. These examples include all of what we may consider the **basic functions**. We collect these formulas in the following table.

Rules for Derivatives of Basic Functions

function	derivative
$mx + b$	m
x^r	rx^{r-1}
$\sin x$	$\cos x$
$\cos x$	$-\sin x$
e^x	e^x
$\ln x$	$1/x$

In the case of the linear function $mx + b$, we obtained the derivative by using its geometric description as the *slope* of the graph of the function. The derivatives of the exponential and logarithm functions came from the definition of the exponential function as the solution of an initial value problem. To find the derivatives of the other functions we will need to start from the definition.

An example: $f(x) = x^3$

We begin by examining the calculation of the derivative of $f(x) = x^3$ using the definition. The change Δy in $y = f(x)$ corresponding to a change Δx in x is given by

$$\begin{aligned}
\Delta y &= f(x + \Delta x) - f(x) \\
&= (x + \Delta x)^3 - x^3 \\
&= 3x^2 \cdot \Delta x + 3x(\Delta x)^2 + (\Delta x)^3.
\end{aligned}$$

From this we get

$$\begin{aligned}
f'(x) &= \lim_{\Delta x \to 0} \frac{\Delta y}{\Delta x} \\
&= \lim_{\Delta x \to 0} 3x^2 + 3x \cdot \Delta x + (\Delta x)^2.
\end{aligned}$$

To see what's happening with this expression, let's consider the specific value $x = 2$ and evaluate the corresponding values of $\Delta y / \Delta x$ for successively smaller Δx.

Δx	$2^2 + 6\Delta x + (\Delta x)^2$	$\Delta y / \Delta x$
.1	$12 + .6 + .01$	12.61
.01	$12 + .06 + .0001$	12.0601
.001	$12 + .006 + .000001$	12.006001
.0001	$12 + .0006 + .00000001$	12.00060001
.00001	$12 + .00006 + .0000000001$	12.0000600001

The value of $\Delta y / \Delta x$ gets closer and closer to 12 as Δx gets smaller and smaller

It is clear from this table that we can make $\Delta y / \Delta x$ as close to 12 as we like by making Δx small enough. Therefore $f'(2) = 12$.

In general, for any given x, the second and third terms in the expansion for $\Delta y / \Delta x$ become vanishingly small as $\Delta x \to 0$, so that $\Delta y / \Delta x$ can be made as close to $3x^2$ as we like by making Δx small enough. For this reason, we say that the derivative $f'(x)$ is *exactly* $3x^2$:

$$f'(x) = \lim_{\Delta x \to 0} 3x^2 + 3x \cdot \Delta x + (\Delta x)^2 = 3x^2.$$

In other words, given the function f specified by the formula $f(x) = x^3$ we have found the formula for its derivative function f': $f'(x) = 3x^2$. Note that this general formula agrees with the specific value $f'(2) = 12$ we have already obtained.

Notice the difference between the statements

$$f'(x) \approx \Delta y / \Delta x \qquad \text{and} \qquad f'(x) = 3x^2.$$

For a particular value of Δx, the corresponding value of $\Delta y/\Delta x$ is an approximation of $f'(x)$. We can obtain another, better approximation by computing $\Delta y/\Delta x$ for a smaller Δx. The successively better approximations differ from one another by less and less. In particular, they differ less and less from the *limit value* $3x^2$. The value of the derivative $f'(x)$ is *exactly* $3x^2$.

More generally, for any function $y = f(x)$, a particular difference quotient $\Delta y/\Delta x$ is an approximation of $f'(x)$. Successively smaller values of Δx give successively better approximations of $f'(x)$. Again $f'(x)$ *exactly* equals the limiting value of these successive approximations. In some cases, however, we are only able to approximate that limiting value, as we often did in chapter 3, and for many purposes the approximation is entirely satisfactory. In this chapter we will concentrate on the exact statements that are possible for functions given by formulas.

The other basic functions

Our formula for the derivative of the function $f(x) = x^3$ is one instance of the general rule for the derivative of $f(x) = x^r$.

For every real number r , the derivative of $\ f(x) = x^r\ $ is $\ f'(x) = r\,x^{r-1}$.

The rule for the
derivative of a
power function

We can prove this rule for the case when r is a positive integer using algebraic manipulations very like the ones carried out for x^3; see the exercises for verifications of this and the other differentiation rules in this section. Using a rule for quotients of functions (coming later in this section), we can show that this rule also holds for *negative* integer exponents. Further arguments using the chain rule show that the pattern still holds for *rational* exponents. We can eliminate this case-by-case approach, though, by recalling the approach developed in chapter 4. We saw that we can give meaning to b^r for any positive base b and any real number r by defining

$$b^r = e^{r\ln(b)}.$$

Using the formulas for the derivatives of e^x and $\ln x$ together with the chain rule, we can prove the rule for $x > 0$ and for arbitrary real exponent r directly, without first proving the special cases for integer or rational exponents. See the exercises for details. Arguments justifying the formulas for the derivatives of the trigonometric functions are also in the exercises.

Combining Functions

We can form new functions by combining functions. We have already studied one of the most useful ways of doing this in chapter 3 when we looked at forming "chains" of functions and developed the **chain rule** for taking the derivative of such a chain. Suppose $u = f(x)$ and $y = g(u)$. Chaining these two functions together we have y as a function of x:

$$y = h(x) = g(f(x)).$$

The chain rule tells us how to find the derivative of y with respect to x. In function notation it takes the form

$$h'(x) = g'(f(x)) \cdot f'(x).$$

In Leibniz notation, using $f(x) = u$ we can write the chain rule as

$$\frac{dy}{dx} = \frac{dy}{du} \cdot \frac{du}{dx}.$$

Functions combined by chains...

The chain rule

We also saw in chapter 3 that the polynomial $5x^3 - 7x^2 + 3$ can be thought of as an *algebraic* combination of simple functions. We can build an even more complicated function by forming a quotient with this polynomial in the numerator and the difference of the functions $\sin x$ and e^x in the denominator. The result is

$$\frac{5x^3 - 7x^2 + 3}{\sin x - e^x}.$$

...and algebraically

The derivative of this function, as well as of other functions formed by adding, subtracting, multiplying and dividing simpler functions, is obtained by use of the following rules for the derivatives of algebraic combinations of differentiable functions.

Rules for Algebraic Combinations of Functions

function	derivative
$f(x) + g(x)$	$f'(x) + g'(x)$
$f(x) - g(x)$	$f'(x) - g'(x)$
$cf(x)$	$cf'(x)$
$f(x) \cdot g(x)$	$f'(x) \cdot g(x) + f(x) \cdot g'(x)$
$\dfrac{f(x)}{g(x)}$	$\dfrac{g(x) \cdot f'(x) - f(x) \cdot g'(x)}{[g(x)]^2}$

Combining functions by adding, subtracting, multiplying and dividing

Notice the signs in the rules

Notice carefully that the product rule has a plus sign but the quotient rule has a *minus* sign. You can remember these formulas better if you think about where these signs come from. Increasing *either* factor increases a (positive) product, so the derivative of *each* factor appears with a plus sign in the formula for the derivative of a product. Similarly, increasing the numerator increases a positive quotient, so the derivative of the numerator appears with a plus sign in the formula for the derivative of a quotient. However, increasing the denominator *decreases* a positive quotient, so the derivative of the denominator appears with a minus sign.

Let's now use the rules to differentiate the quotient

$$\frac{5x^3 - 7x^2 + 3}{\sin x - e^x}.$$

First, the derivative of the numerator $5x^3 - 7x^2 + 3$ is

$$5(3x^2) - 7(2x) + 0 = 15x^2 - 14x.$$

Similarly the derivative of $\sin x - e^x$ is $\cos x - e^x$. Finally, the derivative of the quotient function is obtained by using the rule for quotients:

$$\frac{(\sin x - e^x)(15x^2 - 14x) - (5x^3 - 7x^2 + 3)(\cos x - e^x)}{(\sin x - e^x)^2}.$$

The following examples further illustrate the use of the rules for algebraic combinations of functions.

function	derivative
$-3e^t + \sqrt[3]{t}$	$-3e^t + (1/3)t^{-2/3}$
$\dfrac{5}{x^3} - 7x^4 + \ln x$	$5(-3)x^{-4} - 7(4x^3) + 1/x$
$7\sqrt{x}\cos x$	$7(\dfrac{1}{2\sqrt{x}})\cos x + 7\sqrt{x}(-\sin x)$
$\left(\dfrac{4}{3}\right)\pi r^3$	$\left(\dfrac{4}{3}\right)\pi 3r^2$
$\dfrac{3s^6}{s^2 - s}$	$\dfrac{(s^2 - s)3(6s^5) - 3s^6(2s - 1)}{(s^2 - s)^2}$

For another kind of example, suppose the per capita daily energy consumption in a country is currently 800,000 BTU, and, due to energy conservation efforts, it is falling at the rate of 1,000 BTU per year. Suppose too that the population of the country is currently 200,000,000 people and is rising at the rate of 1,000,000 people per year. Is the total daily energy consumption of this country rising or falling? By how much?

Three different quantities vary with time in this example: daily per capita energy consumption, population and total daily energy consumption. We can model this situation with three functions $C(t)$, $P(t)$ and $E(t)$.

$C(t)$:   per capita consumption at time t
$P(t)$:   population at time t
$E(t)$:   total energy consumption at time t

Since the per capita consumption times the number of people in the population gives the total energy consumption, these three functions are related algebraically:

$$E(t) = C(t) \cdot P(t).$$

If $t = 0$ represents today, then we are given the two rates of change

$$C'(0) = -1,000 = -10^3 \text{ BTU per person per year, and}$$
$$P'(0) = 1,000,000 = 10^6 \text{ persons per year.}$$

Using the product rule we can compute the current rate of change of the total daily energy consumption:

$$
\begin{aligned}
E'(0) &= C(0) \cdot P'(0) + C'(0) \cdot P(0) \\
&= (8 \times 10^5) \cdot (10^6) + (-10^3) \cdot (2 \times 10^8) \\
&= (8 \times 10^{11}) - (2 \times 10^{11}) \\
&= 6 \times 10^{11} \text{ BTU per year.}
\end{aligned}
$$

So the total daily energy consumption is currently rising at the rate of 6×10^{11} BTU per year. Thus the growth in the population more than offsets the efforts to conserve energy.

Finally, it is a useful exercise to check that the units make sense in this computation. Recall that $C(t)$ represents *per capita* daily energy consumption, so the units for $C(0) \cdot P'(0)$ are

Checking units

$$
\frac{\text{BTU}}{\text{person}} \cdot \frac{\text{persons}}{\text{year}} = \frac{\text{BTU}}{\text{year}},
$$

and, similarly, the units for $C'(0) \cdot P(0)$ are

$$
\frac{\text{BTU}}{\text{person}} \cdot \frac{1}{\text{year}} \cdot \text{persons} = \frac{\text{BTU}}{\text{year}}.
$$

Informal Arguments

All of the rules for differentiating algebraic combinations of functions can be proved by using the algebraic definition of the derivative as a limit of a difference quotient. In fact, we will examine such a formal proof below. However, informal arguments based on geometric ideas or other intuitive understandings are also valuable aids to understanding. Here are three examples of such arguments.

Stretching *y*-coordinates

- If a new function g is obtained from f by multiplying by a positive constant c, so $g(x) = cf(x)$, what is the relationship between the graphs of $y = f(x)$ and of $y = g(x)$? Stretching (or compressing, if c is less than 1) the y-coordinates of the points of the graph of f by a factor of c yields the graph of g.

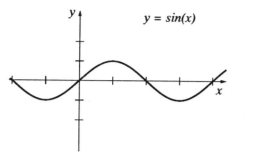

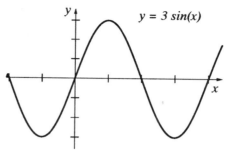

What then is the relationship between the slopes $f'(x)$ and $g'(x)$? If the y-coordinates are tripled, the slope will be three times as great. If they are halved, the slope will also be half as much. More generally, the elongated (or compressed) graph of g has a slope equal to c times the slope of the original graph of f. In other words, $g(x) = cf(x)$ implies $g'(x) = cf'(x)$.

- Now suppose instead that g is obtained from f by adding a constant b, so $g(x) = f(x) + b$. This time the graph of $y = g(x)$ is obtained from the graph of $y = f(x)$ by shifting up or down (according to the sign of c) by $|c|$ units. What is the relationship between the slopes $f'(x)$ and $g'(x)$? The shifted graph has exactly the same slope as the original graph, so in this case, $g(x) = f(x) + b$ implies $g'(x) = f'(x)$.

Shifting y-coordinates

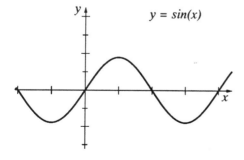

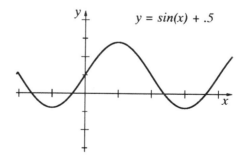

There is a similar pattern when the coordinates of the input variable are stretched or shifted—that is when $y = f(u)$ and u is *rescaled* by the linear relation $u = mx + b$. These results depend on the chain rule and appear in the exercises.

The fact that the derivative of $f(x) + b$ is the same as the derivative of $f(x)$ is a special case of the general *addition rule*, which says *the derivative of a sum is the sum of the derivatives.* In the special case, the derivative of the constant function b is zero, so adding a constant leaves the derivative unchanged. To see that how natural it is to add rates in the general case, consider the following example.

Adding flows

- Suppose we are diluting wine by mixing it with water in a big tub. We may let $f(t)$ be the amount (in gallons) of wine in the tub and $g(t)$ be the amount of water in the tub at time t. Then $f'(t)$ is the rate at which wine is being added at time t (measured in gallons per minute), and $g'(t)$ is the rate at which water is flowing into the tub. $F(t) = f(t) + g(t)$ then gives the total amount of liquid in the tub at time t, and $F'(t)$ is the rate by which that total amount of liquid is changing at time t. Clearly that rate is the sum of the rates of flow of wine and water into the tank. If at some particular moment we are adding wine at the rate of 3.2 gal/min and water at the rate of 1.1 gal/min, the liquid in the tub is increasing by 4.3 gal/min at that moment.

A Formal Proof: the Product Rule

We include here the algebraic calculations yielding the rule for the derivative of the product of two arbitrary functions—just to give the flavor of these arguments. Algebraic arguments for the rest of these rules may be found in the exercises.

The Product Rule:

$$F(x) = f(x) \cdot g(x) \text{ implies } F'(x) = f'(x) \cdot g(x) + f(x) \cdot g'(x)$$

To save some writing, let

$$
\begin{aligned}
\Delta F &= F(x + \Delta x) - F(x), \\
\Delta f &= f(x + \Delta x) - f(x), \\
\text{and} \quad \Delta g &= g(x + \Delta x) - g(x).
\end{aligned}
$$

Rewrite the last two equations as

$$
\begin{aligned}
f(x + \Delta x) &= f(x) + \Delta f \\
g(x + \Delta x) &= g(x) + \Delta g.
\end{aligned}
$$

Now we can write

$$\begin{aligned}
F(x + \Delta x) &= f(x + \Delta x) \cdot g(x + \Delta x) \\
&= (f(x) + \Delta f) \cdot (g(x) + \Delta g) \\
&= f(x) \cdot g(x) + f(x) \cdot \Delta g + \Delta f \cdot g(x) + \Delta f \cdot \Delta g
\end{aligned}$$

This gives us a simple expression for

$$\Delta F = F(x + \Delta x) - F(x)$$

namely,

$$\Delta F = f(x) \cdot \Delta g + \Delta f \cdot g(x) + \Delta f \cdot \Delta g$$

A simple expression for ΔF

These quantities all have nice geometric interpretations. First, think of the numbers $f(x)$ and $g(x)$ as lengths that depend on x; then $F(x)$ naturally stands for the area of the rectangle whose sides are $f(x)$ and $g(x)$. If the sides of the rectangle grow by the amounts Δf and Δg, then the area F grows by ΔF. As the following diagram shows, ΔF has three parts, corresponding to the three terms in the expression we derived algebraically for ΔF.

Interpret Δf, Δg and ΔF as areas

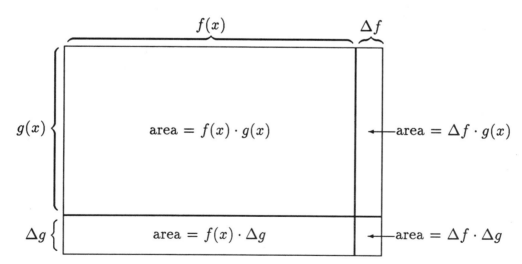

Now we divide ΔF by Δx and finish the argument:

$$\begin{aligned}
\frac{\Delta F}{\Delta x} &= \frac{f(x) \cdot \Delta g + \Delta f \cdot g(x) + \Delta f \cdot \Delta g}{\Delta x} \\
&= f(x) \cdot \frac{\Delta g}{\Delta x} + \frac{\Delta f}{\Delta x} \cdot g(x) + \frac{\Delta f \cdot \Delta g}{\Delta x}
\end{aligned}$$

Consider what happens to each of the three terms as Δx gets smaller and smaller. In the first term, the second factor $\Delta g/\Delta x$ approaches $g'(x)$—by the *definition* of the derivative. The first factor, $f(x)$ doesn't change at all as Δx shrinks. So the first term approaches $f(x) \cdot g'(x)$. Similarly, in the second term, the quotient $\Delta f/\Delta x$ approaches $f'(x)$, and the second term approaches $f'(x) \cdot g(x)$.

Finally, look at the third term. We would know what to expect if we had another factor of Δx in the denominator. We can put ourselves in familiar territory by the "trick" of multiplying the third term by $\Delta x/\Delta x$:

$$\frac{\Delta f \cdot \Delta g}{\Delta x} = \frac{\Delta f}{\Delta x} \cdot \frac{\Delta g}{\Delta x} \cdot \Delta x$$

Thus we can see that as Δx approaches zero, the third term itself approaches $f'(x) \cdot g'(x) \cdot 0 = 0$.

We may summarize our calculation by writing

$$\lim_{\Delta x \to 0} \frac{\Delta F}{\Delta x} = f(x) \cdot \left(\lim_{\Delta x \to 0} \frac{\Delta g}{\Delta x} \right) + \left(\lim_{\Delta x \to 0} \frac{\Delta f}{\Delta x} \right) \cdot g(x)$$
$$+ \left(\lim_{\Delta x \to 0} \frac{\Delta f}{\Delta x} \right) \cdot \left(\lim_{\Delta x \to 0} \frac{\Delta g}{\Delta x} \right) \cdot \left(\lim_{\Delta x \to 0} \Delta x \right)$$

from which we have

$$\lim_{\Delta x \to 0} \frac{\Delta F}{\Delta x} = f(x)g'(x) + f'(x) \cdot g(x) + f'(x) \cdot g'(x) \cdot 0$$
$$= f(x) \cdot g'(x) + f'(x) \cdot g(x).$$

This completes the proof of the product rule. Other formal arguments are left to the exercises.

Exercises

Finding Derivatives

1. Find the derivative of each of the following functions.

a) $3x^5 - 10x^2 + 8$

b) $(5x^{12} + 2)(\pi - \pi^2 x^4)$

c) $\sqrt{u} - 3/u^3 + 2u^7$

d) $mx + b$ (m, b constant)

e) $.5 \sin x + \sqrt[3]{x} + \pi^2$

f) $\dfrac{\pi - \pi^2 x^4}{5x^{12} + 2}$

g) $2\sqrt{x} - \dfrac{1}{\sqrt{x}}$

h) $\tan z \, (\sin z - 5)$

i) $\dfrac{\sin x}{x^2}$

j) $x^2 e^x$

k) $\cos x + e^x$

l) $\sin x / \cos x$

m) $e^x \ln x$

n) $\dfrac{2^x}{10 + \sin x}$

o) $\sin(e^x \cos x)$

p) $6e^{\cos t} / 5\sqrt[3]{t}$

q) $\ln(x^2 + xe^x)$

r) $\dfrac{5x^2 + \ln x}{7\sqrt{x} + 5}$

2. In this problem we examine the effect of stretching or shifting the coordinates of the input variable of a function.

a) Suppose $f(x) = \sin(x)$ and $g(x) = \sin(mx)$, where m is a constant stretching factor. What is the relation between $f'(x)$ and $g'(x)$?

b) As in (a), suppose $f(x) = \sin(x)$, but this time $g(x) = \sin(x + b)$ where b is the size of a (constant) shift. What is the relation between $f'(x)$ and $g'(x)$ this time?

c) Now consider the general case: $f(x)$ in an unspecified differentiable function and $g(x) = f(mx + b)$, where the input variable is stretched by the constant factor m and shifted by the constant amount b. What is the relation between $f'(x)$ and $g'(x)$ in this general case?

3. Which of the following functions has a derivative which is always positive (except at $x = 0$, where neither the function nor its derivative is defined)?

$$1/x \qquad -1/x \qquad 1/x^2 \qquad -1/x^2$$

4. As a function of its radius r, the volume of a sphere is given by the formula $V(r) = \frac{4}{3}\pi r^3$. If r is measured in centimeters, what are the units for $V'(r)$: square cm; cubic cm/hr; hrs/square cm ; cubic cm/cm; cubic cm?

5. Do the following.

a) Show that $\dfrac{1}{1-x^2}$ and $\dfrac{x^2}{1-x^2}$ have the same derivative.

b) If $f'(x) = g'(x)$ for every x, what can be concluded about the relationship between f and g? (Hint: What is $(f(x) - g(x))'$?)

6. Suppose that the current total daily energy consumption in a particular country is 16×10^{13} BTU and is rising at the rate of 6×10^{11} BTU per year. Suppose that the current population is 2×10^8 people and is rising at the rate of 10^6 people per year. What is the current daily per capita energy consumption? Is it rising or falling? By how much?

7. The population of a particular country is 15,000,000 people and is growing at the rate of 10,000 people per year. In the same country the per capita yearly expenditure for energy is $1,000 per person and is growing at the rate of $8 per year. What is the country's current total yearly energy expenditure? How fast is the country's total yearly energy expenditure growing?

8. The population of a particular country is 30 million and is rising at the rate of 4,000 people per year. The total yearly personal income in the country is 20 billion dollars, and it is rising at the rate of 500 million dollars per year. What is the current per capita personal income? Is it rising or falling? By how much?

9. An explorer is marooned on an iceberg. The top of the iceberg is shaped like a square with sides of length 100 feet. The length of the sides is shrinking at the rate of two feet per day. How fast is the area of the top of the iceberg shrinking? Assuming the sides continue to shrink at the rate of two feet per day, what will be the dimensions of the top of the iceberg in five days? How fast will the area of the top of the iceberg be shrinking then?

10. Suppose the iceberg of problem 9 is shaped like a cube. How fast is the volume of the cube shrinking when the sides have length 100 feet? How fast after five days?

Deriving Differentiation Rules

11. In this problem we calculate the derivative of $f(x) = x^4$.

a) Expand $f(x+\Delta x) = (x+\Delta x)^4 = (x+\Delta x)(x+\Delta x)(x+\Delta x)(x+\Delta x)$ as a sum of 12 terms. (Don't collect "like" terms yet.)

b) How many terms in part a involve *no* Δx's? What form do such terms have?

c) How many terms in part a involve exactly *one* Δx? What form do such terms have?

d) Group the terms in part a so that $f(x + \Delta x)$ has the form

$$Ax^4 + B\Delta x + R(\Delta x)^2 \, ,$$

where there are no Δx's among the terms in A or B, but R has several terms, some involving Δx. Use part b to check your value of A; use part c to check your value of B.

e) Compute the quotient $\dfrac{f(x + \Delta x) - f(x)}{\Delta x}$, taking advantage of part d.

f) Now find

$$\lim_{n \to 0} \frac{f(x + \Delta x) - f(x)}{\Delta x} \, .$$

This is the derivative of x^4. Is your result here compatible with the rule for the derivative of x^n ?

12. In this problem we calculate the derivative of $f(x) = x^n$, where n is any positive integer.

a) First show that you can write

$$f(x + \Delta x) = x^n + nx^{n-1}\Delta x + R(\Delta x)^2$$

by developing the following line of argument. Write $(x + \Delta x)^n$ as a product of n identical factors:

$$(x + \Delta x)^n = \underbrace{(x + \Delta x)}_{\text{1-st}}\underbrace{(x + \Delta x)}_{\text{2-nd}}\underbrace{(x + \Delta x)}_{\text{3-rd}}\ldots\underbrace{(x + \Delta x)}_{n\text{-th}}$$

But now, before tackling this general case, look at the following examples. In the examples we use notation to help us keep track of which factors are contributing to the final result.

i) Consider the product $(a + b)(\underline{a} + \underline{b}) = a\underline{a} + a\underline{b} + b\underline{a} + b\underline{b}$. There are four individual terms. Each term contains one of the entries in the first factor (namely a or b) and one of the entries in the second factor (namely $\underline{a}$ or $\underline{b}$). The four terms represent thereby all possible ways of choosing one entry in the first factor and one entry in the second factor.

ii) Multiply out the product $(a + b)(\underline{a} + \underline{b})(A + B)$. (Don't combine like terms yet.) Does each term contain one entry from the first factor, one from the second, and one from the third? How many terms did you get? In fact there are two ways to choose an entry from the first factor, two ways to choose an entry from the second factor, and two ways to choose an entry from the third factor. Therefore, how many ways can you make a choice consisting of one entry from the first, one from the second, and one from the third?

Now return to the general case:

$$(x + \Delta x)^n = \underbrace{(x + \Delta x)}_{\text{1-st}}\underbrace{(x + \Delta x)}_{\text{2-nd}}\underbrace{(x + \Delta x)}_{\text{3-rd}}\ldots\underbrace{(x + \Delta x)}_{n\text{-th}}$$

How many ways can you choose an entry from each factor and *not* get any Δx's? Multiply these chosen entries together; what does the product look like (apart from having no Δx's in it)?

How many ways can you choose an entry from each factor in such a way that the resulting product has *precisely one* Δx? Describe all the various choices which give that result. What does a product that contains precisely one Δx factor look like? What do you obtain for the sum of *all* such terms with precisely one Δx factor?

What is the minimum number of Δx factors in any of the remaining terms in the full expansion of $(x + \Delta x)^n$?

Do your calculations agree with this summary:

$$(x + \Delta x)^n = x^n + nx^{n-1}\Delta x + R(\Delta x)^2 \ ?$$

b) Now find the value of $\dfrac{f(x + \Delta x) - f(x)}{\Delta x}$.

c) Finally, find

$$\lim_{\Delta x \to 0} \frac{f(x + \Delta x) - f(x)}{\Delta x} \ .$$

Do you get nx^{n-1}?

13.   In this problem we give another derivation of the power rule based on writing

$$x^r = e^{r \ln(x)}.$$

Use the chain rule to differentiate $e^{r \ln(x)}$. Explain why your answer equals rx^{r-1}.

14.   Does the rule for the derivative of x^r hold for $r = 0$? Why or why not?

15.   In this exercise we prove the Addition Rule: $F(x) = f(x) + g(x)$ implies $F'(x) = f'(x) + g'(x)$.
a) Show $F(x + \Delta x) - F(x) = f(x + \Delta x) - f(x) + g(x + \Delta x) - g(x)$
b) Divide by Δx and finish the argument.

16.   In this exercise we prove the Quotient Rule: $F(x) = f(x)/g(x)$ implies

$$F'(x) = \frac{g(x)f'(x) - f(x)g'(x)}{(g(x))^2}$$

a) Rewrite $F(x) = f(x)/g(x)$ as $f(x) = g(x)F(x)$. Pretend for the moment that you know what $F'(x)$ is and apply the Product Rule to find $f'(x)$ in terms of $F(x)$, $g(x)$, $F'(x)$, $g'(x)$.
b) Replace $F(x)$ by $f(x)/g(x)$ in your expression for $f'(x)$ in part a.
c) Solve the equation in part b for $F'(x)$ in terms of $f(x)$, $g(x)$, $f'(x)$ and $g'(x)$.

17.   In this problem we calculate the derivative of $f(x) = x^n$ when n is a negative integer. First write $n = -m$, so m is a positive integer. Then $f(x) = x^{-m} = 1/x^m$.
a) Use the Quotient Rule and this new expression for f to find $f'(x)$.
b) Do the algebra to re-express $f'(x)$ as nx^{n-1}.

18.   In this problem we calculate the derivatives of $\sin x$ and $\cos x$. We will need the **addition formulas**:

$$\sin(A + B) = \sin A \cos B + \cos A \sin B$$
$$\cos(A + B) = \cos A \cos B - \sin A \sin B$$

First tackle $f(x) = \sin x$:

a) Use the addition formula for $\sin(A + B)$ to rewrite $f(x + \Delta x)$ in terms of $\sin(x)$, $\cos(x)$, $\sin(\Delta x)$, and $\cos(\Delta x)$.

b) The quotient $\dfrac{f(x + \Delta x) - f(x)}{\Delta x}$ can now be written in the form

$$P \cdot \sin x + Q \cdot \cos x \, ,$$

where P and Q are specific functions of Δx. What are the formulas for those functions?

c) Use a calculator or computer to estimate the limits

$$\lim_{\Delta x \to 0} P \quad \text{and} \quad \lim_{\Delta x \to 0} Q \; .$$

(Try $\Delta x = .1, .01, .001, .0001$. Be sure your calculator is set on radians, not degrees.) Using part b you should now be able to determine the limit

$$\lim_{\Delta x \to 0} \frac{f(x + \Delta x) - f(x)}{\Delta x}$$

by writing it in the form

$$\left(\lim_{\Delta x \to 0} P \right) \cdot \sin x + \left(\lim_{\Delta x \to 0} Q \right) \cdot \cos x \; .$$

d) What is $f'(x)$?

e) Proceed similarly to find the derivative of $g(x) = \cos x$.

19.   Here is a geometric argument showing again (see problem 18) that the derivative of $\sin t$ is $\cos t$ and the derivative of $\cos t$ is $- \sin t$. Imagine a point moving counterclockwise around a unit circle at a speed of one unit per minute. Since the point moves one unit along the circumference per unit time, a radius to the moving point sweeps out one radian per unit time. Assuming the point begins at time $t = 0$ at $(1,0)$, t is the measure in radians of the angle swept out by a radius to the point at time t. The coordinates of the point at time t are thus $(\cos t, \sin t)$. In other words, the x-coordinate of the point is $x = \cos t$ and the y-coordinate is $y = \sin t$. The goal of this problem is to find the rates at which x and y are changing with respect to t.

At a particular moment t, in what direction is the point moving? A tiny person standing on the moving point and facing in the direction of motion would be looking along a line tangent to the circle, pointing counter-clockwise.

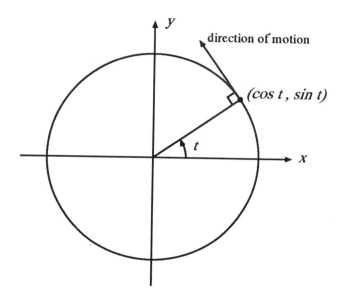

The length of the tangent vector is equal to the speed of the motion, namely one. We also know the tangent to a circle at a point is perpendicular to the radius to that point.

Imagine a right triangle drawn with hypotenuse equal to the tangent vector and legs representing the magnitude and direction of the rate of change of x and y respectively. (See the figure on the next page.)

a) Use geometric arguments to show the length of the horizontal leg is $\sin t$ and the length of the vertical leg is $\cos t$.

b) Finish off with appropriate sign analyses to show

$$\sin' t \;=\; \cos t \text{ and}$$
$$\cos' t \;=\; -\sin t.$$

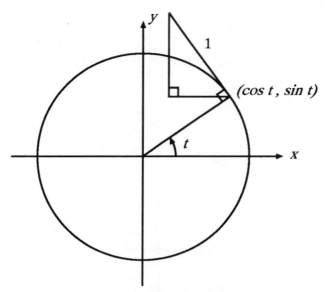

20.   In this problem we calculate the derivatives of the other circular functions. Use the quotient rule together with the derivatives of $\sin x$ and $\cos x$ to verify that the derivatives of the other four circular functions are as given in the table below:

function	derivative
$\tan x = \dfrac{\sin x}{\cos x}$	$\sec^2 x$
$\csc x = \dfrac{1}{\sin x}$	$-\cot x \csc x$
$\sec x = \dfrac{1}{\cos x}$	$\sec x \tan x$
$\cot x = \dfrac{1}{\tan x}$	$-\csc^2 x$

Differential Equations

21.   If $y = f(x)$ then the **second derivative** of f is just the derivative of the derivative of f; it is denoted $f''(x)$ or d^2y/dy^2. Find the second derivative of each of the following functions.

a) $f(x) = e^{3x-2}$

b) $f(x) = \sin \omega x$, where ω is a constant

c) $f(x) = x^2 e^x$

22.   Show that e^{3x} and e^{-3x} both satisfy the (*second order*) differential equation

$$f''(x) = 9f(x).$$

Furthermore, show that *any* function of the form $g(x) = \alpha e^{3x} + \beta e^{-3x}$ satisfies this differential equation. Here α and β are arbitrary constants. Finally, choose α and β so that $g(x)$ also satisfies the two conditions $g(0) = 12$ and $g'(0) = 15$.

23.   Show that $y = \sin x$ satisfies the differential equation $y'' + y = 0$. Show that $y = \cos x$ also satisfies the differential equation. Can you find a function $g(x)$ that satisfies these three conditions:

$$
\begin{aligned}
g''(x) + g(x) &= 0 \\
g(0) &= 1 \\
g'(0) &= 4?
\end{aligned}
$$

24.   Show that $\sin \omega x$ satisfies the differential equation $y'' + \omega^2 y = 0$. What other solutions can you find to this differential equation? Can you find a function $\phi(x)$ that satisfies these three conditions:

$$
\begin{aligned}
\phi''(x) + 4\phi(x) &= 0 \\
\phi(0) &= 36 \\
\phi'(0) &= 64?
\end{aligned}
$$

The Colorado River Problem

·   Make your answer to this sequence of questions an essay. Identify all the variables you consider (e.g., "A stands for the area of the lake"), and indicate the functional relationships between them ("A depends on time t, measured in weeks from the present"). Identify the derivatives of those functions, as necessary.

The Colorado River—which excavated the Grand Canyon, among others—used to empty into the Gulf of California. It no longer does. Instead, it runs into a marshy area some miles from the Gulf and stops. One of the major reasons for this change is the construction of dams— notably the Hoover Dam. Every dam creates a lake behind it, and every lake increases the total surface area of the river. Since the rate at which water evaporates is proportional to the area of the water surface exposed to air, the lakes along the Colorado have increased the loss of river water through evaporation. Over the years, these losses (in conjunction with other factors, like increased usage by a rapidly growing population) have been significant enough to dry up the river at its mouth.

25. Let us analyze the evaporation rate along a river that was recently dammed. Suppose the lake is currently 50 yards wide, and getting wider at a rate of 3 yards per week. As the lake fills, it gets longer, too. Suppose it is currently 950 yards long, and it is extending upstream at a rate of 15 yards per week. Assuming the lake remains approximately rectangular as it grows, find

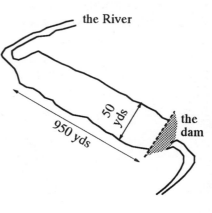

a) the current area of the lake, in square yards;

b) the rate at which the surface of the lake is currently growing, in square yards per week.

26. Suppose the lake continues to spread sideways at the rate of 3 yards per week, and it continues to extend upstream at the rate of 15 yards per week.

a) Express the area of the lake as a (quadratic!) function of time, where time is measured from the present, in weeks, and where the lake's area is as given in problem 19.

b) How many weeks will it take for the lake to cover 30 acres (= 145,200 square yards)?

c) At what rate is the lake surface growing when it covers 30 acres?

27. Compare the rates at which the surface of the lake is growing in problem 24 (which is the "current" rate) and in problem 25 (which is the rate when the lake covers 30 acres). Are these rates the same? If they are not, how do you account for the difference? In particular, the width and length grow at fixed rates, so why doesn't the area? Use what you know about derivatives to answer the question.

28. Suppose the local climate causes water to evaporate from the surface of the lake at the rate of 0.22 cubic yards per week, for each square

yard of surface. Write a formula that expresses total evaporation in terms of area. Use E to denote total evaporation.

29.   The lake is fed by the river, and that in turn is fed by rainwater and groundwater from its watershed. (The **watershed**, or basin, of a river is that part of the countryside containing the ponds and streams which drain into the river.) Suppose the watershed provides the lake, on average, with 25,000 cubic yards of new water each week.

Assuming, as we did in problem 25, that the lake widens at the constant rate of 3 yards per week, and lengthens at the rate of 15 yards per week, will the time ever come that the water being added to the lake from its watershed balances the water being removed by evaporation? In other words, will the lake ever stop filling?

§2.   Finding Partial Derivatives

We know from Chapter 3 that no additional formulas are needed to calculate *partial* derivatives. We simply use the usual differentiation formulas, treating all the variables except one—the one with respect to which the partial derivative is formed—as if they were constants. If we do this we get new techniques for analyzing rate of change in problems that involve functions of several variables.

Some Examples

Here are two examples to illustrate the technique for calculating partial derivatives:

1. Suppose $f(x, y) = x^2 y + 5x^3 - \sqrt{x + y}$. Then

$$f_x(x, y) \; = \; 2xy + 15x^2 - \frac{1}{2\sqrt{x + y}}, \quad \text{and}$$

$$f_y(x, y) \; = \; x^2 - \frac{1}{2\sqrt{x + y}}.$$

Finding formulas for partial derivatives

2. Suppose $g(u, v) = e^{uv} + \dfrac{u}{v}$. Then

$$g_u(u, v) \; = \; v e^{uv} + \frac{1}{v}, \quad \text{and}$$

$$g_v(u, v) \; = \; u e^{uv} - \frac{u}{v^2}.$$

Eradication of Disease

Controlling—or, better still, eradicating—a communicable disease depends first on the development of a vaccine. But even after this step has been accomplished, public health officials must still answer important questions, including:

- What proportion of the population must be vaccinated in order to eliminate the disease?

- At what age should people be vaccinated?

In their 1982 article, "Directly Transmitted Infectious Diseases: Control by Vaccination," (*Science*, Vol. 215, 1053–1060), Roy Anderson and Robert May formulate a model for the spread of disease that permits them to answer these and other questions. For a particular disease in a particular environment, the important variables in their model are

1. The average human life expectancy L, in years;

2. The average age A at which individuals catch the disease, in years;

3. The average age V at which individuals are vaccinated against the disease, in years.

Anderson and May deduce from their model that in order to eradicate the disease, the proportion of the population that is vaccinated must exceed p, where p is given by

$$p = \frac{L + V}{L + A}.$$

Partial derivatives can tell us which variables are most significant

For a disease like measles, public health officials can directly affect the variable V, for example by the recommendations they make to physicians about immunization schedules for children. They may also indirectly affect the variables A and L, because public health policy influences factors which can modify the age at which children catch the disease or the overall life expectancy of the population. (Many other factors affect these variables as well.) Which of these three variables has the greatest effect on the proportion of the population that must be vaccinated?

In other words, which is largest: $\partial p/\partial L$, $\partial p/\partial A$, or $\partial p/\partial V$? Using the rules, we compute:

$$\frac{\partial p}{\partial L} = \frac{(L+A)-(L+V)}{(L+A)^2} = \frac{A-V}{(L+A)^2},$$

$$\frac{\partial p}{\partial A} = \frac{-(L+V)}{(L+A)^2}, \quad \text{and}$$

$$\frac{\partial p}{\partial V} = \frac{1}{L+A}.$$

For measles in the United States, reasonable values of the variables are $L = 70$ years, $A = 5$ years and $V = 1$ year. Using these values, the partial derivatives are

$$\frac{\partial p}{\partial L} = \frac{4}{(75)^2} = .0007,$$

$$\frac{\partial p}{\partial A} = \frac{-71}{(75)^2} = -.0126,$$

$$\frac{\partial p}{\partial V} = \frac{1}{75} = .0133.$$

A comment is in order here on units. While the input variables L, A and V are all measured in years—so the rates are *per year*, the output variable p is dimensionless: it is the ratio of persons vaccinated to persons not vaccinated. It would be reasonable to write p as a percentage. Then we can attach the units *percent per year* to each of the three partial derivatives. Thus we have:

Determining units

$$\frac{\partial p}{\partial L} = .07\% \quad \text{per year}$$

$$\frac{\partial p}{\partial A} = -1.26\% \quad \text{per year}$$

$$\frac{\partial p}{\partial V} = 1.33\% \quad \text{per year.}$$

It is not surprising that a change in average life expectancy has a negligible effect on the proportion p of the population that must be vaccinated in order to eradicate measles. Nor is it surprising that changing the age of vaccination has the greatest effect on p. But it is not obvious ahead of time that changing the age at which children catch the disease has nearly as large an effect on p:

- Decreasing the age of vaccination decreases the proportion p by 1.33% per year of decrease.

- Increasing the age at which children catch measles decreases the proportion p by 1.26% per year of increase.

Changes can also go the "wrong" way. For example, in an area where use of communal child care facilities is growing, contact among very young children increases, and the age at which children are exposed to—and can catch—communicable diseases like measles falls. The Anderson-May model tells us that immunization practices must change to compensate: either the age of vaccination must drop a like amount, or the fraction of the population that is vaccinated must grow by 1.26% per year of decrease in the average age at infection.

Exercises

Finding Partial Derivatives

1. Find the partial derivatives of the following functions.

a) $x^2 y$.

b) $\sqrt{x + y}$

c) e^{xy}

d) $\dfrac{y}{x}$

e) $\dfrac{x + y}{y + z}$

f) $\sin \dfrac{y}{x}$

2. Suppose $f(x, y) = x^2 y + 5x^3 - \sqrt{x + y}$. Find $f_x(x, y)$ and $f_y(x, y)$.

3. Suppose $g(u, v) = e^{uv} + \dfrac{u}{v}$. Find $g_u(u, v)$ and $g_v(u, v)$.

4. The **second partial derivatives** of $z = f(x, y)$ are the partial derivatives of $\partial f / \partial x$ and $\partial f / \partial y$, namely:

$$\frac{\partial^2 f}{\partial x^2} = \frac{\partial}{\partial x}\left(\frac{\partial f}{\partial x}\right)$$

$$\frac{\partial^2 f}{\partial x \partial y} = \frac{\partial}{\partial x}\left(\frac{\partial f}{\partial y}\right)$$

$$\frac{\partial^2 f}{\partial y^2} = \frac{\partial}{\partial y}\left(\frac{\partial f}{\partial y}\right)$$

Find the three second partial derivatives of the following functions.

a) $x^2 y$.

b) $\sqrt{x+y}$

c) e^{xy}

d) $\dfrac{y}{x}$

e) $\sin \dfrac{y}{x}$

Partial differential equations

5. Show that the function $z = \dfrac{1}{\sqrt{t}} \exp \dfrac{-x^2}{4t}$ satisfies the **partial differential equation**

$$\frac{\partial^2 z}{\partial x^2} = \frac{\partial z}{\partial t}.$$

6. Show that *every* linear function of the form $z = px + qy + c$ satisfies the partial differential equation

$$\frac{\partial^2 z}{\partial x^2} + \frac{\partial^2 z}{\partial y^2} = 0.$$

Here p, q, and c are arbitrary constants.

7. Show that the function $z = e^x \sin y$ also satisfies the partial differential equation

$$\frac{\partial^2 z}{\partial x^2} + \frac{\partial^2 z}{\partial y^2} = 0.$$

§3.   The Shape of the Graph of a Function

We know from chapter 3 that the derivative gives us qualitative information about the shape of the graph of a differentiable function.

function	derivative
increasing	positive
decreasing	negative
level	zero
steep (rising or falling)	large (positive or negative)
gradual (rising or falling)	small (positive or negative)
straight	constant

Having a formula for the derivative of a function will thus give us a great deal of information about the behavior of the function itself. In particular we will be interested in using the derivative to solve **optimization problems**—finding maximum or minimum values of a function. Such problems occur frequently in many fields.

Contexts for optimization problems

- Economists actually define human rationality in terms of optimization. Each person is assumed to have a *utility function*, a function that assigns to each of many possible outcomes its *utility*, a numerical measure of its value to her. (Different people may have different utility functions, depending on their personal value sistems.) A rational person is one who acts to maximize her utility. Some utilities are expressed in terms of money. For example, a rational manufacturer will seek to maximize her profit (in dollars). Her profit will depend on—that is, be a function of—such variables as the cost of her raw materials and the unit price she charges for her product.

- Many physical laws are expressed as minimum principles. Ordinary soap bubbles exhibit one of these principles. A soap film has a *surface energy* which is proportional to its surface area. For almost any physical system, its stable state is one which minimizes its energy. Stable soap films are thus examples of *minimal surfaces*. Interfaces involving crystals also have surface energies, leading to the study of crystaline minimal surfaces.

- Statisticians develop mathematical summaries for data—in other words, mathematical models. For example, a relationship between two numerical variables may be summarized by a linear function, say $y = mx + b$, where x and y are the variables of interest. It would be very rare to find data that were exactly linear. In a particular case, the statistician chooses the linear model that minimizes the *discrepancy* between the actual values of y and the theoretical values obtained from the linear function. Statisticians frequently measure this discrepancy by summing the squares of the differences between the actual and the theoretical values of y for each data point. The *best-fitting line* or *regression line* is the graph of the linear function which is optimal in this sense.

- Psychologists who study decision-making have found that some people are "risk averse"; they make their decisions primarily to avoid risks. If we regard risk as a function of the various outcomes under consideration (a bit like a utility function), such a person acts to minimize this function.

The derivative is the key tool here. We will develop a general procedure for using the derivative of a function to locate its extremes.

Language

Here is a graph of what we might consider a "generic" differentiable function.

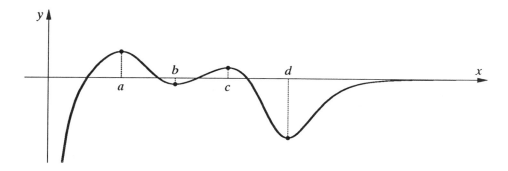

The most distinctive features are the hill tops and valley bottoms, points where the graph levels and the derivative is zero. We distinguish between **local** extremes, like those occurring at the points $x = b$, $x = c$, and $x = d$ and a **global** extreme, like the global maximum at the point $x = a$. The function has a **local minimum** at $x = b$ because $f(x) \geq f(b)$ for all x *sufficiently near* b. The function has a **global maximum** at $x = a$ because $f(x) \leq f(a)$ for *all* x. Notice that this particular function does not have a global minimum. What kinds of

Local extremes
and global
extremes

local extremes does the function have at $x = c$ and $x = d$? The convention is to say that *all* extremes are local extremes, and a local extreme may or may not also be a global extreme.

Some functions have no extremes, even if their derivative equals zero

Examining the graph of as simple a function as $f(x) = x^3$ shows us that a function need not have any extremes at all. Moreover, since $f'(0) = 0$ for this function, a zero derivative doesn't necessarily identify a point where an extreme occurs.

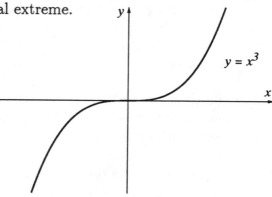

A minimum can occur at a cusp

Can a function have an extreme at a point *other* than where the derivative is zero? Consider the graph of $f(x) = x^{2/3}$ below.

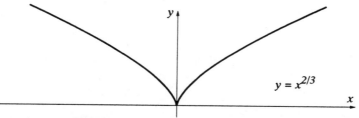

This function is differentiable everywhere except at the point $x = 0$. And it is at this very point, where

$$f'(x) = \frac{2}{3}x^{-1/3} = \frac{2}{3\sqrt[3]{x}}$$

is undefined, that the function has its global minimum. For this reason, points where the derivative fails to exist (or is infinite) are as important as points where the derivative equals zero. All of these kinds of points are called **critical points** for the function.

> A **critical point** for a function f is a point
> where f' equals zero or or infinity or fails to exist.

How can we recognize, from looking at its graph, that a function fails to be differentiable at a point? We know that when the graph has a sharp corner or **cusp** it isn't locally linear at that point, and so has no derivative there. Thus critical points occur at these places. When the graph is locally linear but vertical, the slope cannot be given a finite numerical value. Critical points also occur at these vertical places where the slope is infinite. The graph at the right is an example of such a curve.

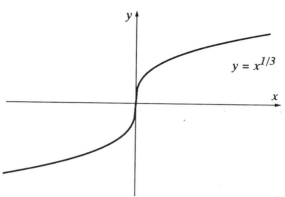

$y = x^{1/3}$

A weaker condition than differentiability, but one that is useful, especially in this context, is **continuity**. We say that a function is **continuous at a point** if

- it is defined at the point, and

- in a neighborhood of the point, small changes in the input produce small changes in the output.

A function is continuous on a set of real numbers if it is continuous at each point of the set. A natural way to think of (and recognize) a continuous function on an interval is by its graph on that interval, which is continuous in the usual sense: it has no gaps or jumps in it. You can draw it without picking up your pencil. Of the four functions whose graphs appear on the next page, f is continuous (and differentiable); g is continuous, but not differentiable (because of the cusp at $x = a$); h is not continuous, because h is undefined at $x = b$; and k is not continuous, because of the "jump" at $x = c$.

Graphs of continuous functions have no gaps or jumps

Luckily, the functions that we are likely to encounter are continuous on their natural domains. Among functions given by formulas, the only exceptions we have to worry about are quotients, like $f(x) = 1/x$, which have gaps in their natural domains at points where the denominator vanishes (at $x = 0$ in the case of $f(x) = 1/x$), so their domains are not single intervals. For convenience, we usually confine our attention to functions continuous on an interval.

A quotient isn't continuous where its denominator is zero

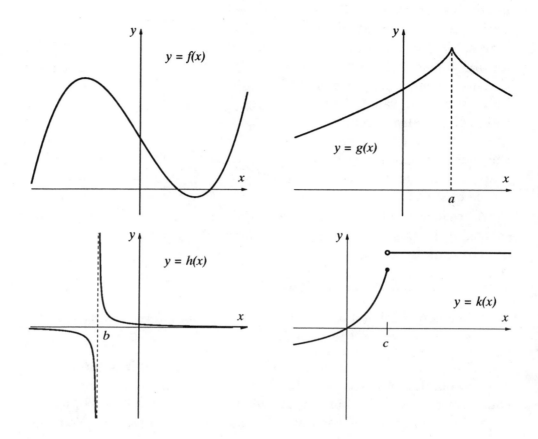

The Existence of Extremes

Not every function has extremes, as the example of $f(x) = x^3$ shows. We are, however, guaranteed extremes for certain functions.

A continuous function on a finite closed interval has extremes

Principle I. We are guaranteed that a function has a global maximum and a global minimum if we know:

- the domain of the function is a *finite closed interval*, and

- the function is *continuous* on this interval.

The domain restriction in most optimization problems is likely to come from the physical constraints of the problem, not the mathematics.

A **finite closed interval**, written $[a, b]$, is the set of all real numbers between a and b, including the endpoints a and b. Other kinds of finite intervals are $(a, b]$ (which excludes the endpoint a), $[a, b)$, and (a, b).

Infinite intervals include open and closed "rays" like $(a, +\infty)$, $[a, +\infty)$, $(-\infty, b)$ and $(-\infty, b]$, and the entire real line.

For example, Principle I does not apply to $f(x) = 1/x$ on the finite closed interval $[-1, 2]$, because this function isn't continuous at every point on this interval—in fact, the function isn't even defined for $x = 0$.

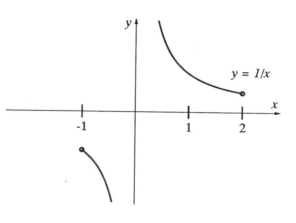

However, Principle I *does* apply to the same function if we change the finite closed interval to one that doesn't include $x = 0$. In the figure below we use $[1, 3]$.

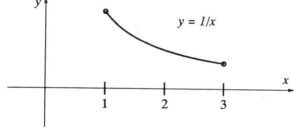

A function which fails to satisfy the first condition of Principle I can still have global extremes. For any function continuous on an interval we can apply the following principle.

Principle II. If a function f is continuous on an interval and has a local extreme at a point $x = c$ of the interval, then either $x = c$ is a critical point for the function, or $x = c$ is an endpoint of the interval. In other words, we are guaranteed that one of the following three conditions holds:

Continuous functions have extremes only at endpoints or critical points

- $f'(c) = 0$,

- $f'(c)$ is undefined,

- $x = c$ is an endpoint of the interval.

The graphs on the next page illustrate three local maxima, each satisfying one of these three conditions.

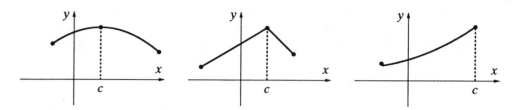

When we apply Principle II to optimization problems, an important part of the task will be ascertaining which, if any, of the critical points or endpoints we find actually gives the extreme we're looking for. We'll examine a variety of techniques, graphical and analytical, for locating critical points and determining what kind of extreme point (if any) they are.

Finding Extremes

Using a graphical approach

Computer graphing can be easy if the general location of the extremes is known

If we can use a computer to examine the graph of the function of interest, we can determine the existence and location of extremes by inspection. However, every graphing utility requires the user to specify the interval on which the function will be graphed, and careful analysis may be required in order to choose an interval that contains all the extremes of interest.

For functions given by data, whose graphs have only finitely many points, we can zoom in to find the exact coordinates of the extreme datapoints. For a function given by a formula, we can estimate the coordinates of an extreme to arbitrary accuracy by zooming in on the point as closely as desired. This is the method we used in some of the exercises of Chapter 1, and it is quite satisfactory in many situations.

Using the formula for the derivative.

Formulas can give exact answers and can handle parameters

In this chapter we are concentrating on functions given by formulas. For these functions we may want a method other than the approximation using a graphing utility described above.

- For some functions, the determination of extremes using a formula for the derivative is at least as easy as using the computer.

- Some functions are described in terms of a *parameter*, a constant whose value may vary from one problem to another. For example, the rate equation from the S-I-R model for change in the number of infected, $I' = aSI - bI$, involves two parameters a and b, the transmission and recovery coefficients. For such a function, we cannot use the computer unless we specify a numerical value for each parameter, thus limiting the generality of our results.

- The computer gives only an approximation, while a precise answer may convey important additional information.

We will assume that the function we are studying is continuous on the domain of interest and that its domain is an interval. Our procedure for finding extremes of a function given by a formula $y = f(x)$ is thus a direct application of Principle II.

The search for local extremes

1. Determine the domain of the function and identify its endpoints, if any. Keep in mind that in an applied context, the domain may be determined by physical or other restrictions.

2. Find a formula for $f'(x)$.

3. Find any roots of $f'(x) = 0$ in the domain. An estimation procedure, for example Newton's Method (see the end of this chapter), may be used for complicated equations.

4. Find any points in the domain where $f'(x)$ is undefined.

5. Determine the shape of the graph to locate any local extremes. Find the shape either by looking directly at the graph of $y = f(x)$ or by analyzing the sign of $f'(x)$ on either side of each critical point $x = c$. (Sometimes the second derivative $f''(c)$ aids the analysis; see the exercises for details.)

The search for global extremes

1. Find the local extremes, as above.

2. If the domain is a finite closed interval, it is only necessary to compare the values of f at each of the critical points and at the endpoints to determine which is the global maximum and which is the global minimum.

3. More generally, use the shape of the graph to ascertain whether the desired global extreme exists and to identify it.

In the succeeding sections and in the exercises we will carry out these search procedures in a variety of situations. Since there are often substantial algebraic difficulties in analyzing the sign of the derivative, or even determining when it is equal to zero, in realistic problems, in section 6 we will develop some numerical methods for handling such complications.

Exercises

Describing functions

1.   For each of the following graphs, is the function continuous? Is the function differentiable?

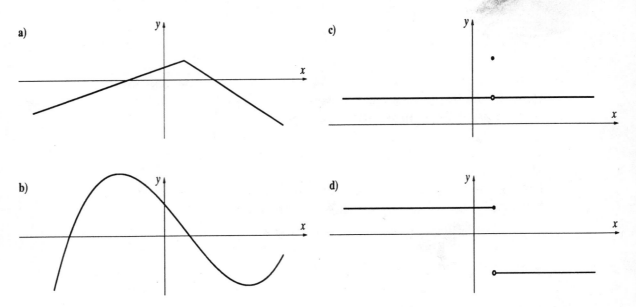

2.   For each of the following graphs of a function $y = f(x)$, is f' increasing or decreasing? At the indicated point, what is the sign of f''? What is the sign of f'?

a)
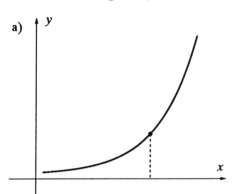

b)
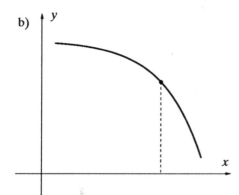

c)
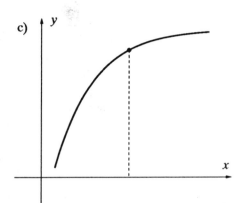

d)
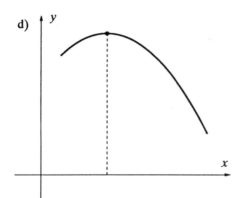

e)
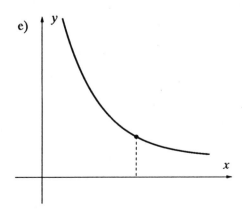

f)
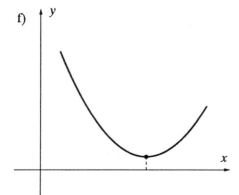

3.   For each of the following, sketch a graph of $y = f(x)$ that is consistent with the given information. On each graph, mark any critical points or extremes.

a) $f'(x) > 0$ for $x < 1$; $f'(1) = 0$; $f'(x) < 0$ for $1 < x < 2$; $f'(2) = 0$; $f'(x) > 0$ for $x > 2$.

b) $f'(x) > 0$ for $x < 2$; $f'(2) = 0$; $f'(x) > 0$ for $x > 2$.

c) $f'(x) > 0$ for $x < 2$; $f'(2) = 0$; $f'(x) < 0$ for $x > 2$.

d) $f'(3) = 0$; $f''(3) > 0$.

4.   **Second derivative test**

- If $f'(c) = 0$ and $f''(c) > 0$, then f has a *local minimum* at $x = c$.

- If $f'(c) = 0$ and $f''(c) < 0$, then f has a *local maximum* at $x = c$.

Using problem 2 as a guide, explain why the second derivative test works.

Finding critical points

5.   For each of the following functions, find the critical points, if any.

a) $f(x) = x^{1/3}$

b) $f(x) = x^3 + \dfrac{3}{2}x^2 - 6x + 5$

c) $f(x) = \dfrac{2x + 1}{x - 1}$

d) $f(x) = \sin x$

e) $f(x) = \sqrt{1 - x^2}$

f) $f(x) = \dfrac{e^x}{x}$

g) $f(x) = x \ln x$

h) $f(x) = x^c + \dfrac{1}{x^c}$ where c is some constant

6. For each of the following graphs, mark any critical points or extremes. Indicate which extremes are local and which are global.

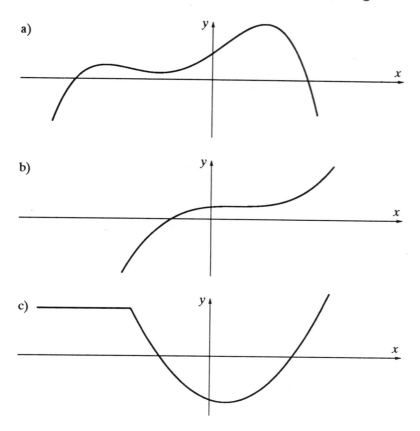

a)

b)

c)

Finding extremes

7.   For what positive value of x does $f(x) = x + \dfrac{7}{x}$ attain its minimum value? Explain how you found this value.

8.   For what value of x in the interval $[1, 2]$ does $f(x) = x + \dfrac{7}{x}$ attain its minimum value? Explain how you found this value.

9.   Use a graphing program to make a sketch of the function $y = f(x) = x^2 2^{-x}$ on the interval $0 \le x \le 10$. From the graph, estimate the value of x which makes y largest, accurately to four decimal places. Then find where y takes on its maximum by setting the derivative $f'(x)$ equal to 0. You should find that the maximum occurs at $x = 2/\ln 2$.

Finally, what is the numerical value of this estimate of $2/\ln 2$, accurate to seven decimal places?

10.   Use a graphing program to sketch the graph of $y = \dfrac{1}{x^2} + x$ on the interval $0 < x < 4$ and estimate the value of x that makes y smallest on this interval. Then use the derivative of y to find the exact value of x that makes y smallest on the same interval. Compare the estimated and exact values.

11.   What is the smallest value $y = \dfrac{4}{x^2} + x$ takes on when x is a positive number? Explain how you found this value.

12.   The function $y = x^4 - 42x^2 - 80x$ has two local minima. Where are they (that is, what are their x coordinates), and what are their values? Which is the lower of the two minima? In this problem you will have to solve a cubic equation. You can use an estimation procedure, but it is also possible to solve by factoring the cubic. (The roots are integers.)

13.   The function $y = x^4 - 6x^2 + 7$ has two local minima. Where are they, and which of the two is lower? In this problem you will have to solve a cubic equation. Do this by factoring and then approximating the x values to 4 decimal places.

14.   Sketch the graph of $y = x \ln x$ on the interval $0 < x < 1$. Where does this function have its minimum, and what is the minimum value?

15.   Let x be a positive number; if $x > 1$ then the cube of x is larger than its square. However, if $0 < x < 1$ then the square is larger than the cube. How *much* larger can it be; that is, what is the greatest amount by which x^2 can exceed x^3? For which x does this happen?

16.   By how much can x^p exceed x^q, when $0 < p < q$ and $0 < x$? For which x does this happen?

§4. Optimal Shapes

The Problem of the Optimal Tin Can

Suppose you are a tin can manufacturer. You must make a can to hold a certain volume V of canned tomatoes. Naturally, the can will be a cylinder, but the proportions, the height h and the radius r, can vary. Your task is to choose the proportions so that you use the least amount of tin to make the can.

What is the minimum surface area?

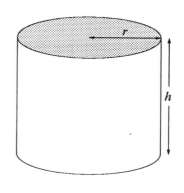

In other words, you want the surface area of the can to be as small as possible.

The Solution

The surface area is the sum of the areas of the two circles at the top and bottom of the can, plus the area of the rectangle that would be obtained if the top and bottom were removed and the side cut vertically.

Thus A depends on r and h:

$$A = 2\pi r^2 + 2\pi r h.$$

However, r and h cannot vary independently. Because the volume V is fixed, r and h are related by

Finding A as a function of r

$$V = \pi r^2 h.$$

Solving the equation above for h in terms of r,

$$h = \frac{V}{\pi r^2},$$

we may express A as a function of r alone:

$$A = f(r) = 2\pi r^2 + 2\pi r \frac{V}{\pi r^2} = 2\pi r^2 + \frac{2V}{r}$$

V is a parameter   Notice that the formula for this function involves the parameter V. The mathematical description of our task is to find the value of r that makes $A = f(r)$ a minimum.

Finding the domain of $f(r)$ Following the procedure of the previous section, we first determine the domain of the function. Clearly this problem makes physical sense only for $r > 0$. Looking at the equation

$$h = \frac{V}{\pi r^2},$$

we see that although V is fixed, r can be arbitrarily large provided h is sufficiently small (resulting in a can that looks like an elephant stepped on it). Thus the domain of our function is $r > 0$, which is not a closed interval, so we have no guarantee that a minimum exists.

Next we compute $f'(r)$, keeping in mind that the symbols V and π represent constants and that we are differentiating with respect to the variable r.

$$f'(r) = 4\pi r - \frac{2V}{r^2} = \frac{4\pi r^3 - 2V}{r^2}$$

Looking for critical points   The derivative is undefined at $r = 0$, which is outside the domain under consideration. So now we set the derivative equal to zero and solve for any possible critical points.

$$
\begin{aligned}
f'(r) &= \frac{4\pi r^3 - 2V}{r^2} \\
0 &= \frac{4\pi r^3 - 2V}{r^2} \\
0 &= 4\pi r^3 - 2V \\
r &= \sqrt[3]{V/2\pi}
\end{aligned}
$$

Thus $r = \sqrt[3]{V/2\pi}$ is the only critical point.

Finding the shape of the graph of A   We can actually sketch the shape of the graph of A versus r based on this analysis of $f'(r)$. The sign of $f'(r)$ is determined by its numerator, since the denominator r^2 is always positive.

- When $r < \sqrt[3]{V/2\pi}$, the numerator is negative, so the graph of A is falling.

- When $r > \sqrt[3]{V/2\pi}$, the numerator is positive, so the graph is rising.

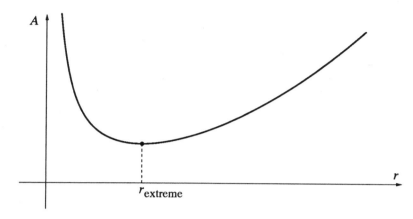

Obviously, A has a *minimum* at

$$r_{\text{extreme}} = \sqrt[3]{V/2\pi}.$$

It is also obvious that there is no *maximum* area, since

$$\lim_{r \to 0} A = \infty \quad \text{and} \quad \lim_{r \to \infty} A = \infty.$$

A has a minimum but no maximum

Thus we see again that not *every* optimization problem has a solution.

The Mathematical Context: Optimal Shapes

It is interesting to find the value of the height $h = h_{\text{extreme}}$ when the area is a minimum,

$$h_{\text{extreme}} = \frac{V}{\pi r_{\text{extreme}}^2}.$$

When we replace r_{extreme} by $\sqrt[3]{V/2\pi}$ and simplify this expression, we obtain

$$h_{\text{extreme}} = 2r_{\text{extreme}}.$$

In other words, the height of the optimal tin can exactly equals its diameter. Campbell soup cans are far from optimal, but a can of Progresso plum tomatoes has diameter 4 inches and height 4.5 inches. Does someone at Progresso know calculus?

It is important to acknowledge that the actual context of this example is not saving money for food canners. In a practical situation, we would probably have had a numerical value for V, so we could have used a graphing utility to approximate $r_{extreme}$ as accurately as our needs warranted. We might have noticed that the optimal radius was about half the height, but we wouldn't have known the relationship is exact. Nor would we have recognized that the relationship holds for cylinders of arbitrary volume. The real context of this example is geometry. In the exercises you will have several more opportunities to observe that, often, the geometric regularity that pleases the eye is also optimal.

Symmetric shapes are often optimal

Exercises

1. Show that the rectangle of perimeter P whose area is a maximum is a square. Use a graphing utility to check your answer for the special case when $P=100$ feet.

2. An open rectangular box is to be made from a piece of cardboard 8 inches wide and 15 inches long by cutting a square from each corner and bending up the sides. Find the dimensions of the box of largest volume. Use a graphing utility to check your answer.

3. One side of an open field is bounded by a straight river. A farmer has L feet of fencing. How should the farmer proportion a rectangular plot along the river in order to enclose as great an area as possible? Use a graphing utility to check your answer for the special case when $L = 100$ feet.

4. An open storage bin with a square base and vertical sides is to be constructed from A square feet of wood. Determine the dimensions of the bin if its volume is to be a maximum. (Neglect the thickness of the wood and any waste in construction.) Use a graphing utility to check your answer for the special case when $A=100$ square feet.

5. A roman window is shaped like a rectangle surmounted by a semicircle. If the perimeter of the window is L feet, what are the dimensions of the window of maximum area? Use a graphing utility to check your answer for the special case when $L = 100$ feet.

6.   Suppose the roman window of problem 5 has clear glass in its rectangular part and colored glass in its semicircular part. If the colored glass transmits only half as much light per square foot as the clear glass does, what are the dimensions of the window that transmits the most light? Use a graphing utility to check your answer for the special case when $L = 100$ feet.

7.   A cylindrical oil can with radius r inches and height h inches is made with a steel top and bottom and cardboard sides. The steel costs 3 cents per square inch, the cardboard costs 1 cent per square inch, and rolling the crimp around the top and bottom edges costs 1/2 cent per linear inch. (Both crimps are done at the same time, so only count the contribution of one circumference.)

a)   Express the cost C of the can as a function of r and h.

b)   Find the dimensions of the cheapest can holding 100 cubic inches of oil. (You'll need to solve a cubic equation to find the critical point. Use a graphing utility or an estimation procedure to approximate the critical point to 3 decimal places.)

§5.   Newton's Method

Finding Critical Points

When we solve optimization problems for functions given by formulas, we begin by calculating the derivative and using the derivative formula to find critical points. Almost always the derivative is defined for all elements of the domain, and we find the critical points by determining the roots of the equation obtained by setting the derivative equal to zero.

Finding the roots of this equation often requires an estimation procedure. For example, consider the function

$$f(x) = x^4 + x^3 + x^2 + x + 1.$$

The derivative of f is

$$f'(x) = 4x^3 + 3x^2 + 2x + 1,$$

which is certainly defined for all x.

Solving
$f'(x) = 0$

In order to use a graphing utility to find the roots of $f'(x) = 0$, we need to choose an interval that will contain the roots we seek. Since $f'(0) = 1 > 0$ and $f'(-1) = -2 < 0$, we know f' has at least one root on $[-1, 0]$. (Why is this?) But might there be other roots outside this interval?

It is easy to see that $f'(x)$ is positive for all $x > 0$, so there are no roots to the right of $[-1, 0]$. What about $x < -1$? Rewriting the derivative as

$$f'(x) = (2x + 1)(2x^2 + 1) + x^2$$

lets us see that $f'(x)$ is negative for all $x < -1$ (check this for yourself), so there are no roots to the left of $[-1, 0]$ either.

Examining the graph of $y = 4x^3 + 3x^2 + 2x + 1$ on $[-1, 0]$, we see that it crosses the x-axis exactly once, so there is a unique critical point. Progressively shrinking the interval, we find that, to eight decimal places, this critical point is

$$c = -.60582958\ldots.$$

There is, however, another estimation procedure we can use, called Newton's method. This has wide applicability and often converges very rapidly to solutions. Newton's method is also an interesting application of both local linearity and successive approximation, two important themes of this course.

Local Linearity and the Tangent Line

The root is an
x-intercept of
the graph

Let's give the derivative function the new name g to emphasize that we now want to consider it as a function in its own right, out of the context of the function f from which it was derived. Look at the graph of the function

$$y = g(x) = 4x^3 + 3x^2 + 2x + 1$$

on the next page. We are seeking the number r such that $g(r) = 0$. The root r is the x-coordinate of the point where the graph crosses the x-axis.

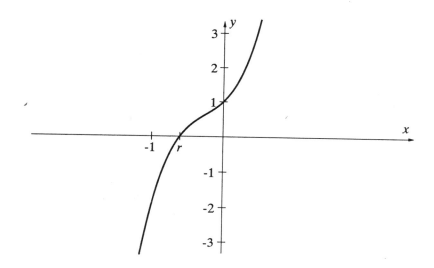

The basic plan of attack in Newton's method is to replace the graph of $y = g(x)$ by a straight line that looks reasonably like the graph near the root r. Then, the x-intercept of that line will be a reasonably good estimate for r. The graph below includes such a line.

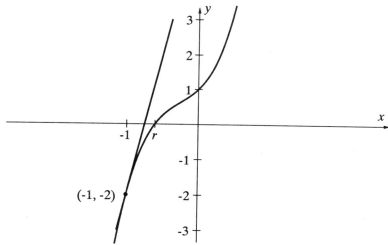

The function g is locally linear at $x = -1$, and we have drawn an extension of the local linear approximation of g at this point. This line is called the **tangent line** to the graph of $y = g(x)$ at $x = -1$, by analogy to the tangent line to a circle. To find the x-intercept of the tangent line, we must know its equation. Clearly the line passes through the **point of tangency** $(-1, g(-1)) = (-1, -2)$. What is its slope? It is the same as the slope of the local linear approximation at

The tangent line extends the local linear approximation

$(-1, g(-1))$, namely

$$g'(-1) = 12(-1)^2 + 6(-1) + 2 = 8.$$

The equation of the tangent line   Thus the equation of the tangent line is

$$y + 2 = 8(x + 1),$$

which we can rewrite as

$$y = -2 + 8(x + 1).$$

Finding the x-intercept of the tangent line   To find the x-intercept of this line, we must set y equal to zero and solve for x: $0 = -2 + 8(x + 1)$ gives us $x = -0.75$. Of course, this x-intercept is not equal to r, but it's a better approximation than, say, -1. To get an even better approximation, we repeat this process, starting with the line tangent to the graph of g at $x = -0.75$ instead of at $x = -1$.

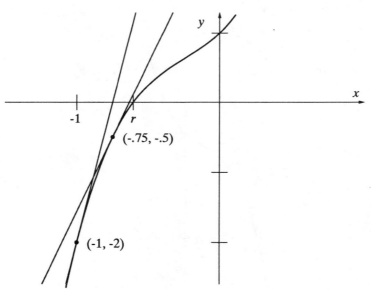

The slope of this new tangent line is $g'(-0.75) = 4.25$, and it passes through $(-0.75, -0.5)$, so its equation is

$$y + 0.5 = 4.25(x + 0.75).$$

Setting $y = 0$ and solving for x gives a new x-intercept equal to -0.6323529, closer still to r.

It seems reasonable to repeat the process yet again, using the tangent line at $x = -0.6323529$, but first we need to introduce some notation to keep track of our computations. Call the original point of tangency x_0, so $x_0 = -1$. Let x_1 be the x-intercept of the tangent line at $x = x_0$. Draw the tangent line at $x = x_1$, and call its x-intercept x_2. Continuing in this way, we get a sequence of points $x_0, x_1, x_2, x_3, \ldots$ which appear to approach nearer and nearer to r. That is, they appear to approach r as their limit.

Successive approximations get closer to the root r

The Algorithm

This process of using one number to determine the next number in the sequence is the heart of Newton's method—it is an iterative method. Moreover, it turns out to be quite simple to calculate each new estimate in terms of the previous one. To see how this works, let's compute x_1 in terms of x_0. We know x_1 is the x-intercept of the line tangent to the graph of g at $(x_0, g(x_0))$. The slope of this line is $g'(x_0)$, so

The general equation of the tangent line

$$y - g(x_0) = g'(x_0)(x - x_0)$$

is the **equation of the tangent line**. Since this line crosses the x-axis at the point $(x_1, 0)$, we set $x = x_1$ and $y = 0$ in the equation to obtain

$$0 - g(x_0) = g'(x_0)(x_1 - x_0).$$

Now it is easy to solve for x_1:

$$
\begin{aligned}
g'(x_0)(x_1 - x_0) &= -g(x_0) \\
x_1 - x_0 &= \frac{-g(x_0)}{g'(x_0)} \\
x_1 &= x_0 - \frac{g(x_0)}{g'(x_0)}.
\end{aligned}
$$

In the same way we get

$$
\begin{aligned}
x_2 &= x_1 - \frac{g(x_1)}{g'(x_1)}, \\
x_3 &= x_2 - \frac{g(x_2)}{g'(x_2)},
\end{aligned}
$$

and so on.

To summarize, suppose that x_0 is given some value START. Then Newton's method is the computation of the sequence of numbers determined by

$$x_0 = \text{START}$$
$$x_{n+1} = x_n - \frac{g(x_n)}{g'(x_n)}, \quad n = 0, 1, 2, 3, \ldots$$

As we have seen many times, the sequence

$$x_1, x_2, x_3, \ldots, x_n, \ldots$$

The limit of the successive approximations is the root

is a list of numbers to which we can always add a new x value—by iterating our method yet again. For most functions, if we begin with an appropriate starting value of x_0, there is another number r that is the *limit* of this list of numbers, in the sense that the difference between x_n and r becomes as small as we wish as n increases without bound,

$$r = \lim_{n \to \infty} x_n.$$

The numbers $x_1, x_2, x_3, \ldots, x_n, \ldots$ constitute a sequence of *successive approximations* for the root r of the equation $g(x) = 0$. We can write a computer program to carry out this algorithm for as many steps as we choose. The program NEWTON does just that for $g(x) = 4x^3 + 3x^2 + 2x + 1$.

Program: NEWTON
Newton's method for solving $g(x) = 4x^3 + 3x^2 + 2x + 1 = 0$

```
start = -1
numberofsteps = 8
x = start
FOR n = 0 to numberofsteps
    print n,x                    {This prints x_n}
    g = 4*x^3+3*x^2+2*x+1
    gprime = 12*x^2+6*x+2
    x = x - g/gprime
NEXT n
```

If we program a computer using this algorithm with START $= -1$, then we get

$$
\begin{aligned}
x_0 &= -1.00000000000 \\
x_1 &= -0.75000000000 \\
x_2 &= -0.63235294118 \\
x_3 &= -0.60687911790 \\
x_4 &= -0.60583128240 \\
x_5 &= -0.60582958619 \\
x_6 &= -0.60582958619 \\
x_7 &= -0.60582958619 \\
x_8 &= -0.60582958619 \, .
\end{aligned}
$$

Thus we have found the root of $g(x) = 0$—the critical point we were looking for. In fact, after only 6 steps we could see that the value of the critical point was specified to at least ten decimal places. Also at the sixth step, we had the eight decimal places obtained with the use of the graphing utility. With the use of the program NEWTON, we will see that in most cases we can obtain results more quickly and to a higher degree of accuracy with Newton's method than by using a graphing utility.

Examples

Example 1. Start with $\cos x = x$. The solution(s) to this equation (if any) will be the x-coordinates of any points of intersection of the graphs of $y = \cos x$ and $y = x$. Sketch these two graphs and convince yourself that there is one solution, between 0 and $\pi/2$. The equation $\cos x = x$ is not in the form $g(x) = 0$, so rewrite it as $\cos x - x = 0$. Now we can apply Newton's method with $g(x) = \cos x - x$. Try starting with $x_0 = 1$. This gives the iteration scheme

$$
\begin{aligned}
x_0 &= 1 \\
x_{n+1} &= x_n - \frac{\cos x_n - x_n}{-\sin x_n - 1}, \quad n = 0, 1, 2, \ldots
\end{aligned}
$$

The numbers we get are

$$
x_0 = 1.000000000
$$

$$x_1 = .750363868\ldots$$
$$x_2 = .739112891\ldots$$
$$x_3 = .739085133\ldots$$
$$x_4 = .739085133\ldots$$

We have the solution to 9 decimal places in only 4 steps. Not only does Newton's method work, it works fast!

Example 2. Suppose we continue with the equation $\cos x = x$, but this time choose $x_0 = 0$. What will we find? The numbers we get are

$$x_0 = 0.000000000$$
$$x_1 = 1.000000000$$
$$x_2 = 0.750363868$$

There's no need to continue; we can see that we will obtain $r = 0.739085133\ldots$ as in Example 1. Look again at your sketch and see why you might have predicted this result.

Example 3. Next, let's find the roots of the polynomial $x^5 - 3x + 1$. This means solving the equation $x^5 - 3x + 1 = 0$. We know the necessary derivative, so we're ready to apply Newton's method, except for one thing: which starting value x_0 do we pick? This is the part of Newton's method that leaves us on our own.

Finding the starting value x_0 can be hard

Assuming that some graphing software is available, the best thing to do is graph the function. But for most graphing utilities, we need to choose an interval. How do we choose one which is sure to include all the roots of the polynomial? The derivative $5x^4 - 3$ of this polynomial is simple enough that we can use it to get an idea of the shape of the graph of $y = x^5 - 3x + 1$ before we turn to the computer. Clearly the derivative is zero only for $x = \pm\sqrt[4]{3/5}$, and the derivative is positive except between these two values of x. In other words, we know the shape of the graph of $y = g(x)$—it is increasing, then decreases for a bit, then increases from there on out. This still is not enough information to tell us how many roots g has, though; its graph might lie in any one of the following configurations and so have 1, 2, or 3 roots (there are two other possibilities not shown—one has 2 roots and one has 1 root).

Using the derivative to find the shape of the graph

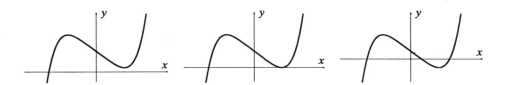

We can thus say that $g(x) = 0$ has at least one and at most three real roots. However, if we further observe that $g(-2) = -25, g(-1) = 3, g(0) = 1, g(1) = -1$, and $g(2) = 27$, we see that the graph of g must cross the x–axis at some value of x between -2 and -1, between 0 and 1, and again between 1 and 2. Therefore the right–hand sketch above must be the correct one.

Or we can almost as easily turn to a graphing utility. If we try the interval $[-5, 5]$, we see again that the graph crosses the x-axis in exactly three points.

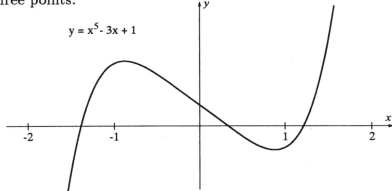

$y = x^5 - 3x + 1$

One of the roots is between -24 and -1, one is between 0 and 1, and the third is between 1 and 2. To find the first, we apply Newton's method with $x_0 = -2$. Then we get

Finding the root between -2 and -1

$$
\begin{aligned}
x_0 &= -2.000000000 \\
x_1 &= -1.67532467\ldots \\
x_2 &= -1.47823803\ldots \\
x_3 &= -1.40044537\ldots \\
x_4 &= -1.39890198\ldots \\
x_5 &= -1.39887920\ldots \\
x_6 &= -1.39887919\ldots \\
x_7 &= -1.39887919\ldots
\end{aligned}
$$

This took a little longer than the other examples, but not a lot. In the exercises you will be asked to compute the other two roots.

Example 4. Let's use Newton's method to find the obvious solution $r = 0$ of $x^3 - 5x = 0$. If we choose x_0 sufficiently close to 0, Newton's method should work just fine. But what does "sufficiently close" mean? Suppose we try $x_0 = 1$. Then we get

$$
\begin{aligned}
x_0 &= 1 \\
x_1 &= -1 \\
x_2 &= 1 \\
x_3 &= -1 \\
x_4 &= 1
\end{aligned}
$$

$$\vdots$$

Newton's
method can fail
The x_n's oscillate endlessly, never getting close to 0. Going back to the geometric interpretation of Newton's method, this oscillation can be explained by the graph below.

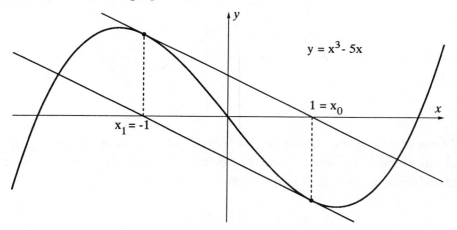

Using more advanced methods, it is possible to get precise estimates for how close x_0 needs to be to r in order for Newton's method to succeed. For now, we'll just have to rely on common sense and trial and error.

One important thing to note is the relation between algebra and Newton's method. Although we can now solve many more equations than we could earlier, this doesn't mean that we can abandon algebra. In fact, given a new equation, you should first try to solve it algebraically, for exact solutions are often better. Only when this fails should you look for approximate solutions using Newton's method. So don't forget algebra—you'll still need it!

Exercises

1. **The Babylonian algorithm** Show that the Babylonian algorithm of chapter 2 is the same as Newton's method applied to the equation $x^2 - a = 0$.

2. When Newton introduced his method, he did so with the example $x^3 - 2x - 5 = 0$. Show that this equation has only one root, and find it.

> This example appeared in 1669 in an unpublished manuscript of Newton's (a published version came later, in 1711). The interesting fact is that Newton's method *differs* from the one presented here: his scheme was more complicated, requiring a different formula to get each approximation. In 1690, Joseph Raphson transformed Newton's scheme into the one used above. Thus, "Newton's method" is more properly called the "Newton–Raphson method", and many modern texts use this more accurate name.

3. Use Newton's method to find a solution of $x^3 + 2x^2 + 10x = 20$ near the point $x = 1$.

> The approximate solution 1;22,7,42,33,4,40 of this equation appears in a book written in 1228 by Leonardo of Pisa (also known as Fibonacci). This number looks odd because it's written in sexagesimal notation: it translates into
>
> $$1 + \frac{22}{60} + \frac{7}{60^2} + \frac{42}{60^3} + \frac{33}{60^4} + \frac{4}{60^5} + \frac{40}{60^6}.$$
>
> This solution is accurate to 10 decimal places, which is not bad for 750 years ago. In the Middle Ages, there was a lot of interest in solving equations. There were even contests, with a prize going to the person who could solve the most. The quadratic formula, which expresses algebraically the roots of any second degree equation, had been known for thousands of years, but there were no general methods for finding roots of higher degree equations. We don't know how Leonardo found his solution—why give away your secrets to your competitors!

4. Use Newton's method to find a solution of $x^3 + 3x^2 = 5$.

In 1530, Nicolo Tartaglia was challenged to solve this equation algebraically. Five years later, in 1535, he found the solution

$$x = \sqrt[3]{\frac{3 + \sqrt{5}}{2}} + \sqrt[3]{\frac{3 - \sqrt{5}}{2}} - 1\,.$$

Initially, Tartaglia could only solve certain types of cubic equations, but this was enough to let him win some famous contests with other mathematicians of the time. By 1541, he knew the general solution, but he made the mistake of telling Geronimo Cardano. Cardano published the solution in 1545 and the resulting formulas are called "Cardan's Formulas".

The above solution of $x^3 + 3x^2 = 5$ is called a **solution by radicals** because it is obtained by extracting various roots or radicals. Similarly, some time before 1545, Luigi Ferrari showed that any fourth degree equation can be solved by radicals. This led to an intense interest in the fifth degree equation. To see what happens in this case, read the next problem.

5.   In Example 3, we saw that one root of $x^5 - 3x + 1$ was $-1.39887919\ldots$. Use Newton's method to find the other two roots.

In 1826, Niels Henrik Abel proved that the general polynomial of degree 5 or greater cannot be solved by radicals. Using the work of Evariste Galois (done around 1830, but not understood until many years later), it can be shown that the equation $x^5 - 3x + 1 = 0$ cannot be solved by radicals. So it can be proved that algebra can't solve this equation!

6.   One of the more surprising applications of Newton's method is to compute reciprocals. To make things more concrete, we will compute $1/3.4567$. Note that this number is the root of the equation $1/x = 3.4567$.

a) Show that the formula of Newton's method gives us

$$x_{n+1} = 2x_n - 3.4567x_n^2$$

b) Using $x_0 = 1$ and the formula from (a), compute $1/3.4567$ to a high degree of accuracy.

This method for computing reciprocals is important because it involves only *multiplication* and *subtraction*. Since $a/b = a \cdot (1/b)$, this implies that division can likewise be built from multiplication and subtraction. Thus, when designing a computer, the division routine doesn't need to be built from scratch—the designer can use the method illustrated here. There are some computers that do division this way.

7. In this problem we will determine the maximum value of the function

$$f(x) = \frac{x+1}{x^4+1}.$$

a) Graph $f(x)$ and convince yourself that the maximum value occurs somewhere around $x = .5$. Of course, the exact location is where the slope of the graph is zero, i.e., where $f'(x) = 0$. So we need to solve this equation.

b) Compute $f'(x)$.

c) Since the answer to (b) is a fraction, it vanishes when its numerator does. Setting the numerator equal to 0 gives a fourth degree equation. Use Newton's method to find a solution near $x = .5$.

d) Compute the maximum value of $f(x)$.

8.   Consider the hyperbola $y = 1/x$ and the circle $x^2 - 4x + y^2 + 3 = 0$.

a) By graphing the circle and the hyperbola, convince yourself that there are two points of intersection.

b) By substituting $y = 1/x$ into the equation of the circle, obtain a fourth degree equation satisfied by the x-coordinate of the points of intersection.

c) Solve the equation from (b) by Newton's method, and then determine the points of intersection.

9.   Sometimes Newton's method doesn't work so nicely. For example, consider the equation $\sin x = 0$.

a) Compute x_1 using Newton's method for each of the four starting values $x_0 = 1.55, 1.56, 1.57$ and 1.58.

b) The answers you get are wildly different. Using the basic formula

$$x_{n+1} = x_n - \frac{g(x_n)}{g'(x_n)}$$

explain why.

The epidemic runs its course

We return to the epidemiology example we have studied since chapter 1. Recall that our *S-I-R* model keeps track of three subgroups of the population: the susceptible, the infected, and the recovered. One of the interesting features of the model is that the larger the initial susceptible population, the more rapidly the epidemic runs its course. We observe this by choosing fixed values of $R_0 = R(0)$ and $I_0 = I(0)$ and looking at graphs of $S(t)$ versus t for various values of $S_0 = S(0)$.

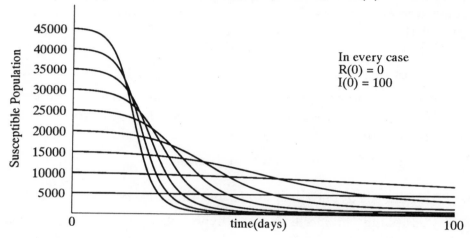

We see in each case that for sufficiently large t the graph of S levels off, approaching a value we'll call S_∞:

$$S_\infty = \lim_{t \to \infty} S(t)$$

What we mean by the epidemic "running its course" is that $S(t)$ reaches this limit value. We can see from the graphs that the value of S_0 affects the number S_∞ of individuals who escape the disease entirely. It turns out that we can actually find the value of S_∞ if we know the values of S_0, I_0, and the parameters a and b.

Recall that a is the *transmission coefficient*, and b is the *recovery coefficient* for the disease. The differential equations of the $S - I - R$ model are

$$\begin{aligned} S' &= -aSI \\ I' &= aSI - bI \\ R' &= bI \end{aligned}$$

10.   Use the differentiation rules together with these differential equations to show that

$$(I + S - (b/a) \cdot \ln S)' = 0$$

11.   Explain why the result of problem 10 means that $I + S - (b/a) \cdot \ln S$ has the same value—call it C—for every value of t.

12.   Look at the graphs of the solutions $I(t)$ for various values of $S(0)$ below.

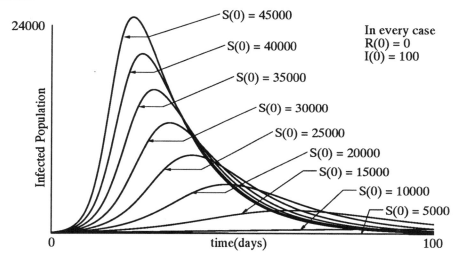

Write $\lim_{t \to \infty} I(t) = I_\infty$. What is the value of I_∞ for all values of $S(0)$?

13.   Use the results of problems 11 and 12 to explain why

$$S_\infty - \frac{b}{a} \ln(S_\infty) = I_0 + S_0 - \frac{b}{a} \ln(S_0) .$$

This equation determines S_∞ implicitly as a function of I_0 and S_0. For particular values of I_0 and S_0 (and of the parameters), you can use Newton's method to find S_∞.

14.   Use the values

$$
\begin{aligned}
a &= .00001 \text{ (person-days)}^{-1} \\
b &= .08 \text{ day}^{-1} \\
I_0 &= 100 \text{ persons} \\
S_0 &= 35,000 \text{ persons}
\end{aligned}
$$

Writing x instead of S_∞ gives

$$x - 8000 \ln(x) = -48,605 \,.$$

Apply Newton's method to find S_∞. Judging from the graph for $S_0 = 35000$, it looks like a reasonable first estimate for S_∞ might be 100.

15.   Using the same values of a, b, and I_0 as in problem 14, determine the value of S_∞ for each of the following initial population sizes:

a) $S_0 = 45,000$.

b) $S_0 = 25,000$.

c) $S_0 = 5,000$.

§6.   Chapter Summary

The Main Ideas

- **Formulas** for **derivatives** can be calculated for functions given by formulas using the definition of the derivative as the limit of difference quotients

$$\lim_{\Delta x \to 0} \frac{f(x + \Delta x) - f(x)}{\Delta x} \,.$$

- Each particular difference quotient is an **approximation** to the derivative. Successive approximations, for smaller and smaller Δx, approach the derivative as a limit. The formula for the derivative gives its *exact* value.

- Formulas for **partial derivatives** are obtained using the formulas for derivatives of functions of a single variable by simply treating all variables other than the one of interest as if they were constants.

- **Optimization** problems occur in many different contexts. For instance, we seek to maximize benefits, minimize energy, and minimize error.

- The **sign** of the derivative indicates where a graph rises and where it falls.

- Functions which are **continuous** on a **finite closed interval** have **global extremes**.

- For a function continuous on an interval, its **local extremes** occur at **critical points**—points where the derivative equals zero or fails to exist—or at endpoints.

- Local linearity permits us to replace the graph of a function $y = g(x)$ by its **tangent line** at a point near a root of $g(x) = 0$, and then the x-intercept of the tangent line is a better approximation to the root. Successive approximations, obtained by iterating this procedure, yield **Newton's method** for solving the equation $g(x) = 0$.

Self-Testing

- You should be able to **differentiate** a function given by a formula.

- You should be able to use differentiation formulas to **calculate** partial derivatives.

- In most cases, you should be able to determine from a graph of a function on an interval whether that function is **continuous** and/or **differentiable** on that interval.

- You should be able to find **critical points** for a function of one or several variables given by a formula.

- You should be able to use the formula for the derivative of a function of a single variable to find **local and global extremes**.

- You should be able to use **Newton's method** to solve an equation of the form $g(x) = 0$.

Chapter Exercises

Prices, demand and profit

Suppose the demand D (in units sold) for a particular product is determined by its price p (in dollars), $D = f(p)$. It is reasonable to assume that when the price is low, the demand will be high, but as the price rises, the demand will fall. In other words, we assume that the slope of the demand function is negative. If the manufacturing cost for each unit of the product is c dollars, then the profit per unit at price p is $p-c$. Finally the total profit T gained at the unit price p will be the number of units sold at the price p (that is, the demand $D(p)$) multiplied by the profit per unit $p - c$.

$$T = g(p) = D(p) \text{ units} \times (p - c) \, \frac{\text{dollars}}{\text{unit}}$$

In this series of problems we will determine the effect of the demand function and of the unit manufacturing cost on the maximum total profit.

1.   Suppose the demand function is linear

$$D = f(p) = 1000 - 500p \text{ units,}$$

and the unit manufacturing cost is .20, so the total profit is

$$T = g(p) = (1000 - 500p)(p - .20) \text{ dollars.}$$

Find the "best" price – that is, find the price that yields the maximum total profit.

2.   Suppose the demand function is the same as in problem 1, but the unit manufacturing cost rises to .30. What is the "best" price now? How much of the rise in the unit manufacturing cost is passed on to the consumer if the manufacturer charges this best price?

3.   Suppose the demand function for a particular product is $D(p) = 2000 - 500p$, and that the unit manufacturing cost is .30? What price should the manufacturer charge to maximize her profit? Suppose the unit manufacturing cost rises to .50. What price should she charge to maximize her profit now? How much of the rise in the unit manufacturing cost should she pass on to the consumer?

4.   If the demand function for a product is $D(p) = 1500 - 100p$, compare the "best" price for unit manufacturing costs of .30 and .50. How much of the rise in cost should the manufacturer pass on to the consumer?

5.   As you may have noticed, problems 2–4 illustrate an interesting phenomenon. In each case, exactly half of the rise in the unit manufacturing cost should be passed on to the consumer. Is this a coincidence?

a)   Consider the most general case of a linear demand function

$$D = f(p) = a - mp.$$

and unit cost c. What is the "best" price? Is exactly half the unit manufacturing cost passed on to the consumer? Explain your answer.

b)   Now consider a non-linear demand function

$$D = f(p) = \frac{1000}{1 + p^2}.$$

Find the "best" price for unit costs
   i) .50 dollar per unit;
   ii) 1.00 dollar per unit;
   iii) 1.50 dollars per unit.
How much of the price increase is passed on to the consumer in cases (ii) and (iii)?

Chapter 6

The Integral

There are many contexts—work, energy, area, volume, distance travelled, and profit and loss are just a few—where the quantity in which we are interested is a product of known quantities. For example, the electrical energy needed to burn three 100 watt light bulbs for Δt hours is $300 \cdot \Delta t$ watt-hours. In this example, though, the calculation becomes more complicated if lights are turned off and on during the time interval Δt. We face the same complication in any context in which one of the factors in a product varies. To describe such a product we will introduce the **integral**.

As you will see, the integral itself can be viewed as a variable quantity. By analyzing the rate at which that quantity changes, we will find that every integral can be expressed as the solution to a particular differential equation. We will thus be able to use all our tools for solving differential equations to determine integrals.

§1. Measuring Work

Human Effort

Let's measure the work done by the staff of an office that processes catalog orders. Suppose a typical worker in the office can process 10 orders an hour. Then we would expect 6 people to process 60 orders an hour; in two hours, they could process 120 orders.

Processing catalog orders

$$10 \, \frac{\text{orders per hour}}{\text{person}} \times 6 \text{ persons} \times 2 \text{ hours} = 120 \text{ orders}.$$

Notice that a staff of 4 people working 3 hours could do the same amount of work:

$$10 \, \frac{\text{orders per hour}}{\text{person}} \times 4 \text{ persons} \times 3 \text{ hours} = 120 \text{ orders.}$$

This example suggests that we should use the product

Human effort is measured as a product

$$\text{number of workers} \times \text{elapsed time}$$

to measure **human effort**. In these terms, it takes 12 "person-hours" of human effort to process 120 orders.

Another name that has been used in the past for this unit of effort is the "man-hour." If the task is large, effort can even be measured in "man-months" or "man-years." The term we will use most of the time is "staff-hour."

Productivity rate Notice that we can re-phrase the rate at which orders are processed as 10 orders per staff-hour. This is sometimes called the **productivity rate**. The productivity rate allows us to translate human effort into work:

$$\text{work done} \; = \; \text{productivity rate} \; \times \; \text{human effort}$$
$$120 \text{ orders} \; = \; 10 \, \frac{\text{orders}}{\text{staff-hour}} \; \times \; 12 \text{ staff-hours.}$$

As this equation shows, work is a linear function of effort in which the productivity rate serves as multiplier (see pages 32–33).

Mowing lawns If we modify the productivity rate, we can use this equation for other kinds of jobs. For example, we can use it to predict how much work a lawn mowing crew will do. Suppose the productivity rate is .7 acres per staff-hour. Then we expect that a staff of S working for H hours can mow

$$.7 \, \frac{\text{acres}}{\text{staff-hour}} \times SH \text{ staff-hours} = .7 \, SH \text{ acres}$$

of lawn altogether.

Staff-hours provide a common measure of work in different jobs Work is measured differently in different jobs—as orders processed, or acres mowed, or houses painted. However, in all these jobs human effort can be measured the *same* way—as staff-hours. Because human effort provides a common unit that can be translated from one job to another, managers tend to use *staff-hour* as the unit of work.

From this new point of view, a staff of S working steadily for H hours does SH staff-hours of work. Suppose the staffing level S is not constant, as in the graph below. Can we still find the total amount of work done?

Non-constant staffing

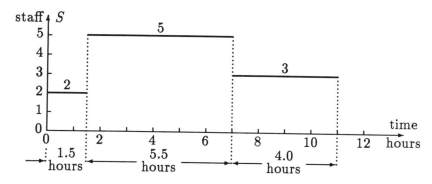

The basic formula works only when the staffing level is constant. But staffing *is* constant over certain time intervals. Thus, to find the total amount of work done, we should simply use the basic formula on each of those intervals, and then add up the individual contributions. These calculations are done in the following table. The total work is 42.5 staff-hours, and since the productivity rate is 10 orders per staff-hour, 425 orders can be processed.

The work done is a sum of products

$$
\begin{array}{rllllll}
2 & \text{staff} & \times & 1.5 & \text{hours} & = & 3.0 & \text{staff-hours} \\
5 & & \times & 5.5 & & = & 27.5 \\
3 & & \times & 4.0 & & = & \underline{12.0} \\
& & & & & & 42.5 & \text{staff-hours}
\end{array}
$$

Accumulated work

The last calculation tells us how much work got done over an entire day. What can we tell an office manager who wants to know how work is progressing *during* the the day?

At the beginning of the day, only two people are working, so after T hours (where $0 \leq T \leq 1.5$)

work done up to time $T = 2$ staff $\times T$ hours $= 2T$ staff-hours.

Even before we consider what happens after 1.5 hours, this expression calls our attention to the fact that *accumulated work is a function*— let's denote it $W(T)$. According to the formula, for the first 1.5 hours

$W(T)$ is a linear function whose multiplier is

$$W' = 2 \; \frac{\text{staff-hours}}{\text{hour}}.$$

Work
accumulates **at**
a rate equal to
the number
of staff

This multiplier is the **rate** at which work is being accumulated. It is also the **slope** of the graph of $W(T)$ over the interval $0 \le T \le 1.5$. With this insight, we can determine the rest of the graph of $W(T)$.

What must $W(T)$ look like on the next time interval $1.5 \le T \le 7$? Here 5 members of staff are working, so work is accumulating at the rate of 5 staff-hours per hour. Therefore, on this interval the graph of W is a straight line segment whose slope is 5 staff-hours per hour. On the third interval, the graph is another straight line segment whose slope is 3 staff-hours per hour. The complete graph of $W(T)$ is shown at the top of the next page.

S is the
derivative
of W, so ...

As the graphs show, the *slope* of the accumulated work function $W(T)$ is the *height* of the staffing function $S(T)$. In other words, S is the *derivative* of W:

$$W'(T) = S(T).$$

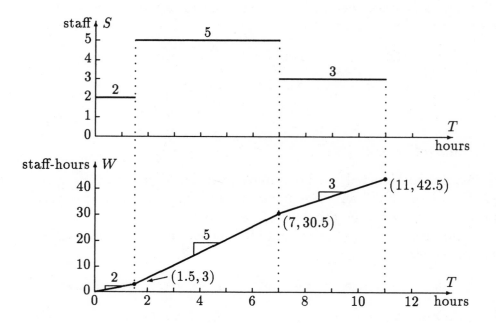

The accumulated work function $W(T)$

Notice that the units for W' and for S are also equivalent:

$$\text{units for } W' = \frac{\text{staff-hours}}{\text{hour}} = \text{staff} = \text{units for } S$$

We can describe the relation between S and W another way. At the moment, we have explained S in terms of W. However, since we started with S, it is really more appropriate to reverse the roles, and explain W in terms of S. Chapter 4, §5 gives us the language to do this: W is an **antiderivative** of S. In other words, $y = W(T)$ is a solution to the differential equation

...W is an
antiderivative
of *S*

$$\frac{dy}{dT} = S(T).$$

As we find accumulation functions in other contexts, this relation will give us crucial information.

Before leaving this example we note some special features of S and W. The staffing function S is said to be **piecewise constant**, or a **step function**. The graphs illustrate the general fact that the derivative of a piecewise *linear* function (W, in this case) is piecewise *constant*.

The derivative of
a piecewise linear
function

Summary

The example of human effort illustrates the key ideas we will meet, again and again, in different contexts in this chapter. Essentially, we have two functions $W(t)$ and $S(t)$ and two different ways of expressing the relation between them: On the one hand,

$W(t)$ is an accumulation function for $S(t)$,

while on the other hand,

$S(t)$ is the derivative of $W(t)$.

Exploring the far-reaching implications of functions connected by such a two-fold relationship will occupy the rest of this chapter.

Electrical Energy

Human energy is one source of work; electricity is another. A power company charges customers for the work done by the electricity it supplies, and it measures that work in a way that is strictly analogous to the way we measure human effort.

For example, suppose we illuminate two light bulbs—one rated at 100 watts, the other at 60 watts. Then it will take the same amount of electrical energy to burn the 100-watt bulb for 3 hours as it will to burn the 60-watt bulb for 5 hours. Both will use 300 watt-hours of electricity. *The analogy between electrical and human energy* The power of the light bulb—measured in watts—is analogous to the number of staff working (and, in fact, workers have sometimes been called man*power*). The time the bulb burns is analogous to the time the staff work. Finally, the product

$$\text{energy} = \text{power} \times \text{elapsed time}$$

for electricity is analogous to the product

$$\text{work} = \text{number of staff} \times \text{elapsed time}$$

for human effort.

Electric *power* is measured in watts, in kilowatts (= 1,000 watts), and in megawatts (= 1,000,000 watts). Electric *energy* is measured in watt-hours, in kilowatt-hours (abbreviated 'kwh') and in megawatt-hours (abbreviated 'mwh'). Since an individual electrical appliance has a power demand of about one kilowatt, kwh are suitable units to use for describing the energy consumption of a house, while mwh are more natural for a whole town.

Suppose the power demand of a town over a 24 hour period is described by the following graph:

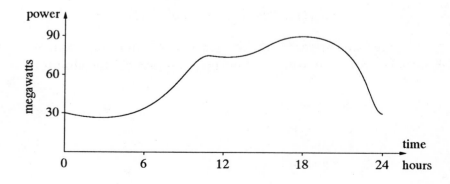

Since this graph decribes *power*, its vertical height over any point t on the time axis tells us the total wattage of the light bulbs, dishwashers, computers, etc. that are turned on in the town at that instant. This demand fluctuates between 30 and 90 megawatts, roughly. The problem is to determine the total amount of *energy* used in a day—how many megawatt-hours are there in this graph? Although the equation

$$\text{energy} = \text{power} \times \text{elapsed time},$$

gives the basic relation between energy and power, we can't use it directly because the power demand isn't constant.

The staffing function $S(t)$ we considered earlier wasn't constant, either, but we were still able to compute staff-hours because $S(t)$ was *piecewise* constant. This suggests that we should replace the power graph by a piecewise constant graph that **approximates** it. Here is one such approximation:

A piecewise constant approximation

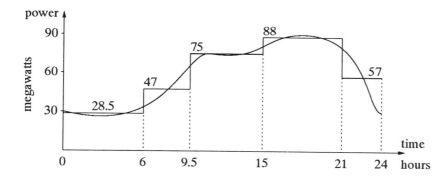

As you can see, the step function has five steps, so our approximation to the total energy consumption of the town will be a sum of five individual products:

$$\text{energy} \approx 28.5 \times 6 + 47 \times 3.5 + \cdots + 57 \times 3 = 1447 \text{ mwh.}$$

This value is only an *estimate*, though. How can we get a better estimate? The answer is clear: start with a step function that approximates the power graph *more closely*. In principle, we can get as good an approximation as we might desire this way. We are limited only by the precision of the power graph itself. As our approximation to the power graph improves, so does the accuracy of the calculation that estimates energy consumption.

Better estimates

In summary, we determine the energy consumption of the town by a sequence of successive approximations. The steps in the sequence are listed in the box below.

---

1. **Approximate** the power demand by a step function.
2. **Estimate** energy consumption from this approximation.
3. **Improve** the energy estimate by choosing a new step function that follows power demand more closely.

Accumulated energy consumption

Energy is being consumed steadily over the entire day; can we determine how much energy has been used through the first T hours of the day? We'll denote this quantity $E(T)$ and call it the **energy accumulation function**. For example, we already have the estimate $E(24) = 1447$ mwh; can we estimate $E(3)$ or $E(17.6)$?

Energy
accumulation

Once again, the earlier example of human effort can guide us. We saw that work accumulates at a rate equal to the number of staff present:

$$W'(T) = S(T).$$

Since $S(T)$ was piecewise constant, this rate equation allowed us to determine $W(T)$ as a piecewise linear function.

We claim that there is analogous relation between accumulated energy consumption and power demand—namely

$$E'(T) = p(T).$$

The function $p(T)$, however, is not piecewise constant, unlike $S(T)$. Therefore, the argument we used to show that $W'(T) = S(T)$ is true will not work here. We need another argument.

Estimating
$E'(T)$

To explain why the differential equation $E'(T) = p(T)$ should be true, we will start by analyzing the derivative $E'(T)$. We have the standard approximation

$$E'(T) \approx \frac{\Delta E}{\Delta T} = \frac{E(T + \Delta T) - E(T)}{\Delta T}.$$

Assume we have made ΔT so small that, to the level of precision we require, the approximation $\Delta E / \Delta T$ agrees with $E'(T)$. The numerator

ΔE is, by definition, the total energy used up to time $T + \Delta T$, minus the total energy used up to time T. This is just the energy used during the time interval ΔT that runs from time T to time $T + \Delta T$:

$$\Delta E = \text{energy used between times } T \text{ and } T + \Delta T.$$

Since the elapsed time ΔT is small, the power demand should be nearly constant, so we can get a good estimate for energy consumption from the basic equation

$$\text{energy used} = \text{power} \times \text{elapsed time}.$$

In particular, if we represent the power by $p(T)$, which is the power demand at the beginning of the time period from T to $T + \Delta T$, then we have

$$\Delta E \approx p(T) \cdot \Delta T.$$

Using this value in our approximation for the derivative $E'(T)$, we get

$$E'(T) \approx \frac{\Delta E}{\Delta T} \approx \frac{p(T) \cdot \Delta T}{\Delta T} = p(T).$$

That is, $E'(T) \approx p(T)$, and the approximation becomes more and more exact as the time interval ΔT shrinks to 0. Thus,

$$E'(T) = \lim_{\Delta T \to 0} \frac{\Delta E}{\Delta T} = p(T).$$

Here is another way to arrive at the same conclusion. Our starting point is the basic formula

A second way to see $E' = p$

$$\Delta E \approx p(T) \cdot \Delta T,$$

which holds over a small time interval ΔT. This formula tells us how E responds to small changes in T. But that is exactly what the **microscope equation** tells us:

$$\Delta E \approx E'(T) \cdot \Delta T.$$

Since these equations give the same information, their multipliers must be the same:

$$p(T) = E'(T).$$

In words, the differential equation $E' = p$ says that *power is the rate at which energy is consumed.* In purely mathematical terms:

> The energy accumulation function $y = E(t)$
> is a solution to the differential equation $dy/dt = p(t)$.

In fact, $y = E(t)$ is *the* solution to the **initial value problem**

$$\frac{dy}{dt} = p(t) \qquad y(0) = 0.$$

We can use all the methods described in chapter 4, §5 to solve this problem.

The relation we have explored between power and energy can be found in an analogous form in many other contexts, as we will see in the next two sections. In §4 we will turn back to accumulation functions and investigate them as solutions to differential equations. Then, in Calculus II, we will look at some special methods for solving the particular differential equations that arise in accumulation problems.

Exercises

Human effort

1.   House-painting is a job that can be done by several people working simultaneously, so we can measure the amount of work done in "staff-hours." Consider a house-painting business run by some students. Because of class schedules, different numbers of students will be painting at different times of the day. Let $S(t)$ be the number of staff present at time t, measured in hours from 8 am, and suppose that during an 8-hour work day, we have

$$S(t) = \begin{cases} 3 & 0 \le t < 2 \\ 2 & 2 \le t < 4.5 \\ 4 & 4.5 \le t \le 8. \end{cases}$$

a)   Draw the graph of the step function defined here, and compute the total number of staff hours.

b)   Draw the graph that shows how staff-hours *accumulate* on this job. This is the graph of the **accumulated work** function $W(T)$. (Compare the graphs of staff and staff-hours on page 334.)

c)   Determine the derivative $W'(T)$. Is $W'(T) = S(T)$?

2. Suppose that there is a house-painting job to be done, and by past experience the students know that four of them could finish it in 6 hours. But for the first 3.5 hours, only two students can show up, and after that, five will be available.

a) How long will the whole job take? [Answer: 6.9 hours.]

b) Draw a graph of the staffing function for this problem. Mark on the graph the time that the job is finished.

c) Draw the graph of the accumulated work function $W(T)$.

d) Determine the derivative $W'(T)$. Is $W'(T) = S(T)$?

Average staffing. Suppose a job can be done in three hours when 6 people work the first hour and 9 work during the last two hours. Then the job takes 24 staff-hours of work, and the **average staffing** is

$$\text{average staffing} = \frac{24 \text{ staff-hours}}{3 \text{ hours}} = 8 \text{ staff}.$$

This means that a *constant* staffing level of 8 persons can accomplish the job in the same time that the given variable staffing level did. Note that the average staffing level (8 persons) is *not* the average of the two numbers 9 and 6!

3. What is the average staffing of the jobs considered in exercises 1 and 2, above?

4. a) Draw the graph that shows how work would accumulate in the job described in exercise 1 if the work-force was kept at the *average* staffing level instead of the varying level described in the exercise. Compare this graph to the graph you drew in exercise 1 b.

b) What is the derivative $W'(T)$ of the work accumulation function whose graph you drew in part (a)?

5. What is the average staffing for the job described by the graph on page 333?

Electrical energy

6. On Monday evening, a 1500 watt space heater is left on from 7 until 11 pm. How many kilowatt-hours of electricity does it consume?

7.   a) That same heater also has settings for 500 and 1000 watts. Suppose that on Tuesday we put it on the 1000 watt setting from 6 to 8 pm, then switch to 1500 watts from 8 till 11 pm, and then on the 500 watt setting through the night until 8 am, Wednesday. How much energy is consumed (in kwh)?

b) Sketch the graphs of power demand $p(t)$ and accumulated energy consumption $E(T)$ for the space heater from Tuesday evening to Wednesday morning. Determine whether $E'(T) = p(T)$ in this case.

c) The **average power demand** of the space heater is defined by:

$$\text{average power demand} = \frac{\text{energy consumption}}{\text{elapsed time}}.$$

If energy consumption is measured in kilowatt-hours, and time in hours, then we can measure average power demand in kilowatts—the same as power itself. (Notice the similarity with average staffing.) What is the average power demand from Tuesday evening to Wednesday morning? If the heater could be set at this average power level, how would the energy consumption compare to the actual energy consumption you determined in part (a)?

8.   The graphs on pages 336 and 337 descibe the power demand of a town over a 24-hour period. Give an estimate of the average power demand of the town during that period. Explain what you did to produce your estimate.   [Answer: 60.29 megawatts is one estimate.]

Work as force × distance

The effort it takes to move an object is also called work. Since it takes *twice* as much effort to move the object twice as far, or to move another object that is twice as heavy, we can see that the work done in moving an object is proportional to both the force applied and to the distance moved. The simplest way to express this fact is to define

$$\text{work} = \text{force} \times \text{distance}.$$

For example, to lift a weight of 20 pounds straight up it takes 20 pounds of force. If the vertical distance is 3 feet then

$$20 \text{ pounds} \times 3 \text{ feet} = 60 \text{ foot-pounds}$$

of work is done. Thus, once again the quantity we are interested in has the form of a product. The *foot-pound* is one of the standard units for measuring work.

9.   Suppose a tractor pulls a loaded wagon over a road whose steepness varies. If the first 150 feet of road are relatively level and the tractor has to exert only 200 pounds of force while the next 400 feet are inclined and the tractor has to exert 550 pounds of force, how much work does the tractor do altogether?

10.   A motor on a large ship is lifting a 2000 pound anchor that is already out of the water at the end of a 30 foot chain. The chain weighs 40 pounds per foot. As the motor lifts the anchor, the part of the chain that is hanging gets shorter and shorter, thereby reducing the weight the motor must lift.

a) What is the combined weight of anchor and hanging chain when the anchor has been lifted x feet above its initial position?

b) Divide the 30-foot distance that the anchor must move into 3 equal intervals of 10 feet each. Estimate how much work the motor does lifting the anchor and chain over each 10-foot interval by multiplying the combined weight at the *bottom* of the interval by the 10-foot height. What is your estimate for the total work done by the motor in raising the anchor and chain 30 feet?

c) Repeat all the steps of part (b), but this time use 6 equal intervals of 5 feet each. Is your new estimate of the work done larger or smaller than your estimate in part (b)? Which estimate is likely to be more accurate? On what do you base your judgment?

d) If you ignore the weight of the chain entirely, what is your estimate of the work done? How much *extra* work do you therefore estimate the motor must do to raise the heavy chain along with the anchor?

§2.   Riemann Sums

In the last section we estimated energy consumption in a town by replacing the power function $p(t)$ by a step function. Let's pause to describe that process in somewhat more general terms that we can adapt to other contexts. The power graph, the approximating step function, and the energy estimate are shown below.

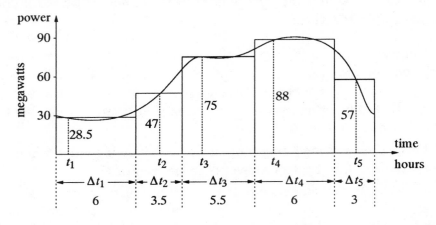

$$\text{energy} \approx 28.5 \times 6 + 47 \times 3.5 + \cdots + 57 \times 3 = 1447 \text{ mwh.}$$

Sampling the power function

The height of the first step is 28.5 megawatts. This is the *actual* power level at the time t_1 indicated on the graph. That is, $p(t_1) = 28.5$ megawatts. We found a power level of 28.5 megawatts by **sampling** the power function at the time t_1. The height of the first step could have been different if we had sampled the power function at a different time. In general, if we sample the power function $p(t)$ at the time t_1 in the interval Δt_1, then we would estimate the energy used during that time to be

$$\text{energy} \approx p(t_1) \cdot \Delta t_1 \text{ mwh.}$$

Notice that t_1 is not in the middle, or at either end, of the first interval. It is simply a time when the power demand is representative of what's happening over the entire interval. Furthermore, t_1 is not even unique; there is another sampling time (near $t = 5$ hours) when the power level is again 28.5 megawatts.

We can describe what happens in the other time intervals the same way. If we sample the k-th interval at the point t_k, then the height of the k-th power step will be $p(t_k)$ and our estimate for the energy used during that time will be

$$\text{energy} \approx p(t_k) \cdot \Delta t_k \text{ mwh.}$$

We now have a general way to construct an approximation for the power function and an estimate for the energy consumed over a 24-hour period. It involves these steps.

1. Choose any number n of subintervals, and let them have arbitrary widths $\Delta t_1, \Delta t_2, \ldots, \Delta t_n$, subject only to the condition

$$\Delta t_1 + \cdots + \Delta t_n = 24 \text{ hours.}$$

A procedure for approximating the power level and energy use

2. Sample the k-th subinterval at any point t_k, and let $p(t_k)$ represent the power level over this subinterval.

3. Estimate the energy used over the 24 hours by the sum

$$\text{energy} \approx p(t_1) \cdot \Delta t_1 + p(t_2) \cdot \Delta t_2 + \cdots + p(t_n) \cdot \Delta t_n \text{ mwh.}$$

The expression on the right is called a **Riemann sum** for the power function $p(t)$ on the interval $0 \le t \le 24$ hours.

> The work of Bernhard Riemann (1826–1866) has had a profound influence on contemporary mathematicians and physicists. His revolutionary ideas about the geometry of space, for example, are the basis for Einstein's theory of general relativity.

The enormous range of choices in this process means there are innumerable ways to construct a Riemann sum for $p(t)$. However, we are not really interested in *arbitrary* Riemanns sums. On the contrary, we want to build Riemann sums that will give us good estimates for energy consumption. Therefore, we will choose each subinterval Δt_k so small that the power demand over that subinterval differs only very little from the sampled value $p(t_k)$. A Riemann sum constructed with *these* choices will then differ only very little from the total energy used during the 24-hour time interval.

Choices that lead to good estimates

Essentially, we use a Riemann sum to resolve a dilemma. We know the basic formula

$$\text{energy} = \text{power} \times \text{time}$$

works when power is constant, but in general power *isn't* constant— that's the dilemma. We resolve the dilemma by using instead a *sum* of

The dilemma

terms of the form *power × time*. With this sum we get an estimate for the energy.

In this section we will explore some other problems that present the same dilemma. In each case we will start with a basic formula that involves a product of two constant factors, and we will need to adapt the formula to the situation where one of the factors varies. The solution will be to construct a Riemann sum of such products, producing an estimate for the quantity we were after in the first place. As we work through each of these problems, you should pause to compare it to the problem of energy consumption.

Calculating Distance Travelled

It is easy to tell how far a car has travelled by reading its odometer. The problem is more complicated for a ship, particularly a sailing ship in the days before electronic navigation was common. The crew always had instruments that could measure—or at least estimate—the velocity of the ship at any time. Then, during any time interval in which the ship's velocity is constant, the distance travelled is given by the familiar formula

Estimating velocity and distance

$$\text{distance} = \text{velocity} \times \text{elapsed time}.$$

If the velocity is *not* constant, then this formula does not work. The remedy is to break up the long time period into several short ones. Suppose their lengths are $\Delta t_1, \Delta t_2, \ldots, \Delta t_n$. By assumption, the velocity is a function of time t; let's denote it $v(t)$. At some time t_k during each time period Δt_k measure the velocity: $v_k = v(t_k)$. Then the Riemann sum

Sampling the velocity function

$$v(t_1) \cdot \Delta t_1 + v(t_2) \cdot \Delta t_2 + \cdots + v(t_n) \cdot \Delta t_n$$

is an estimate for the total distance travelled.

For example, suppose the velocity is measured five times during a 15 hour trip—once every three hours—as shown in the table on the next page. Then the basic formula

$$\text{distance} = \text{velocity} \times \text{elapsed time}.$$

gives us an estimate for the distance travelled during each three-hour period, and the sum of these distances is an estimate of the total distance travelled during the fifteen hours. These calculations appear in the right-hand column of the table. (Note that the first measurement is used to calculate the distance travelled between hours 0 and 3, while the last measurement, taken 12 hours after the start, is used to calculate the distance travelled between hours 12 and 15.)

sampling time (hours)	elapsed time (hours)	velocity (miles/hour)	distance travelled (miles)		
0	3	1.4	3×1.4	=	4.20
3	3	5.25	3×5.25	=	15.75
6	3	4.3	3×4.3	=	12.90
9	3	4.6	3×4.6	=	13.80
12	3	5.0	3×5.0	=	15.00
					61.65

Thus we estimate the ship has travelled 61.65 miles during the fifteen hours. The number 61.65, obtained by adding the numbers in the right-most column, is a Riemann sum for the velocity function.

The estimated distance is a Riemann sum for the velocity function

Consider the specific choices that we made to construct this Riemann sum:

$$\Delta t_1 = \Delta t_2 = \Delta t_3 = \Delta t_4 = \Delta t_5 = 3$$
$$t_1 = 0, \quad t_2 = 3, \quad t_3 = 6, \quad t_4 = 9, \quad t_5 = 12.$$

These choices differ from the choices we made in the energy example in two notable ways. First, all the subintervals here are the same size. This is because it is natural to take velocity readings at regular time intervals. By contrast, in the energy example the subintervals were of different widths. Those widths were chosen in order to make a piecewise constant function that followed the power demand graph closely. Second, all the sampling times lie at the beginning of the subintervals in which they appear. Again, this is natural and convenient for velocity measurements. In the energy example, the sampling times were chosen with an eye to the power graph. Even though we can make arbitrary choices in constructing a Riemann sum, we will do it systematically whenever possible. This means choosing subintervals of equal size and sampling points at the "same" place within each interval.

Intervals and sampling times are chosen in a systematic way

Let's turn back to our estimate for the total distance. Since the velocity of the ship could have fluctuated significantly during each of

the three-hour periods we used, our estimate is rather rough. To improve the estimate we could measure the velocity more frequently—for
example, every 15 minutes. If we did, the Riemann sum would have
60 terms (four distances per hour for 15 hours). The individual terms
in the sum would all be much smaller, though, because they would be
estimates for the distance travelled in 15 minutes instead of in 3 hours.
For instance, the first of the 60 terms would be

$$1.4 \; \frac{\text{miles}}{\text{hour}} \times .25 \text{ hours} = .35 \text{ miles}.$$

Of course it may not make practical sense to do such a precise calculation. Other factors, such as water currents or the inaccuracy of the
velocity measurements themselves, may keep us from getting a good estimate for the distance. Essentially, the Riemann sum is only a *model*
for the distance covered by a ship.

Calculating Areas

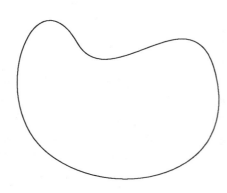

The area of a rectangle is just the product of its
length and its width. How can we measure the
area of a region that has an irregular boundary,
like the one at the left? We would like to use the
basic formula

$$\text{area} = \text{length} \times \text{width}.$$

However, since the region doesn't have straight
sides, there is nothing we can call a "length" or a
"width" to work with.

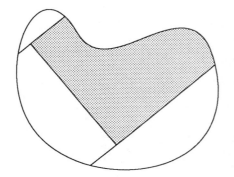

We can begin to deal with this problem by
breaking up the region into smaller regions that
do have straight sides—with, at most, only one
curved side. This can be done many different
ways. The lower figure shows one possibility. The
sum of the areas of all the little regions will be
the area we are looking for. Although we haven't
yet solved the original problem, we have at least
reduced it to another problem that looks simpler
and may be easier to solve. Let's now work on the
reduced problem for the shaded region.

Here is the shaded region, turned so that it sits flat on one of its straight sides. We would like to calculate its area using the formula

width × height,

but this formula applies only to rectangles. We can, however, approximate the region by a collection of rectangles, as shown at the right. The formula *does* apply to the individual rectangles and the sum of their areas will approximate the area of the whole region.

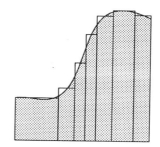

To get the area of a rectangle, we must measure its width and height. Their heights vary with the height of the curved top of the shaded region. To describe that height in a systematic way, we have placed the shaded region in a coordinate plane so that it sits on the x-axis. The other two straight sides lie on the vertical lines $x = a$ and $x = b$.

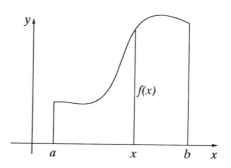

The curved side defines the graph of a function $y = f(x)$. Therefore, at each point x, the vertical height from the axis to the curve is $f(x)$. By introducing a coordinate plane we gain access to mathematical tools—such as the language of functions—to describe the various areas.

The k-th rectangle has been singled out on the left, below. We let Δx_k denote the width of its base. By **sampling** the function f at a properly chosen point x_k in the base, we get the height $f(x_k)$ of the rectangle. Its area is therefore $f(x_k) \cdot \Delta x_k$. If we do the same thing for all n rectangles shown on the right, we can write their total area as

Calculating the areas of the rectangles

$$f(x_1) \cdot \Delta x_1 + f(x_2) \cdot \Delta x_2 + \cdots + f(x_n) \cdot \Delta x_n.$$

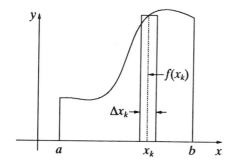
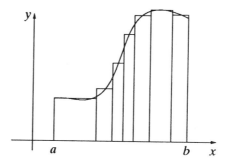

Notice that our estimate for the area has the form of a Riemann sum for the height function $f(x)$ over the interval $a \leq x \leq b$. To get a better estimate, we should use narrower rectangles, and more of them. In other words, we should construct another Riemann sum in which the number of terms, n, is larger and the width Δx_k of every subinterval is smaller. Putting it yet another way, we should *sample* the height more often.

Consider what happens if we apply this procedure to a region whose area we know already. The semicircle of radius $r = 1$ has an area of $\pi r^2/2 = \pi/2 = 1.5707963\dots$. The semicircle is the graph of the function

$$f(x) = \sqrt{1 - x^2},$$

which lies over the interval $-1 \leq x \leq 1$. To get the figure on the left, we sampled the height $f(x)$ at 20 evenly spaced points, starting with $x = -1$. In the better approximation on the right, we increased the number of sample points to 50. The values of the shaded areas were calculated with the program RIEMANN, which we will develop later in this section. Note that with 50 rectangles the Riemann sum is within .005 of $\pi/2$, the exact value of the area.

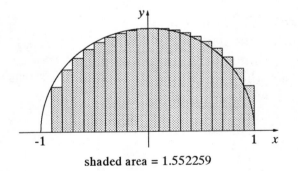

shaded area = 1.552259   shaded area = 1.566098

Calculating Lengths

It is to be expected that products—and ultimately, Riemann sums—will be involved in calculating areas. It is more surprising to find that we can use them to calculate *lengths*, too. In fact, when we are working in a coordinate plane, using a product to describe the length of a straight line is even quite natural.

To see how this can happen, consider a line segment in the x, y-plane that has a known slope m. If we also know the horizontal separation between the two ends, we can find the length of the segment. Call the horizontal separation is Δx and the vertical separation Δy. Then the length of the segment is

$$\sqrt{\Delta x^2 + \Delta y^2}$$

(by the Pythagorean theorem, page 90). Since $\Delta y = m \cdot \Delta x$, we can rewrite this as

$$\sqrt{\Delta x^2 + (m \cdot \Delta x)^2} = \Delta x \cdot \sqrt{1 + m^2}.$$

In other words, if a line has slope m and it is Δx units wide, then its length is the product

$$\sqrt{1 + m^2} \cdot \Delta x.$$

Suppose the line is *curved*, instead of straight. Can we describe its length the same way? We'll assume that the curve is the graph $y = g(x)$. The complication is that the slope $m = g'(x)$ now varies with x.

Suppose $g'(x)$ doesn't vary too much over an interval of length Δx. Then the curve is nearly straight. If we select a single point x_* in the interval and sample the slope $g'(x_*)$ there, we would expect the length of the curve to be approximately

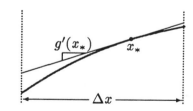

$$\sqrt{1 + (g'(x_*))^2} \cdot \Delta x.$$

As the figure shows, this is the exact length of the straight line segment that lies over the same interval Δx and is tangent to the curve at the point $x = x_*$.

If the slope $g'(x)$ varies appreciably over the interval, we should subdivide the interval into small pieces $\Delta x_1, \Delta x_2, \ldots, \Delta x_n$, over which the curve is nearly straight. Then, if we sample the slope at the point x_k in the k-th subinterval, the sum

$$\sqrt{1 + (g'(x_1))^2} \cdot \Delta x_1 + \cdots + \sqrt{1 + (g'(x_n))^2} \cdot \Delta x_n$$

will give us an estimate for the total length of the curve.

The length of a
curve is
estimated by a
Riemann sum
Once again, we find an expression that has the form of a Riemann sum. There is, however, a new ingredient worth noting. The estimate is a Riemann sum not for the *original* function $g(x)$ but for *another* function

$$f(x) = \sqrt{1 + (g'(x))^2}$$

that we constructed using g. The important thing is that the length is estimated by a Riemann sum for *some* function.

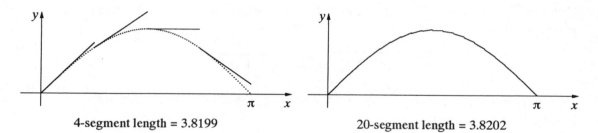

4-segment length = 3.8199   20-segment length = 3.8202

The figure above shows two estimates for the length of the graph of $y = \sin x$ between 0 and π. We used the equal subintervals and we always sampled the slope at the left end of a subinterval. As you can see, the four segments approximate the graph of $y = \sin x$ only very roughly. When we increase the number of segments to 20, on the right, the approximation to the shape of the graph becomes quite good. Notice that the graph itself is not shown on the right; only the 20 segments.

To calculate the two lengths, we constructed Riemann sums for the function $f(x) = \sqrt{1 + \cos^2 x}$. We used the fact that the derivative of $g(x) = \sin x$ is $g'(x) = \cos x$, and we did the calculations using the program RIEMANN. By using the program with still smaller subintervals you can show that

the exact length = 3.820197789 . . . .

Thus, the 20-segment estimate is already accurate to four decimal places.

We have already constructed estimates for the length of a curve, in chapter 2 (pages 89–91). Those estimates were sums, too, but they were not *Riemann* sums. The terms had the form $\sqrt{\Delta x^2 + \Delta y^2}$; they were not *products* of the form $\sqrt{1 + m^2} \cdot \Delta x$. The sums in chapter 2 may seem more straightforward. However, we are developing Riemann sums as a powerful general tool for dealing with many different questions. By expressing lengths as Riemann sums we gain access to that power.

Definition

The Riemann sums that appear in the calculation of power, distance, and length are instances of a general mathematical object that can be constructed for *any* function whatsoever. We pause now to describe that construction apart from any particular context. In what follows it will be convenient for us to write an interval of the form $a \leq x \leq b$ more compactly as $[a, b]$.

Notation: $[a, b]$

Definition. Suppose the function $f(x)$ is defined for x in the interval $[a, b]$. Then a **Riemann sum** for $f(x)$ on $[a, b]$ is an expression of the form

$$f(x_1) \cdot \Delta x_1 + f(x_2) \cdot \Delta x_2 + \cdots + f(x_n) \cdot \Delta x_n.$$

The interval $[a, b]$ has been divided into n subintervals whose lengths are $\Delta x_1, \ldots, \Delta x_n$, and for each k from 1 to n, x_k is some point in the k-th subinterval.

Notice that once the function and the interval have been specified, a Riemann sum is determined by the following data:

Data for a Riemann sum

- A **decomposition** of the original interval into subintervals (which determines the lengths of the subintervals).

- A **sampling point** chosen from each subinterval (which determines a value of the function on each subinterval).

A Riemann sum for $f(x)$ is a sum of products of values of Δx and values of $y = f(x)$. If x and y have units, then so does the Riemann sum; its units are the product of the units for x times the units for y. When a Riemann sum arises in a particular context, the notation may look different from what appears in the definition just given: the variable might not be x, and the function might not be $f(x)$. For example, the energy approximation we considered at the beginning of the section is a Riemann sum for the power demand function $p(t)$ on $[0, 24]$. The length approximation for the graph $y = \sin x$ is a Riemann sum for the function $\sqrt{1 + \cos^2 x}$ on $[0, \pi]$.

Units

It is important to note that, from a mathematical point of view, a Riemann sum is just a number. It's the *context* that provides the

A Riemann sum is just a number

meaning: Riemann sums for a power demand that varies over time approximate total energy consumption; Riemann sums for a velocity that varies over time approximate total distance; and Riemann sums for an length that varies over distance approximate total area.

To illustrate the generality of a Riemann sum, and to stress that it is just a number arrived at through arbitrary choices, let's work through an example without a context. Consider the function

$$f(x) = \sqrt{1 + x^3} \quad \text{on} \quad [1, 3].$$

The data   We will break up the full interval $[1, 3]$ into three subintervals $[1, 1.6]$, $[1.6, 2.3]$ and $[2.3, 3]$. Thus

$$\Delta x_1 = .6 \qquad \Delta x_2 = \Delta x_3 = .7.$$

Next we'll pick a point in each subinterval, say $x_1 = 1.3$, $x_2 = 2$ and $x_3 = 2.8$. Here is the data laid out on the x-axis.

With this data we get the following Riemann sum for $\sqrt{1 + x^3}$ on $[1, 3]$:

$$f(x_1) \cdot \Delta x_1 + f(x_2) \cdot \Delta x_2 + f(x_3) \cdot \Delta x_3$$
$$= \sqrt{1 + 1.3^3} \times .6 + \sqrt{1 + 2^3} \times .7 + \sqrt{1 + 2.8^3} \times .7$$
$$= 6.5263866$$

In this case, the choice of the subintervals, as well as the choice of the point x_k in each subinterval, was haphazard. Different data would produce a different value for the Riemann sum.

Keep in mind that an individual Riemann sum is not especially significant. Ultimately, we are interested in seeing what happens when we recalculate Riemann sums with smaller and smaller subintervals. For that reason, it is helpful to do the calculations systematically.

Calculating a Riemann sum algorithmically. As we have seen with our contextual problems, the data for a Riemann sum is not usually chosen in a haphazard fashion. In fact, when dealing with functions given by formulas, such as the function $f(x) = \sqrt{1 - x^2}$ whose graph

is a semicircle, it pays to be systematic. We use subintervals of equal
size and pick the "same" point from each subinterval (e.g., always pick
the midpoint or always pick the left endpoint). The benefit of systematic choices is that we can write down the computations involved in a
Riemann sum in a simple algorithmic form that can be carried out on
a computer.

Let's illustrate how this strategy applies to the function $\sqrt{1+x^3}$ on
$[1,3]$. Since the whole interval is $3-1=2$ units long, if we construct n
subintervals of equal length Δx, then $\Delta x = 2/n$. For every $k = 1,\ldots,n$,
we choose the sampling point x_k to be the left endpoint of the k-th
subinterval. Here is a picture of the data:

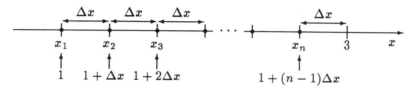

In this systematic approach, the space between one sampling point and
the next is Δx, the same as the width of a subinterval. This puts the
k-th sampling point at $x = 1 + (k-1)\Delta x$.

In the following table, we add up the terms in a Riemann sum S for
$f(x) = \sqrt{1+x^3}$ on the interval $[1,3]$. We used $n = 4$ subintervals and
always sampled f at the left endpoint. Each row shows the following:

1. the current sampling point;

2. the value of f at that point;

3. the current term $\Delta S = f \cdot \Delta x$ in the sum;

4. the accumulated value of S.

left endpoint	current $\sqrt{1+x^3}$	current ΔS	accumulated S
1	1.4142	.7071	.7071
1.5	2.0917	1.0458	1.7529
2	3	1.5	3.2529
2.5	4.0774	2.0387	5.2916

The Riemann sum S appears as the final value 5.2916 in the fourth
column.

The program RIEMANN, below, will generate the last two columns in the table on the previous page. The statement **x = a** on the sixth line determines the position of the first sampling point. Within the FOR–NEXT loop, the statement **x = x + deltax** moves the sampling point to its next position.

Program: RIEMANN
Left endpoint Riemann sums

```
DEF fnf (x) = SQR(1 + x ^ 3)
a = 1
b = 3
numberofsteps = 4
deltax = (b - a) / numberofsteps
x = a
accumulation = 0
FOR k = 1 TO numberofsteps
        deltaS = fnf(x) * deltax
        accumulation = accumulation + deltaS
        x = x + deltax
        PRINT deltaS, accumulation
NEXT k
```

By modifying RIEMANN, you can calculate Riemann sums for other sampling points and for other functions. For example, to sample at midpoints, you must start at the midpoint of the first subinterval. Since the subinterval is Δx units wide, its midpoint is $\Delta x/2$ units from the left endpoint, which is $x = a$. Thus, if you change the statement on the sixth line to **x = a + deltax / 2**, the program will then generate midpoint Riemann sums.

Summation Notation

Because Riemann sums arise frequently and because they are unwieldy to write out in full, we now introduce a method—called **summation notation**—that allows us to write them more compactly. To see how it works, look first at the sum

$$1^2 + 2^2 + 3^2 + \cdots + 50^2.$$

Using summation notation, we can express this as

$$\sum_{k=1}^{50} k^2.$$

For a somewhat more abstract example, consider the sum

$$a_1 + a_2 + a_3 + \cdots + a_n,$$

which we can express as

$$\sum_{k=1}^{n} a_k.$$

we use the capital letter *sigma* $\sum$ from the Greek alphabet to denote a sum. For this reason, summation notation is sometimes referred to as **sigma notation**. You should regard $\sum$ as an instruction telling you **Sigma notation** to **sum** the numbers of the indicated form as the index k runs through the integers, starting at the integer displayed below the $\sum$ and ending at the integer displayed above it. Notice that changing the index k to some other letter has no effect on the sum. For example,

$$\sum_{k=1}^{20} k^3 = \sum_{j=1}^{20} j^3,$$

since both expressions give the sum of the cubes of the first twenty positive integers. Other aspects of summation notation will be covered in the exercises.

Summation notation allows us to write the Riemann sum

$$f(x_1) \cdot \Delta x_1 + \cdots + f(x_n) \cdot \Delta x_n$$

more efficiently as

$$\sum_{k=1}^{n} f(x_k) \cdot \Delta x_k.$$

Be sure not to get tied into one particular way of using these symbols. For example, you should instantly recognize

$$\sum_{i=1}^{m} \Delta t_i \, g(t_i)$$

as a Riemann sum. In what follows we will commonly use summation notation when working with Riemann sums. The important thing to remember is that summation notation is only a "shorthand" to express a Riemann sum in a more compact form.

Exercises

Making approximations

1.   Estimate the average velocity of the ship whose motion is described on page 347. The voyage lasts 15 hours.

2.   The aim of this question is to determine how much electrical energy was consumed in a house over a 24-hour period, when the power demand p was measured at different times to have these values:

time (24-hour clock)	power (watts)
1:30	275
5:00	240
8:00	730
9:30	300
11:00	150
15:00	225
18:30	1880
20:00	950
22:30	700
23:00	350

Notice that the time interval is from $t = 0$ hours to $t = 24$ hours, but the power demand was not sampled at either of those times.

a) Set up an estimate for the energy consumption in the form of a Riemann sum $p(t_1)\Delta t_1 + \cdots + p(t_n)\Delta t_n$ for the power function $p(t)$. To do this, you must identify explicitly the value of n, the sampling times t_k, and the time intervals Δt_k that you used in constructing your estimate. [Note: the sampling times come from the table, but there is wide latitude in how you choose the subintervals Δt_k.]

b) What is the estimated energy consumption, using your choice of data? There is no single "correct" answer to this question. Your estimate depends on the choices you made in setting up the Riemann sum.

c) Plot the data given in the table in part (a) on a (t, p)-coordinate plane. Then draw on the same coordinate plane the step function that represents your estimate of the power function $p(t)$. The width of the

k-th step should be the time interval Δt_k that you specified in part (a); is it?

d) Estimate the *average* power demand in the house during the 24-hour period.

Waste production. A colony of living yeast cells in a vat of fermenting grape juice produces waste products—mainly alcohol and carbon dioxide—as it consumes the sugar in the grape juice. It is reasonable to expect that another yeast colony, twice as large as this one, would produce twice as much waste over the same time period. Moreover, if we double the time period we would expect our colony to produce twice as much waste.

These observations suggest that waste production is proportional to both the size of the colony and the amount of time that passes. If P is the size of the colony, in grams, and Δt is a short time interval, then we can express waste production W as a function of P and Δt:

$$W = k \cdot P \cdot \Delta t \text{ grams.}$$

If Δt is measured in hours, then the multiplier k has to be measured in units of grams of waste per hour per gram of yeast.

The preceding formula is useful only over a time interval Δt in which the population size P does not vary significantly. If the time interval is large, and the population size can be expressed as a function $P(t)$ of the time t, then we can estimate waste production by breaking up the whole time interval into a succession of smaller intervals $\Delta t_1, \Delta t_2, \ldots, \Delta t_n$ and forming a Riemann sum

$$k\, P(t_1)\, \Delta t_1 + \cdots + k\, P(t_n)\, \Delta t_n \approx W \text{ grams.}$$

The time t_k must lie within the time interval Δt_k, and $P(t_k)$ must be a good approximation to the population size $P(t)$ throughout that time interval.

3. Suppose the colony starts with 300 grams of yeast (i.e., at time $t = 0$ hours) and it grows exponentially according to the formula

$$P(t) = 300\, e^{0.2\, t}.$$

If the waste production constant k is 0.1 grams per hour per gram of yeast, estimate how much waste is produced in the first four hours.

Use a Riemann sum with four hour-long time intervals and measure the population size of the yeast in the middle of each interval—that is, "on the half-hour."

Using RIEMANN

4.   a) Calculate left endpoint Riemann sums for the function $\sqrt{1 + x^3}$ on the interval $[1, 3]$ using 40, 400, 4000, and 40000 equally-spaced subintervals. How many digits in this sequence have stabilized?

b) The left endpoint Riemann sums for $\sqrt{1 + x^3}$ on the interval $[1, 3]$ seem to be approaching a limit as the number of subintervals increases without bound. Give the numerical value of that limit, accurate to four decimal places.

c) Calculate left endpoint Riemann sums for the function $\sqrt{1 + x^3}$ on the interval $[3, 7]$. Construct a sequence of Riemann sums using more and more subintervals, until you can determine the limiting value of these sums, accurate to four decimal places. What is that limit?

d) Calculate left endpoint Riemann sums for the function $\sqrt{1 + x^3}$ on the interval $[1, 7]$ in order to determine the limiting value of the sums to four decimal place accuracy. What is that value? How are the limiting values in parts (b), (c), and (d) related? How are the corresponding *intervals* related?

5.   Modify RIEMANN so it will calculate a Riemann sum by sampling the given function at the *midpoint* of each subinterval, instead of the left endpoint. Describe exactly how you changed the program to do this.

6.   a) Calculate *midpoint* Riemann sums for the function $\sqrt{1 + x^3}$ on the interval $[1, 3]$ using 40, 400, 4000, and 40000 equally-spaced subintervals. How many digits in this sequence have stabilized?

b) Roughly how many subintervals are needed to make the midpoint Riemann sums for $\sqrt{1 + x^3}$ on the interval $[1, 3]$ stabilize out to the first four digits? What is the stable value? Compare this to the limiting value you found earlier for left endpoint Riemann sums. Is one value larger than the other; could they be the same?

c) Comment on the relative "efficiency" of midpoint Riemann sums versus left endpoint Riemann sums (at least for the function $\sqrt{1 + x^3}$

on the interval $[1, 3]$). To get the same level of accuracy, an *efficient* calculation will take fewer steps than an *inefficient* one.

7.   a) Modify RIEMANN to calculate *right endpoint* Riemann sums, and use it to calculate right endpoint Riemann sums for the function $\sqrt{1 + x^3}$ on the interval $[1, 3]$ using 40, 400, 4000, and 40000 equally-spaced subintervals. How many digits in this sequence have stabilized?

b) Comment on the efficiency of right endpoint Riemann sums as compared to left endpoint and to midpoint Riemann sums—at least as far as the function $\sqrt{1 + x^3}$ is concerned.

8.   Calculate left endpoint Riemann sums for the function

$$f(x) = \sqrt{1 - x^2} \quad \text{on the interval } [-1, 1].$$

Use 20 and 50 equally-spaced subintervals. Compare your values with the estimates for the area of a semicircle given on page 350.

9.   a) Calculate left endpoint Riemann sums for the function

$$f(x) = \sqrt{1 + \cos^2 x} \quad \text{on the interval } [0, \pi].$$

Use 4 and 20 equally-spaced subintervals. Compare your values with the estimates for the length of the graph of $y = \sin x$ between 0 and π, given on page 352.

b) What is the limiting value of the Riemann sums, as the number of subintervals becomes infinite? Find the limit to 11 decimal places accuracy.

10.   Calculate left endpoint Riemann sums for the function

$$f(x) = \cos(x^2) \quad \text{on the interval } [0, 4],$$

using 100, 1000, and 10000 equally-spaced subintervals.

[Answer: With 10000 equally-spaced intervals, the left endpoint Riemann sum has the value .59485189.]

11.   Calculate left endpoint Riemann sums for the function

$$f(x) = \frac{\cos x}{1 + x^2} \quad \text{on the interval } [2, 3],$$

using 10, 100, and 1000 equally-spaced subintervals. The Riemann sums are all negative; why? (A suggestion: sketch the graph of f. What does that tell you about the signs of the terms in a Riemann sum for f?)

12. a) Calculate midpoint Riemann sums for the function

$$H(z) = z^3 \quad \text{on the interval } [-2, 2],$$

using 10, 100, and 1000 equally-spaced subintervals. The Riemann sums are all zero; why?

b) Repeat part (a) using *left endpoint* Riemann sums. Are the results still zero? Can you explain the difference, if any, between these two results?

Volume as a Riemann sum

If you slice a rectangular parallelepiped (e.g., a brick or a shoebox) parallel to a face, the area A of a **cross-section** does not vary. The same is true for a cylinder (e.g., a can of spinach or a coin). For *any* solid that has a constant cross-section (e.g., the object on the right, below), its volume is just the product of its cross-sectional area with its thickness.

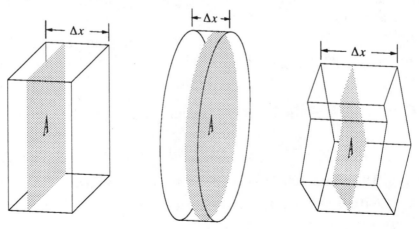

volume = area of cross-section × thickness = $A \cdot \Delta x$

Most solids don't have such a regular shape. They are more like the one shown below. If you take cross-sectional slices perpendicular to some fixed line (which will become our x-axis), the slices will not generally have a regular shape. They may be roughly oval, as shown below, but they will generally vary in area. Suppose the area of the cross-section x inches along the axis is $A(x)$ square inches. Because $A(x)$ varies with x, you cannot calculate the volume of this solid using the simple formula above. However, you can *estimate* the volume as a Riemann sum for A.

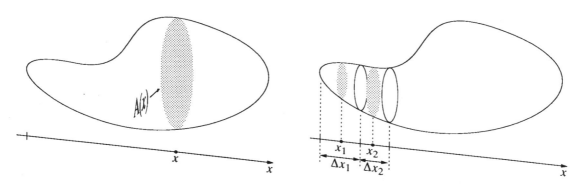

The procedure should now be familiar to you. Subdivide the x-axis into segments of length $\Delta x_1, \Delta x_2, \ldots, \Delta x_n$ inches, respectively. The solid piece that lies over the first segment has a thickness of Δx_1 inches. If you slice this piece at a point x_1 inches along the x-axis, the area of the slice is $A(x_1)$ square inches, and the volume of the piece is approximately $A(x_1) \cdot \Delta x_1$ cubic inches. The second piece is Δx_2 inches thick. If you slice it x_2 inches along the x-axis, the slice has an area of $A(x_2)$ square inches, so the second piece has an approximate volume of $A(x_2) \cdot \Delta x_2$ cubic inches. If you continue in this way and add up the n volumes, you get an estimate for the total volume that has the form of a Riemann sum for the area function $A(x)$:

$$\text{volume} \approx A(x_1)\,\Delta x_1 + A(x_2)\,\Delta x_2 + \cdots + A(x_n)\,\Delta x_n \text{ cubic inches.}$$

One place where this approach can be used is in medical diagnosis. The X-ray technique known as a CAT scan provides a sequence of precisely-spaced cross-sectional views of a patient. From these views much information about the state of the patient's internal organs can be gained without invasive surgery. In particular, the volume of a specific piece of tissue can be estimated, as a Riemann sum, from the areas of individual slices and the spacing between them. The next exercise gives an example.

13.   A CAT scan of a human liver shows us X-ray "slices" spaced 2 centimeters apart. If the areas of the slices are 72, 145, 139, 127, 111, 89, 63, and 22 square centimeters, estimate the volume of the liver.

14.   The volume of a sphere whose radius is r is exactly $V = 4\pi r^3/3$.

a) Using the formula, determine the volume of the sphere whose radius is 3. Give the numerical value to four decimal places accuracy.

One way to get a sphere of radius 3 is to rotate the graph of the semi-circle

$$r(x) = \sqrt{9 - x^2} \qquad -3 \leq x \leq 3$$

around the x-axis. Every cross-section perpendicular to the x-axis is a circle. At the point x, the radius of the circle is $r(x)$, and its area is

$$A = \pi r^2 = \pi(r(x))^2 = A(x).$$

You can thus get estimates for the volume of the sphere by constructing Riemann sums for $A(x)$ on the interval $[-3, 3]$.

b) Calculate a sequence of estimates for the volume of the sphere that use more and more slices, until the value of the estimate stabilizes out to four decimal places. Does this value agree with the value given by the formula in part (a)?

15.   a) Rotate the graph of $r(x) = .5\,x$, with $0 \leq x \leq 6$ around the x-axis. What shape do you get? Describe it precisely, and find its volume using an appropriate geometric formula.

b) Calculate a sequence of estimates for the volume of the same object by constructing Riemann sums for the area function $A(x) = \pi(r(x))^2$. Continue until your estimates stabilize out to four decimal places. What value do you get?

Summation notation

16.   Determine the numerical value of each of the following:

a) $\displaystyle\sum_{k=1}^{10} k$ b) $\displaystyle\sum_{k=1}^{5} k^2$ c) $\displaystyle\sum_{j=0}^{4} 2j + 1$

[Answer: $\displaystyle\sum_{k=1}^{5} k^2 = 55.$]

17.   Write "The sum of the first five positive even integers" in summation notation.

18.   Determine the numerical value of

a) $\displaystyle\sum_{n=1}^{5} \left(\frac{1}{n} - \frac{1}{n+1} \right)$
      b) $\displaystyle\sum_{n=1}^{500} \left(\frac{1}{n} - \frac{1}{n+1} \right)$

19.   Express the following sums using summation notation.

a) $1^2 + 2^2 + 3^2 + \cdots + n^2$.

b) $2^1 + 2^2 + 2^3 + \cdots + 2^m$.

c) $f(s_1)\Delta s + f(s_2)\Delta s + \cdots + f(s_{12})\Delta s$.

d) $y_1^2 \Delta y_1 + y_2^2 \Delta y_2 + \cdots + y_n^2 \Delta y_n$.

20.   Express each of the following as a sum written out term-by-term. (There is no need to calculate the numerical value, even when that can be done.)

a) $\displaystyle\sum_{l=3}^{n-1} a_l$
      b) $\displaystyle\sum_{j=0}^{4} \frac{j+1}{j^2+1}$
      c) $\displaystyle\sum_{k=1}^{5} H(x_k)\,\Delta x_k$.

21.   Acquire experimental evidence for the claim

$$\left(\sum_{k=1}^{n} k \right)^2 = \sum_{k=1}^{n} k^3$$

by determining the numerical values of both sides of the equation for $n = 2, 3, 4, 5,$ and 6.

22.   Let $g(u) = 25 - u^2$ and suppose the interval $[0, 2]$ has been divided into 4 equal subintervals Δu and u_j is the left endpoint of the j-th interval. Determine the numerical value of the Riemann sum

$$\sum_{j=1}^{4} g(u_j)\,\Delta u.$$

Length and area

23.   Using Riemann sums with equal subintervals, estimate the length of the parabola $y = x^2$ over the interval $0 \le x \le 1$. Obtain a sequence of estimates that stabilize to four decimal places. How many subintervals

did you need? (Compare your result here with the earlier result on page 92.)

24.   Using Riemann sums, obtain a sequence of estimates for the area under each of the following curves. Continue until the first four decimal places stabilize in your estimates.

a) $y = x^2$ over $[0, 1]$ b) $y = x^2$ over $[0, 3]$ c) $y = x \sin x$ over $[0, \pi]$

25.   What is the area under the curve $y = \exp(-x^2)$ over the interval $[0, 1]$? Give an estimate that is accurate to four decimal places. Sketch the curve and shade the area.

26.   a) Estimate, to four decimal place accuracy, the length of the graph of the natural logarithm function $y = \ln x$ over the interval $[1, e]$.

b) Estimate, to four decimal place accuracy, the length of the graph of the exponential function $y = \exp(x)$ over the interval $[0, 1]$.

27.   a) What is the length of the hyperbola $y = 1/x$ over the interval $[1, 4]$? Obtain an estimate that is accurate to four decimal places.

b) What is the area under the hyperbola over the same interval? Obtain an estimate that is accurate to four decimal places.

28.   The graph of $y = \sqrt{4 - x^2}$ is a semicircle whose radius is 2. The circumference of the whole circle is 4π, so the length of the part of the circle in the first quadrant is exactly π.

a) Using left endpoint Riemann sums, estimate the length of the graph $y = \sqrt{4 - x^2}$ over the interval $[0, 2]$ in the first quadrant. How many subintervals did you need in order to get an estimate that has the value $3.14159\ldots$?

b) There is a technical problem that makes it impossible to use *right* endpoint Riemann sums. What is the problem?

§3. The Integral

Refining Riemann Sums

In the last section, we estimated the electrical energy a town consumed by constructing a Riemann sum for the power demand function $p(t)$. Because we sampled the power function only five times in a 24-hour period, our estimate was fairly rough. We would get a better estimate by sampling more frequently—that is, by constructing a Riemann sum with more terms and shorter subintervals. The process of refining Riemann sums in this way leads to the mathematical object called the **integral**.

To see what an integral is—and how it emerges from this process of refining Riemann sums—let's return to the function

$$\sqrt{1 + x^3} \quad \text{on} \quad [1, 3]$$

we analyzed at the end of the last section. What happens when we refine Riemann sums for this function by using smaller subintervals? If we systematically choose n equal subintervals and evaluate $\sqrt{1 + x^3}$ at the left endpoint of each subinterval, then we can use the program RIEMANN (page 356) to produce the values in the following table. For future reference we record the size of the subinterval $\Delta x = 2/n$ as well.

Refining Riemann sums with equal subintervals

Left endpoint Riemann sums for $\sqrt{1 + x^3}$ on $[1, 3]$

n	Δx	Riemann sum
100	.02	6.191 236 2
1 000	.002	6.226 082 6
10 000	.0002	6.229 571 7
100 000	.00002	6.229 920 6

The first four digits have stabilized, suggesting that these Riemann sums, at least, approach the limit 6.229....

It's too soon to say that *all* the Riemann sums for $\sqrt{1 + x^3}$ on the interval $[1, 3]$ approach this limit, though. There is such an enormous diversity of choices at our disposal when we construct a Riemann sum. We haven't seen what happens, for instance, if we choose midpoints instead of left endpoints, or if we choose subintervals that are not all of the same size. Let's explore the first possibility.

To modify RIEMANN to choose midpoints, we need only change the line

```
x = a
```

that determines the position of the first sampling point to

```
x = a + deltax / 2.
```

With this modification, RIEMANN produces the following data.

Midpoint Riemann sums for $\sqrt{1 + x^3}$ on $[1, 3]$

n	Δx	Riemann sum
10	.2	6.227 476 5
100	.02	6.229 934 5
1 000	.002	6.229 959 1
10 000	.0002	6.229 959 4

This time, the first *seven* digits have stabilized, even though we used only 10,000 subintervals—ten times fewer than we needed to get four digits to stabilize using left endpoints! This is further evidence that the Riemann sums converge to a limit, and we can even specify the limit more precisely as 6.229959....

These tables also suggest that midpoints are more "efficient" than left endpoints in revealing the limiting value of successive Riemann sums. This is indeed true. In Calculus II, we will see why this happens.

We still have another possibility to consider: what happens if we choose subintervals of different sizes—as we did in calculating the energy consumption of a town? By allowing variable subintervals, we make the problem messier to deal with, but it does not become conceptually more difficult. In fact, we can still get all the information we really need in order to understand what happens when we refine Riemann sums.

To see what form this information will take, look back at the two tables for midpoints and left endpoints, and compare the values of the Riemann sums that they report for the same size subinterval. When the subinterval was fairly large, the values differed by a relatively large amount. For example, when $\Delta x = .02$ we got two sums (namely

6.2299345 and 6.1912362) that differ by more than .038. As the subinterval got smaller, the difference between the Riemann sums got smaller too. (When $\Delta x = .0002$, the sums differ by less than .0004.) This is the general pattern. That is to say, for subintervals with a given maximum size, the various Riemann sums that can be produced will still differ from one another, but those sums will all lie within a certain range that gets smaller as the size of the largest subinterval gets smaller.

Refining Riemann sums with unequal subintervals

The connection between the range of Riemann sums and the size of the largest subinterval is subtle and technically complex; this course will not explore it in detail. However, we can at least see what happens concretely to Riemann sums for the function $\sqrt{1 + x^3}$ over the interval $[1, 3]$. The following table shows the smallest and largest possible Riemann sum that can be produced when no subinterval is larger than the maximum size Δx_k given in the first column.

maximum size of Δx_k	Riemann sums range from	to	difference between extremes
.02	6.113 690	6.346 328	.232 638
.002	6.218 328	6.241 592	.023 264
.000 2	6.228 796	6.231 122	.002 326
.000 02	6.229 843	6.230 076	.000 233
.000 002	6.229 948	6.229 971	.000 023

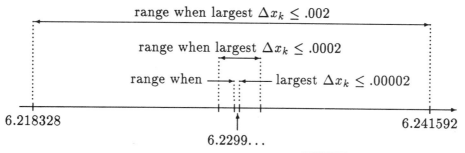

The range of Riemann sums for $\sqrt{1 + x^3}$ on $[1, 3]$

This table provides the most compelling evidence that there is a single number 6.2299... that *all* Riemann sums will be arbitrarily close to, if they are constructed with sufficiently small subintervals Δx_k. This number is called the **integral** of $\sqrt{1 + x^3}$ on the interval $[1, 3]$, and we will express this by writing

The integral

$$\int_1^3 \sqrt{1 + x^3}\, dx = 6.2299\ldots .$$

Each Riemann sum approximates this integral, and in general the approximations get better as the size of the largest subinterval is made smaller. Moreover, as the subintervals get smaller, the location of the sampling points matters less and less.

The unusual symbol $\int$ that appears here reflects the historical origins of the integral. We'll have more to say about it after we consider the definition.

Definition

The purpose of the following definition is to give a name to the number to which the Riemann sums for a function converge, *when those sums do indeed converge.*

> **Definition.** Suppose all the Riemann sums for a function $f(x)$ on an interval $[a, b]$ get arbitrarily close to a single number when the lengths $\Delta x_1, \ldots, \Delta x_n$ are made small enough. Then this number is called the **integral** of $f(x)$ on $[a, b]$ and it is denoted
>
> $$\int_a^b f(x)\, dx.$$

The function f is called the **integrand**. The definition begins with a *Suppose ...* because there are functions whose Riemann sums don't converge. We'll look at an example on page 372. However, that example is quite special. All the functions that typically arise in context, and nearly all the functions we study in calculus, *do* have integrals. In particular, every continuous function has an integral, and so do many non-continuous functions—such as the step functions with which we began this chapter. (Continuous functions are discussed on pages 295–297).

A typical function has an integral

Notice that the definition doesn't speak about the choice of sampling points. The condition that the Riemann sums be close to a single number involves only the subintervals $\Delta x_1, \Delta x_2, \ldots, \Delta x_n$. This is important; it says *once the subintervals are small enough, it doesn't matter which sampling points x_k we choose—all of the Riemann sums will be close to the value of the integral.* (Of course, some will still be closer to the value of the integral than others.)

The integral allows us to resolve the dilemma we stated at the beginning of the chapter. Here is the dilemma: how can we describe How an integral expresses a product the product of two quantities when one of them varies? Consider, for example, how we expressed the energy consumption of a town over a 24-hour period. The basic relation

$$\text{energy} = \text{power} \times \text{elapsed time}$$

cannot be used directly, because power demand varies. Indirectly, though, we can use the relation to build a Riemann sum for power demand p over time. This gives us an *approximation*:

$$\text{energy} \approx \sum_{k=1}^{n} p(t_k) \, \Delta t_k \quad \text{megawatt-hours}.$$

As these sums are refined, two things happen. First, they converge to the true level of energy consumption. Second, they converge to the integral—by the definition of the integral. Thus, energy consumption is described *exactly* by the integral

$$\text{energy} = \int_0^{24} p(t) \, dt \quad \text{megawatt-hours}$$

of the power demand p. In other words, *energy is the integral of power over time.*

On page 346 we asked how far a ship would travel in 15 hours if we knew its velocity was $v(t)$ miles per hour at time t. We saw the distance could be estimated by a Riemann sum for the v. Therefore, reasoning just as we did for energy, we conclude that the *exact* distance is given by the integral

$$\text{distance} = \int_0^{15} v(t) \, dt \quad \text{miles}.$$

The energy integral has the same units as the Riemann sums that The units for an integral approximate it. Its units are the product of the megawatts used to measure p and the hours used to measure dt (or t). The units for the distance integral are the product of the miles per hour used to measure velocity and the hours used to measure time. In general, the units for the integral

$$\int_a^b f(x) \, dx$$

are the product of the units for f and the units for x.

Because the integral is approximated by its Riemann sums, we can use summation notation (introduced in the previous section) to write

$$\int_1^3 \sqrt{1 + x^3}\, dx \approx \sum_{k=1}^n \sqrt{1 + x_k^3}\, \Delta x_k.$$

This expression helps reveal where the rather unusual-looking notation for the integral comes from. In seventeenth century Europe (when calculus was being created), the letter 's' was written two ways: as 's' and as '$\int$'. The $\int$ that appears in the integral and the $\sum$ that appears in the Riemann sum both serve as abbreviations for the word *sum*. While we think of the Riemann sum as a sum of products of the form $\sqrt{1 + x_k^3} \cdot \Delta x_k$, in which the various Δx_k are small quantities, some of the early users of calculus thought of the integral as a sum of products of the form $\sqrt{1 + x^3} \cdot dx$, in which dx is an "infinitesimally" small quantity.

Now we do not use infinitesimals or regard the integral as a sum directly. On the contrary, for us the integral is a *limit* of Riemann sums as the subinterval lengths Δx_k all shrink to 0. In fact, we can express the integral directly as a limit:

$$\int_a^b f(x)\, dx = \lim_{\Delta x_k \to 0} \sum_{k=1}^n f(x_k)\, \Delta x_k.$$

The process of calculating an integral is called **integration**. Integration means "putting together." To see why this name is appropriate, notice that we determine energy consumption over a long time interval by putting together a lot of energy computations $p \cdot \Delta t$ over a succession of short periods.

A function that does not have an integral

Riemann sums converge to a single number for many functions—but not all. For example, Riemann sums for

$$J(x) = \begin{cases} 0 & \text{if } x \text{ rational} \\ 1 & \text{if } x \text{ is irrational} \end{cases}$$

do not converge. Let's see why.

A rational number is the quotient p/q of one integer p by another q. An irrational number is one that is *not* such a quotient; for example, $\sqrt{2}$ is irrational. The values of J are continually jumping between 0 and 1.

Suppose we construct a Riemann sum for J on the interval $[0, 1]$ Why Riemann sums for $J(x)$ do not converge using the subintervals $\Delta x_1, \Delta x_2, \ldots, \Delta x_n$. Every subinterval contains both rational and irrational numbers. Thus, we could choose all the sampling points to be rational numbers $r_1, r_2, \ldots, r_n$. In that case,

$$J(r_1) = J(r_2) = \cdots = J(r_n) = 0,$$

so the Riemann sum has the value

$$\sum_{k=1}^{n} J(r_k)\,\Delta x_k = \sum_{k=1}^{n} 0 \cdot \Delta x_k = 0.$$

But we could also choose all the sampling points to be irrational numbers $s_1, s_2, \ldots, s_n$. In that case,

$$J(s_1) = J(s_2) = \cdots = J(s_n) = 1,$$

and the Riemann sum would have the value

$$\sum_{k=1}^{n} J(s_k)\,\Delta x_k = \sum_{k=1}^{n} 1 \cdot \Delta x_k = \sum_{k=1}^{n} \Delta x_k = 1.$$

(For any subdivision Δx_k, $\Delta x_1 + \Delta x_2 + \cdots + \Delta x_n = 1$ because the subintervals together form the interval $[0, 1]$.) If some sampling points are rational and others irrational, the value of the Riemann sum will lie somewhere between 0 and 1. Thus, the Riemann sums range from 0 to 1, *no matter how small the subintervals Δx_k are chosen*. They cannot converge to any single number.

The function J shows us that not every function has an integral. The definition of the integral (page 370) takes this into account. It doesn't guarantee that an arbitrary function will have an integral. It simply says that *if* the Riemann sums converge to a single number, then we can give that number a name—the integral.

Can you imagine what the graph of $y = J(x)$ would look like? It would consist of two horizontal lines with gaps; in the upper line, the gaps would be at all the rational points, in the lower at the irrational points. This is impossible to draw! Roughly speaking, any graph you can draw you can integrate.

Visualizing the Integral

The eye plays an important role in our thinking. We visualize concepts whenever we can. Given a function $y = f(x)$, we visualize it as a graph in the x, y-coordinate plane. We visualize the derivative of f as the slope of its graph at any point. We can visualize the integral of f, too. We will view it as the area under that graph. Let's see why we can.

We have already made a connection between areas and Riemann sums, in last section. Our starting point was the basic formula

$$\text{area} = \text{height} \times \text{width}.$$

Since the height of the graph $y = f(x)$ at any point x is just $f(x)$, we were tempted to say that

$$\text{area} = f(x) \cdot (b - a).$$

Of course we couldn't do this, because the height $f(x)$ is variable. The remedy was to slice up the interval $[a, b]$ into small pieces Δx_k, and assemble a collection of products $f(x_k) \Delta x_k$:

$$\sum_{k=1}^{n} f(x_k) \Delta x_k = f(x_1) \Delta x_1 + f(x_2) \Delta x_2 + \cdots + f(x_n) \Delta x_n.$$

Riemann sums converge to the area ... This is a Riemann sum. It represents the total area of a row of side-by-side rectangles whose tops approximate the graph of f. As the Riemann sums are refined, the tops of the rectangles approach the shape of the graph, and their areas approach the area under the graph. But the **...and to the integral** process of refining Riemann sums leads to the integral, so the integral must be the area under the graph.

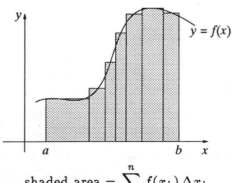

$$\text{shaded area} = \sum_{k=1}^{n} f(x_k) \Delta x_k$$

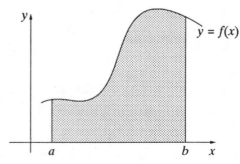

$$\text{shaded area} = \int_{a}^{b} f(x) \, dx$$

Every integral we have encountered can be visualized as the area under a graph. For instance, since

$$\text{energy use} = \int_0^{24} p(t)\, dt,$$

we can now say that the energy used by a town is just the area under its power demand graph.

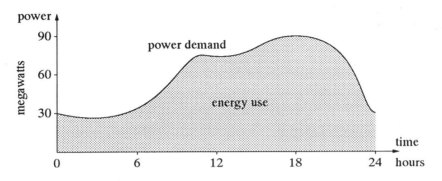

Energy used is the area under the power graph

Although we can always visualize an integral as an area, it may not be very enlightening in particular circumstances. For example, in the last section (page 352) we estimated the length of the graph $y = \sin x$ from $x = 0$ to $x = \pi$. Our estimates came from Riemann sums for the function $f(x) = \sqrt{1 + \cos^2 x}$ over the interval $[0, \pi]$. These Riemann sums converge to the integral

The area interpretation is not always helpful!

$$\int_0^{\pi} \sqrt{1 + \cos^2 x}\, dx,$$

which we can now view as the area under the graph of $\sqrt{1 + \cos^2 x}$.

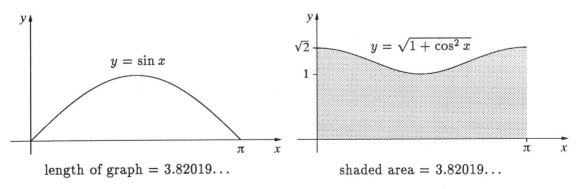

length of graph = 3.82019...   shaded area = 3.82019...

More generally, the length of $y = g(x)$ will always equal the area under $y = \sqrt{1 + (g'(x))^2}$.

The integral of a negative function

Up to this point, we have been dealing with a function $f(x)$ that is never negative on the interval $[a, b]$: $f(x) \geq 0$. Its graph therefore lies entirely above the x-axis. What happens if $f(x)$ *does* take on negative values? We'll first consider an example.

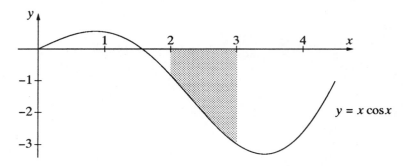

The graph of $f(x) = x \cos x$ is shown above. On the interval $[2, 3]$, it lies entirely below the x-axis. As you can check,

$$\int_2^3 x \cos x \, dx = -1.969080.$$

The integral is negative, but areas are positive. Therefore, it seems we can't interpret *this* integral as an area. But there is more to the story. The shaded region is 1 unit wide and varies in height from 1 to 3 units. If we say the average height is about 2 units, then the area is about 2 square units. Except for the negative sign, our rough estimate for the area is almost exactly the value of the integral.

The integral is the *negative* of the area

In fact, the integral of a *negative* function is always the *negative* of the area between its graph and the x-axis. To see why this is always true, we'll look first at the simplest possibility—a constant function.

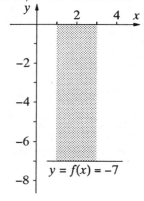

The integral of a constant function is just the product of that constant value by the width of the interval. For example, suppose $f(x) = -7$ on the interval $[1, 3]$. The region between the graph and the x-axis is a rectangle whose area is $7 \times 2 = 14$. However,

$$f(x) \cdot \Delta x = -7 \times 2 = -14.$$

This is the *negative* of the area of the region.

Let's turn now to an arbitrary function $f(x)$ whose values vary but remain negative over the interval $[a, b]$. Each term in the Riemann sum on the left is the *negative* of one of the shaded rectangles. In the process of refinement, the total area of the rectangles approaches the shaded area on the right. At the same time, the Riemann sums approach the integral. Thus, the integral must be the negative of the shaded area.

An arbitrary function that takes only negative values

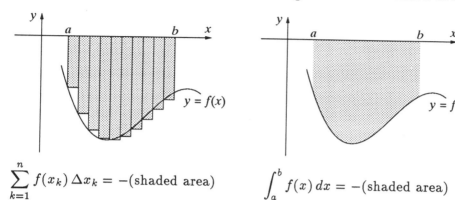

$$\sum_{k=1}^{n} f(x_k)\,\Delta x_k = -(\text{shaded area}) \qquad \int_{a}^{b} f(x)\,dx = -(\text{shaded area})$$

Functions with both positive and negative values

The final possibility to consider is that $f(x)$ takes both positive and negative values on the interval $[a, b]$. In that case its graph lies partly above the x-axis and partly below. By considering these two parts separately we can see that

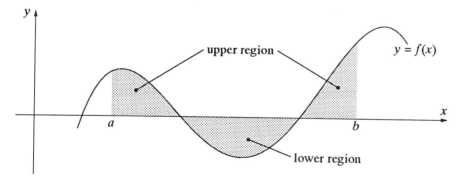

$$\int_{a}^{b} f(x)\,dx = (\text{area of upper region}) - (\text{area of lower region}).$$

The graph of $y = \sin x$ on the interval $[0, \pi]$ is the mirror image of the graph on the interval $[\pi, 2\pi]$. The first half lies above the x-axis, the second half below. Since the upper and lower areas are equal, it follows that

$$\int_0^{2\pi} \sin x \, dx = 0.$$

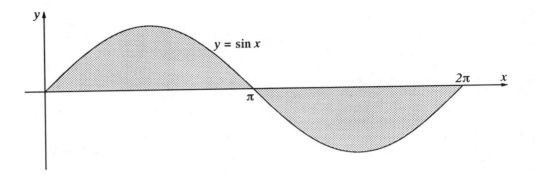

Signed area

There is a way to simplify the geometric interpretation of an integral as an area. It involves introducing the notion of *signed area*, by analogy with the notion of signed length.

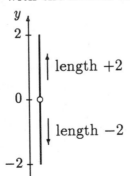

Consider the two points 2 and -2 on the y-axis at the right. Although the line that goes up from 0 to 2 has the same length as the line that goes down from 0 to -2 we customarily attach a sign to those lengths to take into account the *direction* of the line. Specifically, we assign a positive length to a line that goes up and a negative length to a line that goes down. Thus the line from 0 to -2 has **signed length** -2.

A region below the x-axis has negative area

To adapt this pattern to areas, just assign to any area that goes up from the x-axis a positive value and to any area that goes down from the x-axis a negative value. Then the **signed area** of a region that is partly above and partly below the x-axis is just the sum of the areas of the parts—taking the signs of the different parts into account.

Consider, for example, the graph of $y = x$ over the interval $[-2, 3]$. The upper region is a triangle whose area is 4.5. The lower region is another triangle; its area 2, and its *signed* area is -2. Thus, the total signed area is $+2.5$, and it follows that

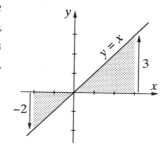

$$\int_{-2}^{3} x \, dx = 2.5.$$

You should confirm that Riemann sums for $f(x) = x$ over the interval $[-2, 3]$ converge to the value 2.5 (see exercise 13).

Now that we can describe the signed area of a region in the x, y-plane, we have a simple and uniform way to visualize the integral of any function:

$$\int_{a}^{b} f(x) \, dx = \text{the } \textbf{signed area}$$

between the graph of $f(x)$ and the x-axis.

Error Bounds

A Riemann sum determines the value of an integral only approximately. For example,

The error in a Riemann sum

$$\int_{1}^{3} \sqrt{1 + x^3} \, dx = 6.229959\ldots,$$

but a left endpoint Riemann sum with 100 equal subintervals Δx gives

$$\sum_{k=1}^{100} \sqrt{1 + x^3} \, \Delta x = 6.191236.$$

If we use this sum as an estimate for the value of the integral, we make an **error** of

$$6.229959 - 6.191236 = .038723.$$

By increasing the number of subintervals, we can reduce the size of the error. For example, with 100,000 subintervals, the error is only .000053. (This information comes from pages 367–369.) The fact that the first four digits in the error are now 0 means, roughly speaking, that the first four digits in the new estimate are correct.

Finding the error
without knowing
the exact value

In this example, we could measure the error in a Riemann sum because we knew the value of the integral. Usually, though, we *don't* know the value of the integral—that's why we're calculating Riemann sums! We will describe here a method to decide how inaccurate a Riemann sum is *without first knowing the value of the integral.* For example, suppose we estimate the value of

$$\int_0^1 e^{-x^2}\, dx$$

using a left endpoint Riemann sum with 1000 equal subintervals. The value we get is .747140. Our method will tell us that this differs from the true value of the integral by *no more than* .000633. So the method does not tell us the exact size of the error. It says only that the error

Error bounds

is not larger than .000633. Such a number is called an **error bound**. The actual error—that is, the true difference between the value of the integral and the value of the Riemann sum—may be a lot less than .000633. (That is indeed the case. In the exercises you are asked to show that the actual error is about half this number.)

We have two ways to indicate that .747140 is an estimate for the value of the integral, with an error bound of .000633. One is to use a "plus-minus" sign ($\pm$):

$$\int_0^1 e^{-x^2}\, dx = .747140 \pm .000633.$$

Since $.747140 - .000633 = .746507$ and $.747140 + .000633 = .747773$, this is the same as

$$.746507 \le \int_0^1 e^{-x^2}\, dx \le .747773.$$

Upper and lower
bounds

The number .746507 is called a **lower bound** for the integral, and .747773 is called an **upper bound**. Thus, the true value of the integral is .74..., and the third digit is either a 6 or a 7.

Our method will tell us even more. In this case, it will tell us that the original Riemann sum is *already* larger than the integral. In other words, we can drop the upper bound from .747773 to .747140:

$$.746507 \le \int_0^1 e^{-x^2}\, dx \le .747140.$$

The method

We want to get a bound on the difference between the integral of a function and a Riemann sum for that function. By visualizing both the integral and the Riemann sum as areas, we can visualize their difference as an area, too. We'll assume that all subintervals in the Riemann sum have the same width Δx. This will help keep the details simple. Thus

Visualize the error as an area

$$\text{error} = \left| \int_a^b f(x)\, dx - \sum_{k=1}^n f(x_k)\, \Delta x \right|.$$

(The absolute value $|u - v|$ of the difference tells us how far apart u and v are.)

We'll also assume that the function $f(x)$ is *positive* and *increasing* on an interval $[a, b]$. We say $f(x)$ is **increasing** if its graph rises as x goes from left to right.

Let's start with left endpoints for the Riemann sum. Because $f(x)$ is increasing, the rectangles lie entirely below the graph of the function:

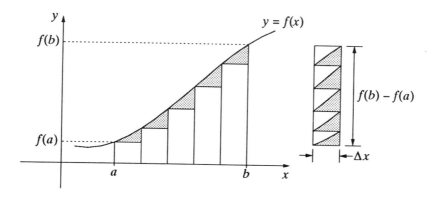

The integral is the area under the graph, and the Riemann sum is the area of the rectangles. Therefore, the error is the area of the shaded region that lies above the rectangles and below the graph. This region consists of a number of separate pieces that sit on top of the individual rectangles. Slide them to the right and stack them on top of one another, as shown in the figure. They fit together inside a single rectangle of width Δx and height $f(b) - f(a)$. Thus the error (which is the area of the shaded region) is no greater than the area of this rectangle:

Slide the little errors into a stack

$$\text{error} \le \Delta x \, (f(b) - f(a)).$$

The number $\Delta x\,(f(b) - f(a))$ is our error bound. It is clear from the figure that this is not the *exact* value of the error. The error is smaller, but it is difficult to say exactly how much smaller. Notice also that we need very little information to find the error bound—just Δx and the function values $f(a)$ and $f(b)$. We do *not* need to know the exact value of the integral!

The error bound is proportional to Δx

The error bound is proportional to Δx. If we cut Δx in half, that will cut the error bound in half. If we make Δx a tenth of what it had been, that will make the error bound a tenth of what *it* had been. In the figure below, Δx is 1/5-th its value in the previous figure. The rectangle on the right shows how much smaller the error bound has become as a result. It demonstrates how Riemann sums converge as the size of the subinterval Δx shrinks to zero.

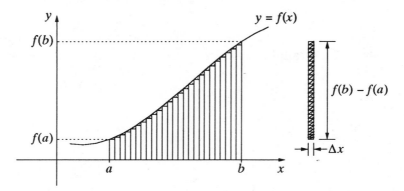

Right endpoints

Let's go back to the original Δx, but switch to *right* endpoints. Then the tops of the rectangles lie above the graph. The error is the vertically hatched region between the graph and the tops of the rectangles. Once again, we can slide the little errors to the right and stack

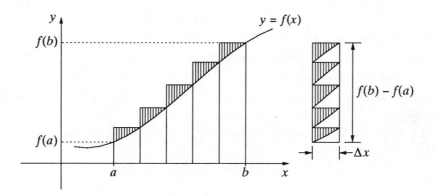

them on top of one another. They fit inside the same rectangle we had before. Thus, whether the Riemann sum is constructed with left endpoints or right endpoints, we find the same error bound:

$$\text{error} \le \Delta x \left(f(b) - f(a) \right).$$

Because $f(x)$ is increasing, the left endpoint Riemann sum is smaller than the integral, while the right endpoint Riemann sum is larger. On a number line, these three values are arranged as follows:

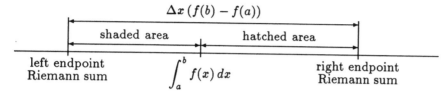

The distance from the left endpoint Riemann sum to the integral is represented by the shaded area, and the distance from the right endpoint by the hatched area. Notice that these two areas exactly fill the rectangle that gives us the error bound. Thus the distance between the two Riemann sums on the number line is exactly $\Delta x \left(f(b) - f(a) \right)$, as shown.

Finally, suppose that the Riemann sum has *arbitrary* sampling points x_k:

Arbitrary sampling points

$$\sum_{k=1}^{n} f(x_k) \, \Delta x.$$

Since f is increasing on the interval $[a, b]$, its values get larger as x goes from left to right. Therefore, on the k-th subinterval,

$$f(\text{left endpoint}) \le f(x_k) \le f(\text{right endpoint}).$$

In other words, the rectangle built over the left endpoint is the shortest, and the one built over the right endpoint is the tallest. The one built over the sampling point x_k lies somewhere in between.

The areas of these three rectangles are arranged in the same order:

$$f(\text{left endpoint}) \, \Delta x \le f(x_k) \, \Delta x \le f(\text{right endpoint}) \, \Delta x.$$

If we add up these areas, we get Riemann sums. The Riemann sums are arranged in the same order as their individual terms:

$$\left\{ \begin{matrix} \text{left endpoint} \\ \text{Riemann sum} \end{matrix} \right\} \le \sum_{k=1}^{n} f(x_k) \, \Delta x \le \left\{ \begin{matrix} \text{right endpoint} \\ \text{Riemann sum} \end{matrix} \right\}.$$

Thus, the left endpoint and the right endpoint Riemann sums are extremes: *every* Riemann sum for f that uses a subinterval size of Δx lies between these two.

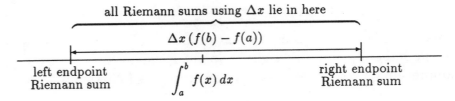

It follows that $\Delta x \left(f(b) - f(a) \right)$ is an error bound for all Riemann sums whose subintervals are Δx units wide.

Error bounds for a decreasing function

If $f(x)$ is a positive function but is *decreasing* on the interval $[a, b]$, we get essentially the same result. Any Riemann sum for f that uses a subinterval of size Δx differs from the integral by no more than $\Delta x \left(f(a) - f(b) \right)$. When f is decreasing, however, $f(a)$ is larger than $f(b)$, so we write the height of the rectangle as $f(a) - f(b)$. To avoid having to pay attention to this distinction, we can use absolute values to describe the error bound:

$$\text{error} \leq \Delta x \left| f(b) - f(a) \right|.$$

Furthermore, when f is decreasing, the left endpoint Riemann sum is larger than the integral, while the right endpoint Riemann sum is smaller. The difference between the right and the left endpoint Riemann sums is still $\Delta x \left| f(b) - f(a) \right|$.

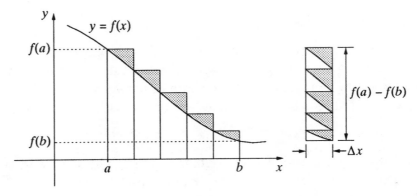

Monotonic functions

Up to this point, we have assumed $f(x)$ was either always increasing, or else always decreasing, on the interval $[a, b]$. Such a function is said to be **monotonic**. If $f(x)$ is *not* monotonic, the process of getting an error bound for Riemann sums is only slightly more complicated.

Here is how to get an error bound for the Riemann sums constructed for a non-monotonic function. First break up the interval $[a, b]$ into smaller pieces on which the function *is* monotonic.

Error bounds for non-monotonic functions

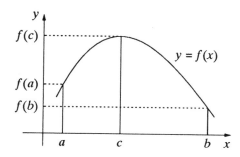

In this figure, there are two such intervals: $[a, c]$ and $[c, b]$. Suppose we construct a Riemann sum for f by using rectangles of width Δx_1 on the first interval, and Δx_2 on the second. Then the total error for this sum will be no larger than the sum of the error bounds on the two intervals:

The monotonic pieces

$$\text{total error} \le \Delta x_1 \, |f(c) - f(a)| + \Delta x_2 \, |f(b) - f(c)|.$$

By making Δx_1 and Δx_2 sufficiently small, we can make the error as small as we wish.

This method can be applied to any non-monotonic function that can be broken up into monotonic pieces. For other functions, more than two pieces may be needed.

Using the method

Earlier we said that when we use a left endpoint Riemann sum with 1000 equal subintervals to estimate the value of the integral

An example

$$\int_0^1 e^{-x^2} \, dx,$$

the error is no larger than .000633. Let's see how our method would lead to this conclusion.

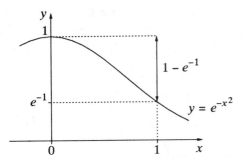

On the interval $[0, 1]$, $f(x) = e^{-x^2}$ is a decreasing function. Furthermore,

$$f(0) = 1 \qquad f(1) = e^{-1} \approx .3679.$$

If we divide $[0, 1]$ into 1000 equal subintervals Δx, then $\Delta x = 1/1000 = .001$. The error bound is therefore

$$.001 \times |.3679 - 1| = .001 \times .6321 = .0006321.$$

Any number *larger* than this one will also be an error bound. By "rounding up," we get .000633. This is slightly shorter to write, and it is the bound we claimed earlier. An even shorter bound is .00064.

Furthermore, since $f(x) = e^{-x^2}$ is decreasing, any Riemann sum constructed with *left* endpoints is larger than the actual value of the integral. Since the left endpoint Riemann sum with 1000 equal subdivisions has the value .747140, upper and lower bounds for the integral are

$$.746507 = .747140 - .000633 \le \int_0^1 e^{-x^2}\, dx \le .747140.$$

Integration Rules

Just as there are rules that tell us how to find the derivative of various combinations of functions, there are other rules that tell us how to find the integral. Here are three that are exactly analogous to differentiation rules.

$$\int_a^b f(x) + g(x)\, dx = \int_a^b f(x)\, dx + \int_a^b g(x)\, dx$$

$$\int_a^b f(x) - g(x)\, dx = \int_a^b f(x)\, dx - \int_a^b g(x)\, dx$$

$$\int_a^b c\, f(x)\, dx = c \int_a^b f(x)\, dx$$

Let's see why the third rule is true. A Riemann sum for the integral on the left looks like

Compare
Riemann sums

$$\sum_{k=1}^n c\, f(x_k)\, \Delta x_k = c\, f(x_1)\, \Delta x_1 + \cdots + c\, f(x_n)\, \Delta x_n.$$

Since the factor c appears in every term in the sum, we can move it outside the summation:

$$c\left(f(x_1)\, \Delta x_1 + \cdots + f(x_n)\, \Delta x_n\right) = c\left(\sum_{k=1}^n f(x_k)\, \Delta x_k\right).$$

The new expression is c times a Riemann sum for

$$\int_a^b f(x)\, dx.$$

Since the Riemann sum expressions are equal, the integral expressions they converge to must be equal, as well. You can use similar arguments to show why the other two rules are true.

Here is one example of the way we can use these rules:

Using the rules

$$\int_1^3 4\sqrt{1 + x^3}\, dx = 4 \int_1^3 \sqrt{1 + x^3}\, dx = 4 \times 6.229959 = 24.919836.$$

(The value of the second integral is given on page 379.) Here is another example:

$$\int_2^9 5x^7 - 2x^3 + 24x\, dx = 5 \int_2^9 x^7\, dx - 2 \int_2^9 x^3\, dx + 24 \int_2^9 x\, dx.$$

Of course, we must still determine the value of various integrals of the form

$$\int_a^b x^n\, dx.$$

However, the example shows us that, once we know the value of these special integrals, we can determine the value of the integral of any polynomial.

Here are two more rules that have no direct analogue in differentiation. The first says that if $f(x) \leq g(x)$ for every x in the interval $[a, b]$, then

$$\int_a^b f(x)\, dx \leq \int_a^b g(x)\, dx.$$

In the second, c is a point somewhere in the interval $[a, b]$:

$$\int_a^b f(x)\, dx = \int_a^c f(x)\, dx + \int_c^b f(x)\, dx.$$

If you visualize an integral as an area, it is clear why these rules are true.

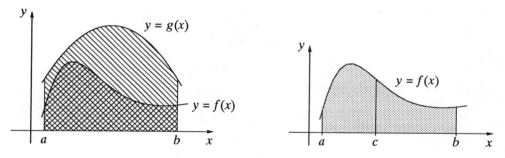

The second rule can be used to understand the results of exercise 4 on page 360. It concerns the three integrals

$$\int_1^3 \sqrt{1 + x^3}\, dx = 6.229959, \qquad \int_3^7 \sqrt{1 + x^3}\, dx = 45.820012,$$

and

$$\int_1^7 \sqrt{1 + x^3}\, dx = 52.049971.$$

Since the third interval $[1, 7]$ is just the first $[1, 3]$ combined the second $[3, 7]$, the third integral is the sum of the first and the second. The numerical values confirm this.

Exercises

1. Determine the values of the following integrals.

a) $\int_2^{15} 3\, dx$ b) $\int_{-5}^{-3} 7\, dx$ c) $\int_{-4}^{9} -2\, dz$

2. a) Sketch the graph of

$$g(x) = \begin{cases} 7 & \text{if } 1 \le x < 5, \\ -3 & \text{if } 5 \le x \le 10 \end{cases}$$

b) Determine $\int_1^7 g(x)\, dx$, $\int_7^{10} g(x)\, dx$, and $\int_1^{10} g(x)\, dx$.

Refining Riemann sums

3. a) By refining Riemann sums, find the value of the following integral to four decimal places accuracy. Do the computations twice: first, using left endpoints; second, using midpoints.

$$\int_0^1 \frac{1}{1 + x^3}\, dx.$$

b) How many subintervals did you need to get four decimal places accuracy when you used left endpoints and when you used midpoints? Which sampling points gave more efficient computations—left endpoints or midpoints?

4. By refining appropriate Riemann sums, determine the value of each of the following integrals, accurate to four decimal places. Use whatever sampling points you wish, but justify your claim that your answer is accurate to four decimal places.

a) $\int_1^4 \sqrt{1 + x^3}\, dx$ b) $\int_4^7 \sqrt{1 + x^3}\, dx$ c) $\int_0^3 \frac{\cos x}{1 + x^2}\, dx$

[Answer: $\int_0^3 \frac{\cos x}{1 + x^2}\, dx = .6244\ldots$.]

5. Determine the value of the following integrals to four decimal places accuracy.

a) $\int_1^2 e^{-x^2}\, dx$ c) $\int_0^4 \sin(x^2)\, dx$

b) $\int_0^4 \cos(x^2)\, dx$ d) $\int_0^1 \frac{4}{1 + x^2}\, dx$

6.   a) What is the length of the graph of $y = \sqrt{x}$ from $x = 1$ to $x = 4$?

b)   What is the length of the graph of $y = x^2$ from $x = 1$ to $x = 2$?

c)   Why are the answers in parts (a) and (b) the same?

7.   Both of the curves $y = 2^x$ and $y = 1 + x^{3/2}$ pass through $(0, 1)$ and $(1, 2)$. Which is the shorter one? Can you decide simply by looking at the graphs?

8.   A pyramid is 30 feet tall. The area of a horizontal cross-section x feet from the top of the pyramid measures $2x^2$ square feet. What is the area of the base? What is the volume of the pyramid, to the nearest cubic foot?

Error bounds

9.   A left endpoint Riemann sum with 1000 equally spaced subintervals gives the estimate .135432 for the value of the integral

$$\int_1^2 e^{-x^2} \, dx.$$

a)   Is the true value of the integral larger or smaller than this estimate? Explain.

b)   Find an error bound for this estimate.

c)   Using the information you have already assembled, find lower and upper bounds A and B:

$$A \le \int_1^2 e^{-x^2} \, dx \le B.$$

d)   The lower and upper bounds allow you to determine a certain number of digits in the *exact* value of the integral. How many digits do you know, and what are they?

10.   A left endpoint Riemann sum with 100 equally spaced subintervals gives the estimate .342652 for the value of the integral

$$\int_0^{\pi/4} \tan x \, dx.$$

a) Is the true value of the integral larger or smaller than this estimate? Explain.

b) Find an error bound for this estimate.

c) Using the information you have already assembled, find lower and upper bounds A and B:

$$A \le \int_0^{\pi/4} \tan x \, dx \le B.$$

d) The lower and upper bounds allow you to determine a certain number of digits in the *exact* value of the integral. How many digits do you know, and what are they?

11. a) In the next section you will see that

$$\int_0^{\pi/2} \sin x \, dx = 1$$

exactly. Here, estimate the value by a Riemann sum using the left endpoints of 100 equal subintervals.

b) Find an error bound for this estimate, and use it to construct the best possible lower and upper bounds

$$A \le \int_0^{\pi/2} \sin x \, dx \le B.$$

12. In the text (page '367), a Riemann sum using left endpoints on 1000 equal subintervals produces an estimate of 6.226083 for the value of

$$\int_1^3 \sqrt{1 + x^3} \, dx.$$

a) Is the true value of the integral larger or smaller than this estimate? Explain your answer, and do so without referring to the fact that the true value of the integral is known to be 6.229959....

b) Find an error bound for this estimate.

c) Find the upper and lower bounds for the value of the integral that are determined by this estimate.

d) According to these bounds, how many digits of the value of the integral are now known for certain?

The average value of a function

In the exercises for §1 we saw that the average staffing level for a job is

$$\text{average staffing} = \frac{\text{total staff-hours}}{\text{hours worked}}.$$

If $S(t)$ represents the number of staff working at time t, then the total staff-hours accumulated between $t = a$ and $t = b$ hours is

$$\text{total staff-hours} = \int_a^b S(t)\, dt \quad \text{staff-hours}.$$

Therefore the average staffing is

$$\text{average staffing} = \frac{1}{b-a} \int_a^b S(t)\, dt \quad \text{staff}.$$

Likewise, if a town's power demand was $p(t)$ megawatts at t hours, then its average power demand between $t = a$ and $t = b$ hours is

$$\text{average power demand} = \frac{1}{b-a} \int_a^b p(t)\, dt \quad \text{megawatts}.$$

Geometric meaning of the average

We can define the **average value** of an arbitrary function $f(x)$ over an interval $a \le x \le b$ by following this pattern:

$$\text{average value of } f = \frac{1}{b-a} \int_a^b f(x)\, dx.$$

The average value of f is sometimes denoted $\overline{f}$. Since

$$\overline{f} \cdot (b - a) = \int_a^b f(x)\, dx,$$

the area under the horizontal line $y = \overline{f}$ between $x = a$ and $x = b$ is the same as the area under the graph $y = f(x)$. See the graph on the opposite page.

13.　a) What is the average value of $f(x) = 5$ on the interval $[1, 7]$?

b) What is the average value of $f(x) = \sin x$ on the interval $[0, 2\pi]$? On $[0, 100\pi]$?

c) What is the average value of $f(x) = \sin^2 x$ on the interval $[0, 2\pi]$? On $[0, 100\pi]$?

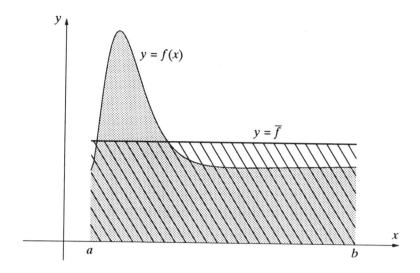

14. a) What is the average value of the function

$$H(x) = \begin{cases} 1 & \text{if} \quad 0 \le x < 4, \\ 12 & \text{if} \quad 4 \le x < 6 \\ 1 & \text{if} \quad 6 \le x \le 20 \end{cases}$$

on the interval $[0, 20]$?

b) Is the average $\overline{H}$ larger or smaller than the average of the two numbers 12 and 1 that represent the largest and smallest values of the function?

c) Sketch the graph $y = H(x)$ along with the horizontal line $y = \overline{H}$, and show directly that the same area lies under each of these two graphs over the interval $[0, 20]$.

15. a) What are the maximum and minimum values of $f(x) = x^2 e^{-x}$ on the interval $[0, 20]$? What is the *average* of the maximum and the minimum?

b) What is the average $\overline{f}$ of $f(x)$ on the interval $[0, 20]$?

c) Why aren't these two averages the same?

The integral as a signed area

16. By refining Riemann sums, confirm that $\displaystyle\int_{-2}^{3} x \, dx = 2.5$.

17.   a) Sketch the graphs of $y = \cos x$ and $y = 5 + \cos x$ over the interval $[0, 4\pi]$.

b) Find $\int_0^{4\pi} \cos x \, dx$ by visualizing the integral as a signed area.

c) Find $\int_0^{4\pi} 5 + \cos x \, dx$. Why does $\int_0^{4\pi} 5 \, dx$ have the same value?

18.   a) By refining appropriate Riemann sums, determine the value of the integral $\int_0^{\pi} \sin^2 x \, dx$ to four decimal places accuracy.

b) Sketch the graph of $y = \sin^2 x$ on the interval $0 \leq x \leq \pi$. Note that your graph lies inside the rectangle formed by the lines $y = 0$, $y = 1$, $x = 0$ and $x = \pi$. (Sketch this rectangle.)

c) Explain why the area under the graph of $y = \sin^2 x$ is exactly *half* of the area of the rectangle you sketched in part (b). What is the area of that rectangle?

d) Using your observations in part (c), explain why $\int_0^{\pi} \sin^2 x \, dx$ is *exactly $\pi/2$*.

19.   a) On what interval $a \leq x \leq b$ does the graph of the function $y = 4 - x^2$ lie *above* the x-axis?

b) Sketch the graph of $y = 4 - x^2$ on the interval $a \leq x \leq b$ you determined in part (a).

c) What is the area of the region that lies above the x-axis and below the graph of $y = 4 - x^2$?

20.   a) What is the signed area (see page 378 in the text) between the graph of $y = x^3 - x$ and the x-axis on the interval $-1 \leq x \leq 2$?

b) Sketch the graph of $y = x^3 - x$ on the interval $-1 \leq x \leq 2$. On the basis of your sketch, support or refute the following claim: the *signed* area between the graph of $y = x^3 - x$ and the x-axis on the interval $-1 \leq x \leq 2$ is exactly the same as the area between the graph of $y = x^3 - x$ and the x-axis on the interval $+1 \leq x \leq 2$.

§4. The Fundamental Theorem of Calculus

Two Views of Power and Energy

In §1 we considered how much energy a town consumed over 24 hours when it was using $p(t)$ megawatts of power at time t. Suppose $E(T)$ megawatt-hours of energy were consumed during the first T hours. Then the integral, introduced in §3, gave us the language to describe how E depends on p:

$$E(T) = \int_0^T p(t)\, dt.$$

Energy is the integral of power...

Because E is the integral of p, we can visualize $E(T)$ as the area under the power graph $y = p(t)$ as the time t sweeps from 0 hours to T hours:

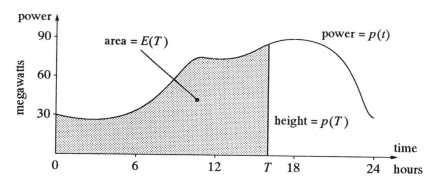

As T increases, so does $E(T)$. The exact relation between E and T is shown in the graph below. The height of the graph of E at any point T is equal to the area under the graph of p from 0 out to the point T.

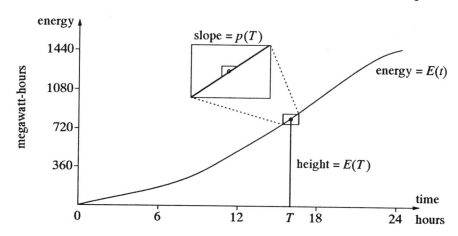

The microscope window on the graph of E reminds us that the slope of the graph at any point T is just $p(T)$:

...and power is the derivative of energy

$$p(T) = E'(T).$$

We discovered this fact in §1, where we stated it in the following form: *power is the rate at which energy is consumed.*

Pause now to study the graphs of power and energy. You should convince yourself that the *height* of the E graph at any time equals the *area* under the p graph up to that time. For example, when $T = 0$ no area has accumulated, so $E(0) = 0$. Furthermore, up to $T = 6$ hours, power demand was almost constant at about 30 megawatts. Therefore, $E(6)$ should be about

$$30 \text{ megawatts} \times 6 \text{ hours} = 180 \text{ megawatt-hours.}$$

It is. You should also convince yourself that the *slope* of the E graph at any point equals the *height* of the p graph at that point. Thus, for example, the graph of E will be steepest where the graph of p is tallest.

Notice that when we write the energy accumulation function $E(T)$ as an integral,

$$E(T) = \int_0^T p(t)\, dt,$$

A new ingredient: variable limits of integration

we have introduced a new ingredient. The time variable T appears as one of the "limits of integration." By definition, the integral is a single number. However, that number depends on the interval of integration $[0, T]$. As soon as we treat T as a variable, the integral itself becomes a variable, too. Here the value of the integral varies with T.

We now have two ways of viewing the relation between power and energy. According to the first view, the energy accumulation function E is the integral of power demand:

$$E(T) = \int_0^T p(t)\, dt.$$

An integral is a solution to a differential equation

According to the second view, the energy accumulation function is the solution $y = E(t)$ to an initial value problem defined by power demand:

$$y' = p(t); \qquad y(0) = 0.$$

If we take the first view, then we find E by refining Riemann sums— because that is the way to determine the value of an integral. If we take the second view, then we can find E by using any of the methods for solving initial value problems that we studied in chapter 4. Thus, the energy integral is a solution to a certain differential equation.

    This is unexpected. Differential equations involve derivatives. At first glance, they have nothing to do with integrals. Nevertheless, the relation between power and energy shows us that there is a deep connection between derivatives and integrals. As we shall see, the connection holds for the integral of *any* function. The connection is so important— because it links together the two basic processes of calculus—that it has been called the **fundamental theorem of calculus**.

The fundamental theorem of calculus

    The fundamental theorem gives us a powerful new tool to calculate integrals. Our aim in this section is to see why the theorem is true, and to begin to explore its use as a tool. In Calculus II we will consider many specific integration techniques that are based on the fundamental theorem.

Integrals and Differential Equations

We begin with a statement of the fundamental theorem for a typical function $f(x)$.

The Fundamental Theorem of Calculus

The solution $y = A(x)$ to the initial value problem

$$y' = f(x) \qquad y(a) = 0$$

is the **accumulation function** $A(X) = \displaystyle\int_a^X f(x)\, dx.$

We have always been able to find the value of the integral

$$\int_a^b f(x)\, dx$$

by refining Riemann sums. The fundamental theorem gives us a new way. It says: First, find the solution $y = A(x)$ to the initial value problem $y' = f(x)$, $y(a) = 0$ using any suitable method for solving the differential equation. Then, once we have $A(x)$, we get the value of the integral by evaluating A at $x = b$.

A test case   To see how all this works, let's find the value of the integral

$$\int_0^4 \cos(x^2)\,dx$$

two ways: by refining Riemann sums and by solving the initial value problem

$$y' = \cos(x^2); \quad y(0) = 0$$

using Euler's method.

RIEMANN   To estimate the value of the integral, we use the program RIEMANN
versus...   from page 356. It produces the sequence of left endpoint sums shown in
the table below on the left. Since the first three digits have stabilized,
the value of the integral is .594 . . . . The integral was deliberately chosen
to lead to the initial value problem we first considered on page 260. We
...TABLE   solved that problem by Euler's method, using the program TABLE,
and produced the table of estimates for the value of $y(4)$ that appear
in the table on the right. We see $y(4) = .594 . . . .$

<table>
<tr><td>Program: RIEMANN
Left endpoint Riemann sums</td><td>Program: TABLE
Euler's method</td></tr>
</table>

```
DEF fnf (x) = COS(x ^ 2)
a = 0
b = 4
numberofsteps = 2 ^ 3
deltax = (b - a) / numberofsteps
x = a
accumulation = 0
FOR k = 1 TO numberofsteps
    deltaS = fnf(x) * deltax
    accumulation = accumulation + deltaS
    x = x + deltax
NEXT k
PRINT accumulation
```

```
DEF fnf (t) = COS(t ^ 2)
tinitial = 0
tfinal = 4
numberofsteps = 2 ^ 3
deltat = (tfinal - tinitial) / numberofsteps
t = tinitial
accumulation = 0
FOR k = 1 TO numberofsteps
    deltay = fnf(t) * deltat
    accumulation = accumulation + deltay
    t = t + deltat
NEXT k
PRINT accumulation
```

number of steps	estimated value of the integral		number of steps	estimated value of $y(4)$
2^3	1.13304		2^3	1.13304
2^6	.65639		2^6	.65639
2^9	.60212		2^9	.60212
2^{12}	.59542		2^{12}	.59542
2^{15}	.59458		2^{15}	.59458
2^{18}	.59448		2^{18}	.59448

RIEMANN estimates the value of an integral by calculating Riemann sums. TABLE solves a differential equation by constructing estimates using Euler's method. These appear to be quite different tasks, but they lead to exactly the same results! But this is no accident. Compare the programs. (In both, the PRINT statement has been put outside the loop. This speeds up the calculations but still gives us the final outcome.) Once you make necessary modifications (e.g., change x to t, a to tinitial, *et cetera*), you can see the two programs are the same.

The two programs are the same

The very fact that these two programs do the same thing gives us one proof of the fundamental theorem of calculus.

Graphing accumulation functions

Let's take a closer look at the solution $y = A(x)$ to the initial value problem

$$y' = \cos(x^2); \qquad y(0) = 0.$$

We can write it as the accumulation function

$$A(X) = \int_0^X \cos(x^2)\, dx.$$

By switching to the graphing version of TABLE (this is the program PLOT, shown on page 261), we can graph A. Here is the result.

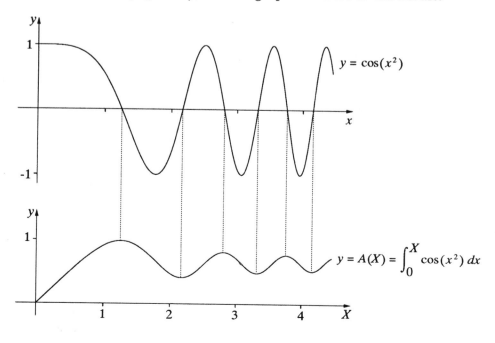

The relation between $y = \cos(x^2)$ and $y = A(x)$ is the same as the relation between power and energy.

- The *height* of the graph $y = A(x)$ at any point $x = X$ is equal to the *signed area* between the graph $y = \cos(x^2)$ and the x-axis over the interval $0 \le x \le X$.

- On the intervals where $\cos(x^2)$ is positive, $A(x)$ is increasing. On the intervals where $\cos(x^2)$ is negative, $A(x)$ is decreasing.

- When $\cos(x^2) = 0$, $A(x)$ has a maximum or a minimum.

- The *slope* of the graph of $y = A(x)$ at any point $x = X$ is equal to the *height* of the graph $y = \cos(x^2)$ at that point.

In summary, the lower curve is the integral of the upper one, and the upper curve is the derivative of the lower one.

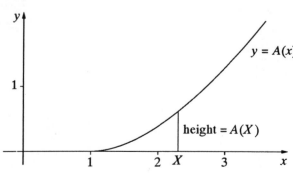

To get a better idea of the simplicity and power of this approach to integrals, let's look at another example:

$$A(X) = \int_1^X \ln x \, dx.$$

We find A, as always, by solving the initial value problem

$$y' = \ln x; \qquad y(1) = 0$$

using Euler's method. The graphs $y = \ln x$ and $y = A(x)$ are shown at the left. Notice once again that the height of the graph of $A(x)$ at any point $x = X$ equals the area under the graph of $y = \ln x$ from $x = 1$ to $x = X$. Also, the graph of A becomes steeper as the height of $\ln x$ increases. In particular, the graph of A is horizontal (at $x = 1$) when $\ln x = 0$.

Antiderivatives

We have developed a novel approach to integration in this section. We start by replacing a given integral by an accumulation function:

Find an integral...

$$\int_a^b f(x)\,dx \qquad \rightsquigarrow \qquad A(X) = \int_a^X f(x)\,dx.$$

Then we try to find $A(X)$. If we do, then the original integral is just the value of A at $X = b$.

At first glance, this doesn't seem to be a sensible approach. We appear to be making the problem harder: instead of searching for a single number, we must now find an entire function. However, we know that $y = A(x)$ solves the initial value problem

$$y' = f(x), \qquad y(a) = 0.$$

This means we can use the complete 'bag of tools' we have for solving differential equations to find A. The real advantage of the new approach is that it reduces integration to the fundamental activity of calculus—solving differential equations.

...by solving a differential equation ...

The differential equation $y' = f(x)$ that arises in integration problems is special. The right hand side depends only on the input variable x. We studied this differential equation in chapter 4, §5, where we developed a special method to solve it—**antidifferentiation.**

...using antidifferentiation

We say that $F(x)$ is an **antiderivative** of $f(x)$ if f is the derivative of F: $F'(x) = f(x)$. Here are some examples (from page 257):

function	antiderivative
$5x^4 - 2x^3$	$x^5 - \frac{1}{2}x^4$
$5x^4 - 2x^3 + 17x$	$x^5 - \frac{1}{2}x^4 + \frac{17}{2}x^2$
$6 \cdot 10^z + 17/z^5$	$6 \cdot 10^z / \ln 10 - 17/6z^6$
$3\sin t - 2t^3$	$-3\cos t - \frac{1}{2}t^4$
$\pi \cos x + \pi^2$	$\pi \sin x + \pi^2 x$

Since $y = A(x)$ solves the differential equation $y' = f(x)$, we have $A'(x) = f(x)$. Thus, $A(x)$ is an antiderivative of $f(x)$, so we can try to find A by antidifferentiating f.

A is an antiderivative of f

Here is an example. Suppose we want to find

$$A(X) = \int_2^X 5x^4 - 2x^3 \, dx.$$

Notice that $f(x) = 5x^4 - 2x^3$ is the first function in the previous table. Since A must be an antiderivative of f, let's try the antiderivative for f that we find in the table:

$$F(x) = x^5 - \tfrac{1}{2}x^4.$$

Check the condition $A(2) = 0$

The problem is that F must also satisfy the initial condition $F(2) = 0$. However,

$$F(2) = 2^5 - \tfrac{1}{2} \cdot 2^4 = 32 - \tfrac{1}{2} \cdot 16 = 24 \neq 0,$$

so the initial condition does not hold *for this particular choice* of antiderivative. But this problem is easy to fix. Let

$$A(x) = F(x) - 24.$$

Since $A(x)$ differs from $F(x)$ only by a constant, it has the same derivative—namely, $5x^4 - 2x^3$. So $A(x)$ is still an antiderivative of $5x^4 - 2x^3$. But it also satisfies the initial condition:

$$A(2) = F(2) - 24 = 24 - 24 = 0.$$

Therefore $A(x)$ solves the problem; it has the right derivative *and* the right value at $x = 2$. Thus we have a formula for the accumulation function

$$\int_2^X 5x^4 - 2x^3 \, dx = X^5 - \tfrac{1}{2}X^4 - 24.$$

The key step in finding the correct accumulation function A was to recognize that a given function f has infinitely many antiderivatives: if F is an antiderivative, then so is $F + C$, for any constant C. The general procedure for finding an accumulation function involves these two steps:

> To find $A(X) = \int_a^X f(x) \, dx$:
>
> 1. first find *an* antiderivative $F(x)$ of $f(x)$,
> 2. then set $A(x) = F(x) - F(a)$.

Comment: Recall that some functions f simply cannot be integrated— the Riemann sums they define may not converge. Although we have not stated it explicitly, you should keep in mind that the procedure just described applies only to functions that can be integrated.

A caution

Example. To illustrate the procedure, let's find the accumulation function

$$A(X) = \int_0^X \sin x\, dx.$$

The first step is to find an antiderivative for $f(x) = \sin x$. A natural choice is $F(x) = -\cos x$. To carry out the second step, note that $a = 0$. Since $F(0) = -\cos 0 = -1$, we set

$$A(x) = F(x) - F(0) = -\cos x - (-1) = 1 - \cos x.$$

The graphs $y = \sin x$ and $y = 1 - \cos x$ are shown below.

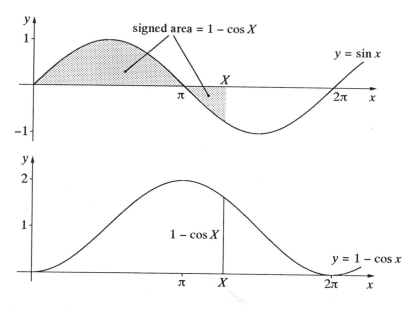

Now that we have a formula for $A(x)$ we can find the exact value of integrals involving $\sin x$. For instance,

Evaluating specific integrals

$$\int_0^\pi \sin x\, dx = A(\pi) = 1 - \cos \pi = 1 - (-1) = 2.$$

Also,

$$\int_0^{\pi/2} \sin x\, dx = A(\pi/2) = 1 - \cos \pi/2 = 1 - (0) = 1,$$

and
$$\int_0^{2\pi} \sin x \, dx = A(2\pi) = 1 - \cos 2\pi = 1 - (1) = 0.$$

These values are *exact*, and we got them without calculating Riemann sums. However, we have already found the value of the third integral. On page 378 we argued from the shape of the graph $y = \sin x$ that the signed area between $x = 0$ and $x = 2\pi$ must be 0. That argument is no help in finding the area to $x = \pi$ or to $x = \pi/2$, though. Before the fundamental theorem showed us we could evaluate integrals by finding antiderivatives, we could only make estimates using Riemann sums. Here, for example, are midpoint Riemann sums for

$$\int_0^{\pi/2} \sin x \, dx :$$

subintervals	Riemann sum
10	1.001 028 868
100	1.000 010 325
1000	1.000 000 147
10000	1.000 000 045

According to the table, the value of the integral is 1.000000..., to six decimal places accuracy. That is valuable information, and is often as accurate as we need. However, the fundamental theorem tells us that the value of the integral is 1 *exactly*! With the new approach, we can achieve absolute precision.

The fundamental theorem makes absolute precision possible

Precision is the result of having a *formula* for the antiderivative. Notice how we used that formula to express the value of the integral. Staring with an arbitrary antiderivative $F(x)$ of $f(x)$, we get

$$A(X) = \int_a^X f(x) \, dx = F(X) - F(a).$$

If we set $X = b$ we find

$$\int_a^b f(x) \, dx = F(b) - F(a).$$

In other words,

> If $F(x)$ is any antiderivative of $f(x)$,
> then $\int_a^b f(x) \, dx = F(b) - F(a)$.

Example.   Evaluate $\int_0^{\ln 2} e^{-u}\, du$.

For the antiderivative, we can choose $F(u) = -e^{-u}$. Then

$$
\begin{aligned}
\int_0^{\ln 2} e^{-u}\, du &= F(\ln 2) - F(0) \\
&= -e^{-\ln 2} - (-e^0) \\
&= -(1/2) - (-1) \\
&= 1/2
\end{aligned}
$$

Example.   Evaluate $\int_2^3 \ln t\, dt$.

For the antiderivative, we can choose $F(t) = t\ln t - t$. Using the product rule, we can show that this is indeed an antiderivative:

$$
\begin{aligned}
F'(t) &= t \cdot \frac{1}{t} + 1 \cdot \ln t - 1 \\
&= 1 + \ln t - 1 \\
&= \ln t
\end{aligned}
$$

Thus,

$$
\begin{aligned}
\int_2^3 \ln t\, dt &= F(3) - F(2) \\
&= 3 \cdot \ln 3 - 3 - (2 \cdot \ln 2 - 2) \\
&= \ln 3^3 - \ln 2^2 - 1 \\
&= \ln 27 - \ln 4 - 1 \\
&= \ln(27/4) - 1
\end{aligned}
$$

While formulas make it possible to get exact values, they do present us with problems of their own. For instance, we need to know that $t\ln t - t$ is an antiderivative of $\ln t$. This is not obvious. In fact, there is no guarantee that the antiderivative of a function given by a formula will have a formula! The antiderivatives of $\cos(x^2)$ and $\sin(x)/x$ do not have formulas, for instance. Many techniques have been devised to find the formula for an antiderivative. In chapter 11 we will survey some of those that are most frequently used.

Parameters

In chapter 4, §2 we considered differential equations that involved parameters (see pages 210–214). It also happens that integrals can involve parameters. However, parameters complicate numerical work. If we calculate the value of an integral numerically, by making estimates with Riemann sums, we must first fix the value of any parameters that appear. This makes it difficult to see how the value of the integral depends on the parameters. We would have to give new values to the parameters and then recalculate the Riemann sums.

The outcome is much simpler and more transparent if we are able to use the fundamental theorem to get a *formula* for the integral. The parameters just appear in the formula, so it is immediately clear how the integral depends on the parameters.

Here is an example that we shall explore further in Calculus II. We want to see how the integrals

$$\int_a^b \sin(\alpha x)\, dx \qquad \text{and} \qquad \int_a^b \cos(\alpha x)\, dx$$

depend on the parameter α, and also on the parameters a and b. To begin, you should check that $F(x) = -\cos(\alpha x)/\alpha$ is an antiderivative of $\sin(\alpha x)$. Therefore,

$$\begin{aligned}
\int_a^b \sin(\alpha x)\, dx &= \frac{-\cos(\alpha b)}{\alpha} - \frac{-\cos(\alpha a)}{\alpha} \\
&= \frac{\cos(\alpha a) - \cos(\alpha b)}{\alpha}.
\end{aligned}$$

In a similar way, you should be able to show that

$$\int_a^b \cos(\alpha x)\, dx = \frac{\sin(\alpha b) - \sin(\alpha a)}{\alpha}.$$

Suppose the interval $[a, b]$ is exactly one-half of a full period: $[0, \pi/\alpha]$. Then

$$\begin{aligned}
\int_0^{\pi/\alpha} \sin(\alpha x)\, dx &= \frac{\cos(\alpha \cdot 0) - \cos(\alpha\pi/\alpha)}{\alpha} \\
&= \frac{\cos 0 - \cos \pi}{\alpha} \\
&= \frac{1 - (-1)}{\alpha} = \frac{2}{\alpha}
\end{aligned}$$

Exercises

Constructing accumulation functions

1. a) Obtain a formula for the accumulation function

$$A(X) = \int_2^X 5\,dx$$

and sketch its graph on the interval $2 \le X \le 6$.
b) Is $A'(X) = 5$?

2. Let $f(x) = 2 + x$ on the interval $0 \le x \le 5$.
a) Sketch the graph of $y = f(x)$.
b) Obtain a formula for the accumulation function

$$A(X) = \int_0^X f(x)\,dx$$

and sketch its graph on the interval $0 \le X \le 5$.
c) Verify that $A'(X) = f(X)$ for every X in $0 \le X \le 5$.
d) By comparing the graphs of f and A, verify that, at any point X, the *slope* of the graph of A is the same as the *height* of the graph of f.

3. a) Consider the accumulation function

$$A(X) = \int_0^X x^3\,dx .$$

Using the fact that $A'(X) = X^3$, obtain a formula that expresses A in terms of X.

b) Modify A so that accumulation begins at the value $x = 1$ instead of $x = 0$ as in part (a). Thus

$$A(X) = \int_1^X x^3\,dx .$$

It is still true that $A'(X) = X^3$, but now $A(1) = 0$. Obtain a formula that expresses this modified A in terms of X. How do the formulas For A in parts (a) and (b) differ?

Using the fundamental theorem

4.   Find $A'(X)$ when

a) $A(X) = \displaystyle\int_0^X \cos(x)\,dx$

e) $A(X) = \displaystyle\int_0^X \sin(x^2)\,dx$

b) $A(X) = \displaystyle\int_0^X \sin(x)\,dx$

f) $A(X) = \displaystyle\int_0^X \sin^2 x\,dx$

c) $A(X) = \displaystyle\int_0^X \cos(x^2)\,dx$

g) $A(X) = \displaystyle\int_0^X \ln t\,dt$

d) $A(X) = \displaystyle\int_0^X \cos(t^2)\,dt$

h) $A(X) = \displaystyle\int_0^X x^2 - 4x^3\,dx$

5.   Find all critical points of the function

$$A(X) = \int_0^X \cos(x^2)\,dx$$

on the interval $0 \le X \le 4$. Indicate which critical points are local maxima and which are local minima. (Critical points and local maxima and minima are discussed on pages 294–300.)

[Answer: There are five critical points in the interval $[0, 4]$. The first is a local maximum at $\sqrt{\pi/2}$.]

6.   Find all critical points of the function

$$A(X) = \int_0^X \sin(x^2)\,dx$$

on the interval $0 \le X \le 4$. Indicate which critical points are local maxima and which are local minima.

7.   Find *all* critical points of the function

$$A(X) = \int_0^X x^2 - 4x^3\,dx.$$

Indicate which critical points are local maxima and which are local minima.

8. Express the solution to each of the following initial value problems as an accumulation function (that is, as an integral with a variable upper limit of integration).

a) $y' = \cos(x^2)$, $\quad y(\sqrt{\pi}) = 0$ $\qquad$ c) $y' = \sin(x^2)$, $\quad y(0) = 5$

b) $y' = \sin(x^2)$, $\quad y(0) = 0$ $\qquad$ d) $y' = e^{-x^2}$, $\quad y(0) = 0$

9. Sketch the graphs of the following accumulation functions over the indicated intervals.

a) $\displaystyle\int_0^X \sin(x^2)\,dx$, $\quad 0 \le X \le 4$

b) $\displaystyle\int_0^X \frac{\sin(x)}{x}\,dx$, $\quad 0 \le X \le 4$

Formulas for integrals

10. Determine the exact value of each of the following integrals.

a) $\displaystyle\int_3^7 2 - 3x + 5x^2\,dx$ $\qquad$ e) $\displaystyle\int_1^6 dx/x$

b) $\displaystyle\int_0^{5\pi} \sin x\,dx$ $\qquad$ f) $\displaystyle\int_0^4 7u - 12u^5\,dx$

c) $\displaystyle\int_0^{5\pi} \sin(2x)\,dx$. $\qquad$ g) $\displaystyle\int_0^1 2^t\,dt$

d) $\displaystyle\int_0^1 e^t\,dt$ $\qquad$ h) $\displaystyle\int_{-1}^1 s^2\,ds$

11. Express the values of the following integrals in terms of the parameters they contain.

a) $\displaystyle\int_3^7 kx\,dx$ $\qquad$ e) $\displaystyle\int_{\ln 2}^{\ln 3} e^{ct}\,dt$

b) $\displaystyle\int_0^\pi \sin(ax)\,dx$ $\qquad$ f) $\displaystyle\int_1^b 5 - x\,dx$

c) $\displaystyle\int_1^4 px^2 - x^3\,dx$ $\qquad$ g) $\displaystyle\int_0^1 a^t\,dt$

d) $\displaystyle\int_0^1 e^{ct}\,dt$ $\qquad$ h) $\displaystyle\int_1^2 u^c\,du$

12.   Find a formula for the solution of each of the following initial value problems.

a) $y' = x^2 - 4x^3$, $y(0) = 0$ c) $y' = x^2 - 4x^3$, $y(0) = 0$
b) $y' = x^2 - 4x^3$, $y(3) = 0$ d) $y' = \cos(3x)$, $y(\pi) = 0$

13.   Find the average value of each of the following functions over the indicated interval.

a) $x^2 - x^3$ over $[0, 1]$

b) $\ln x$ over $[1, e]$

c) $\sin x$ over $[0, \pi]$

14.   a) What is the average value of the function $px - x^2$ on the interval $[0, 1]$? The average depends on the parameter p.

b) For which value of p will that average be zero?

§5.   Chapter Summary

The Main Ideas

- A **Riemann sum** for the function $f(x)$ on the interval $[a, b]$ is a sum of the form

$$f(x_1) \cdot \Delta x_1 + f(x_2) \cdot \Delta x_2 + \cdots + f(x_n) \cdot \Delta x_n,$$

where the interval $[a, b]$ has been subdivided into n subintervals whose lengths are $\Delta x_1, \Delta x_2, \ldots, \Delta x_n$, and each x_k is a sampling point in the k-th subinterval (for each k from 1 to n).

- Riemann sums can be used to approximate a variety of quantities expressed as **products** where one factor varies with the other.

- Riemann sums give more accurate **approximations** as the lengths $\Delta x_1, \Delta x_2, \ldots, \Delta x_n$ are made small.

- If the Riemann sums for a function $f(x)$ on an interval $[a, b]$ converge, the limit is called the **integral** of $f(x)$ on $[a, b]$, and it is denoted

$$\int_a^b f(x) \, dx.$$

- The **units** of $\int_a^b f(x)\,dx$ equal the product of the units of $f(x)$ and the units of x.

- **The Fundamental Theorem of Calculus.**
  The solution $y = A(x)$ of the initial value problem

$$y' = f(x) \qquad y(a) = 0$$

  is the **accumulation function**

$$A(X) = \int_a^X f(x)\,dx.$$

- If $F(x)$ is an **antiderivative** of $f(x)$, then

$$\int_a^b f(x)\,dx = F(b) - F(a).$$

- The integral $\int_a^b f(x)\,dx$ equals the **signed area** between the graph of $f(x)$ and the x-axis.

- If $f(x)$ is **monotonic** on $[a, b]$ and if $\int_a^b f(x)\,dx$ is approximated by a Riemann sum with subintervals of width Δx, then the **error** in the approximation is at most $\Delta x \cdot |f(b) - f(a)|$.

Self-Testing

- You should be able to write down (by hand) a Riemann sum to approximate a quantity expressed as a product (e.g., human effort, electrical energy, work, distance travelled, area).

- You should be able to write down an integral giving the *exact* value of a quantity approximated by a Riemann sum.

- You should be able to use **sigma notation** to abbreviate a sum, and you should be able to read sigma notation to calculate a sum.

- You should be able to use a computer program to compute the value of a Riemann sum.

- You should be able to find an error bound when approximating an integral by a Riemann sum.

- You should know and be able to use the **integration rules**.

- You should be able to use the fundamental theorem of calculus to find the value of an integral.

- You should be able to use an antiderivative to find the value of an integral.

Index

Quick Reference

FORMULAS FROM GEOMETRY

Notation: $A=$ area, $V=$ volume, $b=$ length of base, $h=$ height (perpendicular to base), $r=$ radius, $C=$ circumference.

Rectangle:	$A = bh.$
Triangle:	$A = \frac{1}{2}bh.$
Circle:	$A = \pi r^2, \quad C = 2\pi r.$
Cylinder:	$V = \pi r^2 h.$
Sphere:	$V = \frac{4}{3}\pi r^3, \quad A = 4\pi r^2.$

Pythagorean Theorem: If a right triangle has hypotenuse of length c and legs of lengths a and b, then $c^2 = a^2 + b^2$.

FORMULAS FROM TRIGONOMETRY

Definition. Begin with a unit circle centered at the origin. Given the input number t, locate a point P on the circle by tracing an arc of length t along the circle from the point $(1,0)$. If t is positive, trace the arc counterclockwise; if t is negative, trace it clockwise. Because the circle has radius 1, the arc of length t subtends a central angle of **radian** measure t. The trigonometric functions $\cos t$ and $\sin t$ are defined as the coordinates of the point P,

$$P = (\cos t, \sin t).$$

The other trigonometric functions are defined in terms of the sine and cosine functions:

$$\tan t = \sin t / \cos t \qquad \sec t = 1/\cos t$$
$$\cot t = \cos t / \sin t \qquad \csc t = 1/\sin t.$$

We also have the following identities:

$$\sin^2 t + \cos^2 t = 1$$
$$\sin(-t) = -\sin t$$
$$\cos(-t) = \cos t$$
$$\sin(A+B) = \sin A \cos B + \cos A \sin B$$
$$\cos(A+B) = \cos A \cos B - \sin A \sin B$$

THE DERIVATIVE

- A **locally linear** function has a graph that looks approximately straight when magnified under a computer microscope.

- The **slope of the graph** at any point is the **limit** of the slopes seen under a microscope.

- The **rate of change** of a function at a point is the slope of its graph at that point, and thus is also a **limit**. Its dimensional units are (units of output)/(unit of input).

- The **derivative** of $f(x)$ at $x = a$ is the name given to both the rate of change of f at a and the slope of the graph of f at $(a, f(a))$.

- The derivative of $y = f(x)$ at $x = a$ is written $f'(a)$. The **Leibniz notation** for the derivative is dy/dx.

- $$f'(a) = \lim_{\Delta x \to 0} \frac{\Delta y}{\Delta x}$$
  $$= \lim_{h \to 0} \frac{f(a+h) - f(a-h)}{2h} = \lim_{h \to 0} \frac{f(a+h) - f(a)}{h}.$$

- The **microscope equation** $\Delta y \approx f'(a) \cdot \Delta x$ describes the relation between x and $y = f(x)$ as seen under a microscope.

- The microscope equation says the change in output is proportional to the change in the input. The derivative $f'(a)$ is the **multiplier**, or scaling factor.

- Functions that have more than one input variable have **partial derivatives**. A partial derivative is the rate at which the output changes with respect to one variable when we hold all the others constant.

- A function $z = F(x, y)$ of two variables also has a **microscope equation**:
  $$\Delta z \approx F_x(a, b) \cdot \Delta x + F_y(a, b) \cdot \Delta y.$$

  The partial derivatives are the **multipliers** in the microscope equation.

DIFFERENTIATION RULES

Derivatives of Basic Functions:

function	derivative
$mx + b$	m
x^r	rx^{r-1}
$\sin x$	$\cos x$
$\cos x$	$-\sin x$
$\tan x$	$\sec^2 x$
e^x	e^x
$\ln x$	$1/x$
b^x	$\ln b \cdot b^x$

Derivatives of Combinations of Functions:

function	derivative
$g(f(x))$	$g'(f(x)) \cdot f'(x)$
$f(x) + g(x)$	$f'(x) + g'(x)$
$f(x) - g(x)$	$f'(x) - g'(x)$
$cf(x)$	$cf'(x)$
$f(x) \cdot g(x)$	$f'(x) \cdot g(x) + f(x) \cdot g'(x)$
$\dfrac{f(x)}{g(x)}$	$\dfrac{g(x) \cdot f'(x) - f(x) \cdot g'(x)}{[g(x)]^2}$

DIFFERENTIAL EQUATIONS

- **Euler's method** is a procedure to approximate a function defined by a set of differential equations and initial conditions. The approximation is a piecewise linear function.

- The exact function defined by a set of differential equations and initial conditions can be expressed as a limit of a sequence of successive Euler approximations with smaller and smaller step sizes.

- The programs SIR, SIRVALUE, SIRPLOT, TABLE, PLOT, and SEQUENCE all compute Euler approximations.

- The function $y = Ce^{kt}$ is the solution to the initial value problem

$$\frac{dy}{dt} = ky \qquad y(0) = C.$$

THE INTEGRAL

- A **Riemann sum** for the function $f(x)$ on the interval $[a, b]$ is a sum of the form

$$f(x_1) \cdot \Delta x_1 + f(x_2) \cdot \Delta x_2 + \cdots + f(x_n) \cdot \Delta x_n,$$

where the interval $[a, b]$ has been subdivided into n subintervals whose lengths are $\Delta x_1, \Delta x_2, \ldots, \Delta x_n$, and each x_k is a sampling point in the k-th subinterval (for each k from 1 to n).

- Riemann sums can be used to approximate a variety of quantities expressed as *products* where one factor varies.

- Riemann sums give more accurate approximations as the lengths $\Delta x_1, \Delta x_2, \ldots, \Delta x_n$ are made small.

- If all the Riemann sums for a function $f(x)$ on an interval $[a, b]$ get arbitrarily close to a *single number* when the lengths $\Delta x_1, \Delta x_2, \ldots, \Delta x_n$ are made small enough, then that number is called the **integral** of $f(x)$ on $[a, b]$, and it is denoted

$$\int_a^b f(x) \, dx.$$

- The *exact* value of a quantity approximated by a Riemann sum is given by the corresponding integral. If x and $f(x)$ have units, then $\int_a^b f(x) \, dx$ has units equal to the product of the units of $f(x)$ times the units of x.

- **The Fundamental Theorem of Calculus.**
  The solution $y = A(x)$ of the initial value problem

  $$y' = f(x) \qquad y(a) = 0$$

  is the **accumulation function**

  $$A(X) = \int_a^X f(x)\, dx.$$

- If $F(x)$ is an antiderivative of $f(x)$, then

  $$\int_a^b f(x)\, dx = F(b) - F(a).$$

- The integral $\int_a^b f(x)\, dx$ equals the **signed area** between the graph of $f(x)$ and the x-axis.

INTEGRATION RULES
(C can be any constant)

Integrals of Basic Functions:

$$\int x^r\, dx = \frac{1}{r+1} x^{r+1} + C \qquad (r \text{ any constant} \neq -1)$$

$$\int \frac{1}{x}\, dx = \ln(x) + C \quad \text{or} \quad \ln(-x) + C$$

$$\int e^x\, dx = e^x + C$$

$$\int \sin x\, dx = -\cos x + C$$

$$\int \cos x\, dx = \sin x + C$$

$$\int \sec^2 x\, dx = \tan x + C$$

Integrals of Combinations of Functions:

$$\int c f(x)\, dx = c \int f(x)\, dx \qquad (c \text{ any constant})$$

$$\int f(x) \pm g(x)\, dx = \int f(x)\, dx \pm \int g(x)\, dx$$